Formal Approaches to Compı
Information Technolo

FACIT

Springer
London
Berlin
Heidelberg
New York
Barcelona
Budapest
Hong Kong
Milan
Paris
Santa Clara
Singapore
Tokyo

Also in this series:

Proof in VDM: a Practitioner's Guide
J.C. Bicarregui, J.S. Fitzgerald, P.A. Lindsay, R. Moore and B. Ritchie
ISBN 3-540-19813-X

On the Refinement Calculus
C. Morgan and T. Vickers (eds.)
ISBN 3-540-19931-4

Systems, Models and Measures
A. Kaposi and M. Myers
ISBN 3-540-19753-2

Notations for Software Design
L.M.G. Feijs, H.B.M. Jonkers and C.A. Middelburg
ISBN 3-540-19902-0

Formal Object-Oriented Development
K. Lano
ISBN 3-540-19978-0

S.J. Goldsack and S.J.H. Kent (Eds)

Formal Methods and Object Technology

Springer

S.J. Goldsack, BSc, PhD, FBCS, FIEE, CEng
Department of Computing, Imperial College of Science, Technology and Medicine,
180 Queen's Gate, London SW7 2BZ, UK

S.J.H. Kent, BSc, PhD
Department of Computing, University of Brighton, Watts Building,
Lewes Road, Brighton BN2 4GJ, UK

Series Editor

S.A. Schuman, BSc, DEA, CEng
Department of Mathematical and Computing Sciences
University of Surrey, Guildford, Surrey GU2 5XH, UK

ISBN 3-540-19977-2 Springer-Verlag Berlin Heidelberg New York

British Library Cataloguing in Publication Data
Formal methods and object technology. - (Formal approaches
to computing & information technology)
1.Object-oriented programming (Computer science)
I.Goldsack, Stephen J., 1926- II.Kent, Stuart John Harding
005.1'1
ISBN 3540199772

Library of Congress Cataloging-in-Publication Data
Formal methods and object technology / S.J. Goldsack and S.J.H. Kent.
p. cm. - - (Formal approaches to computing and information
technology)
ISBN 3-540-19977-2 (alk. paper)
1. Object-oriented programming (Computer science) I. Goldsack,
Stephen J., 1926- . II. Kent, S. J. H. (Stuart John Harding),
1966- . II. Series.
QA76.64.F67 1996 96-33762
005.1'1 - - dc20 CIP

Printed in Great Britain

Typesetting: Camera ready by editors
Printed and bound at the Athenæum Press Ltd., Gateshead, Tyne and Wear
34/3830-543210 Printed on acid-free paper

Preface

Rationale

Software engineering aims to develop software by using approaches which enable large and complex program suites to be developed in a systematic way. However, it is well known that it is difficult to obtain the level of assurance of correctness required for safety critical software using old fashioned programming techniques. The level of safety required becomes particularly high in software which is to function without a break for long periods of time, since the software cannot be restarted and errors can accumulate. Consequently programming for mission critical systems, for example, needs to address the requirements of correctness with particular care.

In the search for techniques for making software cheaper and more reliable, two important but largely independent influences have been visible in recent years. These are:

- Object Technology
- Formal Methods

First, it has become evident that *objects* are, and will remain an important concept in software. Experimental languages of the 1970's introduced various concepts of package, cluster, module, etc. giving concrete expression to the importance of modularity and encapsulation, the construction of software components hiding their state representations and algorithmic mechanisms from users, exporting only those features (mainly the procedure calling mechanisms) which were needed in order to use the objects. This gives the software components a level of abstraction, separating the view of what a module does for the system from the details of how it does them. These ideas reached a peak with the development of Ada which it was hoped would provide a vehicle for construction of large scale systems with the necessary degrees of reliability. Indeed, Ada83 is a powerful language which remains a useful vehicle for quality software construction.

However, in a complementary development, it became recognised that certain properties of objects, particularly inheritance and the use of object references as part of the data stored by an object, could be used to construct large systems incrementally, as well as making it possible to reuse component objects in different contexts. These ideas were first given expression in Simula 67 but were not widely appreciated at that time (though the all-important concept of Class came from there). It fell to the developers of SmallTalk to bring the ideas to maturity. C^{++} has become a popular language capturing the object-oriented concepts in a familiar but not very secure medium and Eiffel has provided a carefully thought out strongly typed language supporting object-oriented programming. In recent years there has been an ever increasing use of the concepts, and inevitably Ada had to move with the times and Ada95 has now been defined to make it conform more nearly to the new ethos.

No doubt the progress of research will continue to present new ideas about languages and structuring concepts for reliable software, but we are convinced that the essential object-oriented concepts will have a permanent place in software technology.

The other thread of development is formality. At least for highly critical software systems, it seems essential to give software engineering the same basis in mathematics that is the hall mark of other important engineering disciplines. In this there has been much progress, resulting in three main paradigms: model-based, algebraic and process algebra based. However, except for a few notable exceptions, take up of formal approaches in industry has been disappointing. There are, no doubt, a number of reasons for this, but one may be the lack of flexible structuring mechanisms required to model large systems—an area in which Object Technology is particularly strong.

We believe that approaches which include the best aspects of formal methods and Object Technology may succeed in providing the tools required for secure development of large scale software in a form which industry will find appealing. This book provides a cross-section of recent research contributing to the merging of the two paradigms.

Organisation

The book is organised into four parts. The first part is introductory in nature, comprising two chapters which, respectively, introduce the main concepts of object technology, and provide an overview of research combining formal methods with this technology.

The other parts comprise a series of contributions from leading researchers in the area, organised into three strands.

Formal Methods in Object Technology. This involves the application of formal techniques to Object Technology. The aim of research in this strand is to take ideas from formal methods and apply them in the further development of existing object-oriented methods.

Object Technology in Formal Methods. This involves the extension of formal methods with object-oriented concepts. The aim of research in this strand has been to incorporate object-oriented concepts to provide structuring and reuse in the development software using existing formal methods.

Formal foundations of Object Technology. The third strand includes more fundamental research, whose purpose has been to develop mathematical characterisations of object-oriented concepts. This work underpins the other two strands by providing a proper semantic basis essential for ensuring correctness and for the development of tools to support the use of formal techniques. This can be compared, for example, with the way in which the use of mathematics in characterising the semantics for programming languages has led to techniques for the development of correctand efficient compilers.

A fourth aspect which is of considerable interest to the research community lies in the use of object structures in the development of concurrent systems. In the real world objects coexist, and a realistic model of the world must support concurrent execution of software objects modelling world entities. Concurrency is the default in many application domains. However, concurrency is an issue which is in many ways orthogonal to the three aspects discussed above. Work in any of the domains may or may not discuss topics arising from concurrency.

Nor are the boundaries between the three strands sharply defined. Nevertheless, we feel that this categorisation does serve to identify the main emphasis of each contribution.

Chapter 2, entitled *Formality in Object Technology*, provides an assessment of each contribution, relating it to other current and recent research in the area.

Readership

The book is aimed both at researchers in Object Technology and Formal Methods, and at practioners who wish to obtain an overview of recent research in the merging of these two fields. The introductory part of the book would be suitable for students, on advanced level courses or embarking on a research program, as an introduction to and overview of the area.

Acknowledgements

We would like to thank all contributors for their efforts, and Rosie Kemp at Springer Verlag for her support and patience whilst the manuscript was being prepared.

Stephen Goldsack and Stuart Kent
December 1995.

Contents

List of Contributors

The following lists the authors, with affiliations, for each chapter.

1. S.J. Goldsack
   Department of Computing,
   Imperial College of Science, Technology and Medicine,
   London SW7 2BZ, England.
   e-mail: sjg@doc.ic.ac.uk

2. K. Lano
   Department of Computing,
   Imperial College of Science, Technology and Medicine,
   London SW7 2BZ, England.
   e-mail: kcl@doc.ic.ac.uk

3. A. Moreira
   Departamento de Informática,
   Faculdade de Ciências e Tecnologia,
   Universidade Nova de Lisboa, Portugal.
   e-mail: amm@fct.unl.pt

   R.G.Clark
   Department of Computing Science and Mathematics,
   University of Stirling, Stirling, Scotland.
   e-mail: rgc@cs.stir.ac.uk

4. J. Armstrong
Dependable Computing Systems Centre
University of York,
Heslington,
York, England.
e-mail: jma@minster.york.ac.uk

J. Howse, R. Mitchell
Department of Computing,
University of Brighton,
Lewes Rd.,
Brighton, England.
e-mail: John.Howse@bton.ac.uk
e-mail: Richard.Mitchell@bton.ac.uk

I. Maung
Dept of Computer Science,
University of Warwick,
Coventry CV4 7AL, England.
e-mail: Ian.Maung@dcs.warwick.ac.uk

5. K.L. Clark, T. I. Wang
Department of Computing,
Imperial College of Science, Technology and Medicine,
London SW7 2BZ, England.
e-mail: klc@doc.ic.ac.uk

6. E.H.Dürr
Cap Volmac and Utrecht University,
Utrecht, Netherlands.
e-mail: E.H.Durr@ruunfs.fys.ruu.nl

S.J. Goldsack
Department of Computing,
Imperial College of Science, Technology and Medicine,
London SW7 2BZ, England.
e-mail: sjg@doc.ic.ac.uk

7. K. Lano
Department of Computing,
Imperial College of Science, Technology and Medicine,
London SW7 2BZ, England.
e-mail: kcl@doc.ic.ac.uk

H. Houghton
Trading Research,
J. P. Morgan Corp.
100 New Bridge St.
London EC4, England.

P. Wheeler
Applied Information Engineering
Lloyd's Register of Shipping,
29 Wellesley Rd.,
Croydon CR0 2AJ, England.

8. Y. Yang, N. Treves
TELESYSTEMES,
Paris, France.

Z. Yao
Department of Electrical Engineering,
University of Quebec,
Trois-Rivieres,
Quebec G9A 5H7, Canada.
e-mail: Ziwen_Yao@UQTR.UQuebec.ca

9. J. Fiadeiro
Department of Informatics,
Faculty of Sciences,
University of Lisbon,
Lisbon, Portugal.
e-mail: llf@di.fc.ul.pt

T.S.E. Maibaum
Department of Computing,
Imperial College of Science, Technology and Medicine,
London SW7 2BZ, England.
e-mail: tsem@doc.ic.ac.uk

10. G. Malcolm
Programming Research Group,
Oxford University Computing Laboratory,
Oxford, England.
e-mail Grant.Malcolm@comlab.ox.ac.uk

11. P. Borba, J.A. Goguen
Programming Research Group,
Oxford University Computing Laboratory,
Oxford, England.
e-mail: phmb@di.ufpe.br
e-mail: Joseph.Goguen@comlab.ox.ac.uk

12. S. Drossopolou, S. Karathanos, Dan Yang
Department of Computing,
Imperial College of Science, Technology and Medicine,
London SW7 2BZ, England.
email: sd@doc.ic.ac.uk
email: yd@doc.ic.ac.uk

13. I. Maung
Dept of Computer Science,
University of Warwick,
Coventry CV4 7AL, England.
e-mail: Ian.Maung@dcs.warwick.ac.uk

14. F. Piessens, E. Steegmans
Dept of Computer Science,
Katholieke Universiteit Leuven,
Leuven, Belgium.
email: Frank.Piessens@cs.kuleuven.ac.be
email: Eric.Steegmans@cs.kuleuven.ac.be

15. E. Sekerinski
Department of Computer Science,
Åbo Akademi University,
Turku, Finland.
e-mail: Emil.Sekerinski@abo.fi

Part I

Introduction

Chapter 1

The Object Paradigm

This book is about the approach to software development which we call Object Technology. The first chapter discusses what objects are and why they are important. The treatment here is informal, giving the kind of intuitive account of the semantics of the features such as could be found in a reference manual or in an introductory textbook for a programming language. Some of the ideas are quite difficult to explain and understand at this informal level; particularly, it is difficult to comprehend in full the effects of constructs which involve polymorphism and deferred binding of method calls. It is also especially difficult to know what to expect if the features are misused.

The informal account of OO given in this chapter will be followed in chapter 2 by a review of how a more formal basis can be given to the concepts, and in particular to the aspects covered by the remaining chapters of the book. The treatment of formal matters is not, however, systematic, so not all the OO concepts will be found treated formally in the later chapters.

1.1 The Ubiquitous Object

All software relates to some problem domain, and a programmer sets out to construct programs which are models of his particular application. That it is useful to construct the software in terms of objects is partly a reflection of their importance in the application world. This point of view is illustrated by figure 1.1 which shows how an accurate model of an object-structured world will itself be object-structured. So let us consider first the world around us.

There are objects everywhere. As I look around my office I see many objects. To name just a few, there are books, a table, a desk (which is a kind of table), chairs, a radiator, a word-processor (which is a kind of computer). There are bookshelves which are objects containing books which are themselves also objects. I can continue an endless catalogue of such concrete things. But there are less concrete things which I can also consider to be examples of the object paradigm. The documents on my desk are clearly objects, but there are other documents in electronic form in the memory of the computer. The programs

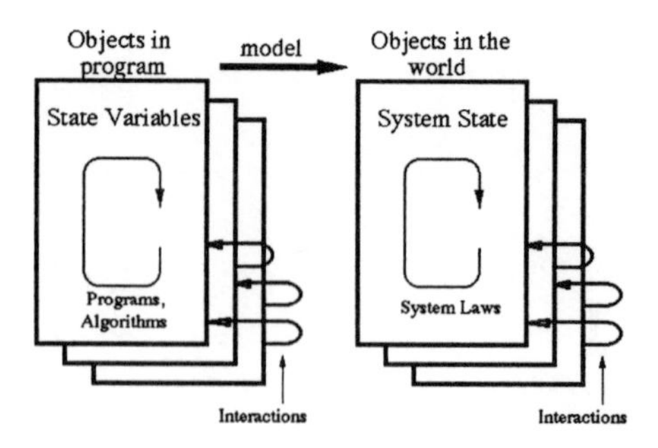

Figure 1.1: An object structured world is modeled by object structured software.

in the machine are objects: they do things in response to my commands. Some do things for me without my command - the system maintains its integrity and reports if I make errors!

People around me are objects: some have special roles, such as bank clerks, train drivers, shop assistants, but they are all people, and they are all objects. Some of the objects are static, immutable things. A book may perhaps be open at some page, or change its position in space in response to an action by a person, but the book itself is a rather immutable thing. Many other objects are more adaptable, and can change their state in response to some command. The file in which I am at present writing this text will change only in response to my keyboard commands. Other files in the file store will (I hope) remain unchanged until I next have occasion to open them and start issuing commands to modify them. The filestore itself is an object.

1.1.1 Active Objects

An object like a file or a book changes its state only when some appropriate action is invoked by a user, requesting or commanding it to do so. There are, however, objects in the world which change their states spontaneously, of their own volition and without external command. Such objects include people, especially people having roles. Often we perceive the actions of persons as actions of the things they control. For example, we may consider that a train is an object which moves from place to place "spontaneously" without command from its passengers; however, on a more careful inspection of its mode of operation, we could see that it really does so in response to commands from its human driver. Nevertheless, there are objects which behave in a spontaneous way and do not reflect the actions of people. Importantly there are automated control systems, which may be computer based. My bank account is a computer file which changes mainly in response to commands to enter or withdraw cash, or make payments to other accounts. However, if interest is paid on a per diem basis, its addition will probably be spontaneous, and it may be appropriate to consider it as if it were an active entity, spontaneously changing its state according to pre-defined laws.

Concurrency of Active Objects

An important feature of active objects is that, in carrying out their activities, they do so concurrently with others in their environment. People live concurrent lives, and trains move about concurrently with other trains and with the people who use them, as well, of course, as with the buses and aeroplanes which also provide transport, and all the many things which go on in every-day life. Active world objects which co-exist involve concurrency in their activities.

To avoid conflicts of interest there have to be means of synchronising interactions between such objects. People may board a train only when it is stopped at a suitable platform, and passengers must form a queue when the train arrives at their platform and enter it by the door one at a time. There will also be problems in my computer file if I try to print it while I continue to change it by typing new text.

Such conflicts, involving contention for the use of a shared resource, are often life threatening, safety concerns. Concurrent movement of trains may be freely permitted until they are about to enter a shared track segment. To avoid accidents, access to such a track segment must be strictly controlled to ensure that only one train is using it at any time. The distinction between passive and active objects, and the means of specifying the synchronisation constraints have been central concerns in the development of VDM^{++} and are described in chapter 6.

1.2 Defining the Object Concept

So objects are ubiquitous: but what *is* an object, and why is object structuring helpful in software development? Generally, we consider an object to be a conceptual entity with a *unique* and *persistent* identity; it has an essential self-hood which distinguishes it from other objects, whether they are the same kind of thing or not. It may be anonymous but more usually has one or more names by which it is known to its users.

Another important feature of an object is its *state*. A detailed description of an object will include the values of certain quantities. Some will be constants, but others will vary and develop with time, often due to interactions with other objects. A description does not have to be exhaustive; which variables are included depends on the problem domain of interest. In describing my office, I might or might not wish to include the colour of the walls or the temperature of the desk as elements in the description. The entities whose values are relevant to the description of an object are called its *state attributes*, and may be constants or variables.

Having a persistent identity means, of course, that the object and its identity remain part of the system over a continuous period of time. Having fixed my attention on some object, I am able to continue to observe it and recognise it as distinct from others, even though it may change its state during the period of observation. An object's state may change dramatically during its life, while its essential identity is preserved. I could select the egg laid by a butterfly, and follow it through all its metamorphoses until it eventually emerges from its pupal cocoon to become a butterfly. Yet there is no doubt that it is the same insect throughout these changes.

An object does not, however, have to be eternal. It can come into existence as a result of some creation event, and later it may be removed from the scope of the user's interest by a destruction event.

Different objects may be in identical states at any time, but that does not make them the same object. I might have two copies of some book, both open at the same page; but they would still be distinct entities, and at a later stage one might become closed and be placed on the bookshelf, while the other is being read by some person - yet with a little care I could continue to know which was which.

An object usually has an owner, or *client*, some other object for which it performs its particular purpose and who issues commands for it to perform its actions. We may say that it is a *server* or *supplier* for that client. However, ownership is not always unique, and it may be owned jointly by more than one client object. Moreover, the ownership pattern may also change in the course of time. A person may change his employer, and a car may be used by different drivers at different times, or a book may be moved to a different bookshelf. This is discussed in more detail in section 1.5

An object usually has actions which it can perform upon request by its clients. Performing such actions may change the state of the object, and requesting such actions will be the only way in which its state can be changed by the clients. The state information is hidden from users unless it has operations which can observe the state and return the value to the client.

1.3 Objects in Software

In software, an object is formed by providing in the language concepts of *encapsulation* and *scope*. Within the scope, only properties of the object which are intended to be accessible to the clients are "visible" to other parts of the program.

Structuring software in terms of objects serves several important purposes which have been recognised as techniques of software engineering for many years. Classic papers by Parnas [Parnas, 1972] discussed modularity, and introduced the concept of information hiding. Yourdon[Yourdon, 1989] originated Structured Analysis and Design, while Liskov[Liskov and Zilles, 1977] had important influence in the area of abstraction and modularity. Objects are modules which possess properties identified as desirable, such as:

- Modularity
- Abstraction
- Information hiding

- Reusability
- Strong cohesion
- Weak coupling

Objects are natural modules, which provide an abstract view which separates the concerns of the designer for *what* an object does for the system from the internal details of *how* it does it. Encapsulation hides from the user of a component object any details of the state representations and internal mechanisms, ensuring that the execution of the object's behaviour cannot be damaged by unexpected or unintended access to the state variables or operations by its clients. Finally, objects are natural units of reuse, providing entities with abstractly defined properties, which may be reused in programs in a different context.

A programming language, such as Ada83, supporting the object concept, but not supporting techniques of inheritance discussed in the following sections, is usually referred to as an *object based* language[Wegner, 1987].

Each object has an interface, sometimes called its protocol, by which it can interact with its clients. This will specify the operations which the object can perform for its clients, and how they can be invoked. Following the SmallTalk tradition[Goldberg and Robson, 1983], the operations are almost always called *methods*. Some methods may change the internal state, in which case they may be referred to as *mutating* methods; others may serve only to enable the client to request information about the current state, in which case they are usually called *enquiry* methods. The essential structure of an object is illustrated in figure 1.2.

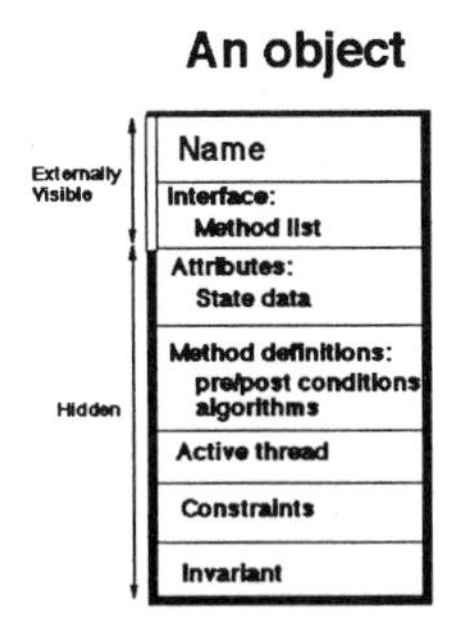

Figure 1.2: The structure of a typical object

1.4 Object Classes

Objects are usually recognised as belonging to *classes*, sets of objects which, while having distinct identities from each other, share identical properties[1].

[1]The concept of a Class in relation to software was first introduced in Simula67[Birtwistle *et al.*, 1979] which was the first Object Oriented programming language. Its importance

Often, in English, we distinguish the class from the individual by the use of the definite or indefinite article. We talk of *a* table and distinguish it from *the* table. In the first case we are considering some table object from the class of all tables, while in the second we are referring to a particular table. We commonly may say that an object is a member, or an *instance*, of a particular class.

An object class may, as remarked in the last paragraph, be considered to be a *set* of objects in the world of discourse, sharing identical properties. However, that is not the only way in which it is often understood. If we try to define the properties shared by all the members of a class, we do so by defining the properties of a typical member of the class. Defining the class in this way effectively provides a template or blue-print, describing the construction of each object of the class, and which can be used in creating new objects belonging to the class. In many object oriented programming languages programmers are encouraged to think of the class in this way, as a template used by the compiler for creating objects of the class.

There is also a third view which may be taken of the class concept. It may be considered as a *type*, of which all the objects are instances, in just the way that variables in a programming language are considered to have well defined types. A module encapsulating its state, and presenting operations to its users, implements an *Abstract Data Type*[Guttag, 1980], [Liskov and Zilles, 1977]. It is abstract, in the sense that it may be represented in different ways, yet if the operations are suitably formulated, the services provided for the clients may be exactly the same. For example, an object implementing the properties of a physical vector may have its state stored in polar or Cartesian coordinates. With suitable external operations to add vectors, compute outer and inner products, rotate vectors etc., the users need not be aware of the different implementation. A class as a type will regard such differences as irrelevant, and the objects would be of the same class. However, the template description defines the actual implementation, and PolarVector and CartesianVector would be different classes. This example is discussed further at the end of section1.5.2.

Class and Instance Variables

Objects are said to be instances of the classes to which they belong. In the class definition attributes, constants and variables, will be defined which are replicated in each object which is an instance of the class. Such variables are usually called *instance variables.* Some languages support also a concept of *class variables.* These represent state information which is known to all the objects of the class, and whose use is shared between them. In a world with concurrency between the actions of different objects, such sharing must be disallowed, or used with extreme caution.

---

was not immediately widely recognised, and it fell to Smalltalk[Goldberg and Robson, 1983] to make the concepts widely known and understood. More recently Meyer developed a strongly typed language with OO features known as Eiffel[Meyer, 1988], and C++, an object structured version of C[Stroustrup, 1991] has become popular.

1.4.1 Class Hierarchies and Inheritance of Properties

Classes of objects can often be considered to belong in hierarchies. Books may be classified according to subject categories, and these into subcategories, and so on. A well known classification scheme is that used by biologists to describe the animal kingdom. But many other objects can be seen to belong in categories. We have already hinted that a desk is a kind of table. The class of desks is a subclass of the class of tables. To buy a table, or a desk, we would go to a furniture store which sells items of a class furniture. There we would find also beds, chairs and cupboards, all objects of the class of furniture; but we would not expect to buy books! Bank clerks and train drivers and students are each special categories of people, and can be seen as inheriting the properties of a person from a class Person.

A class is a set of objects possessing some characteristic properties; an object of a subclass possesses all the properties characteristic of its parent, but other properties too. These additional properties characterise a subset of the parent class. It is often helpful to consider a Venn diagram, as in figure 1.3 where a class of students is seen to be a subclass of the class of persons. All members of the class Student are *ipse facto* members also of the class Person, and possess all the properties required to be a Person object.

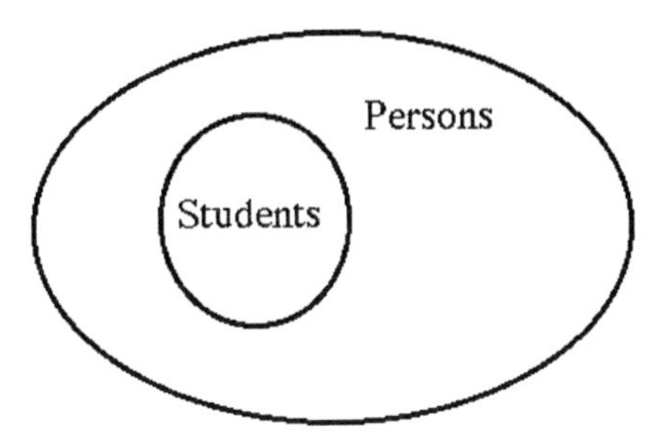

Figure 1.3: A simple view of the subclass relationship

Such subclassing relationships can be repeated at a lower level; students of history form a subclass of students, so forming a hierarchy - a tree. This is shown as part of a hierarchy tree in figure 1.4.

In such a hierarchy, the elements higher in the tree (ancestors) have fewer defined properties than their more concrete descendents. In a true class hierarchy objects of a child (heir) class have *all* the properties of objects of the parent class, and more besides. We say they *inherit* the properties of objects of the parent class. They *specialise* the parent objects in some way either by *extending* them (adding new properties) or by restricting them, by adding constraints. (see figure 1.5

A student is a person, with all the basic properties of persons, but extended with additional properties relevant to students such as a timetable of lectures and a method *study*, for example. But a female student may be considered

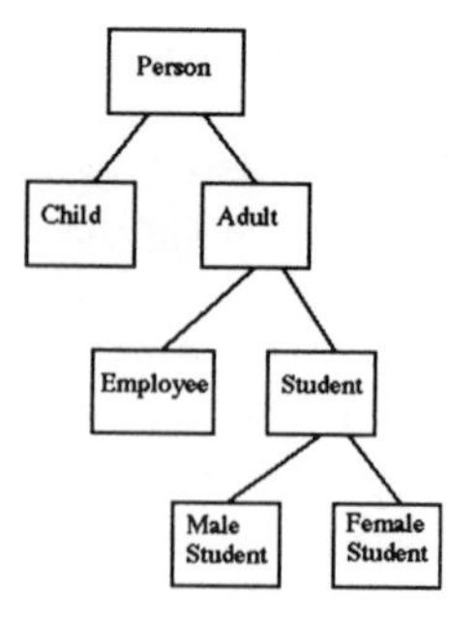

Figure 1.4: An example of a class hierarchy

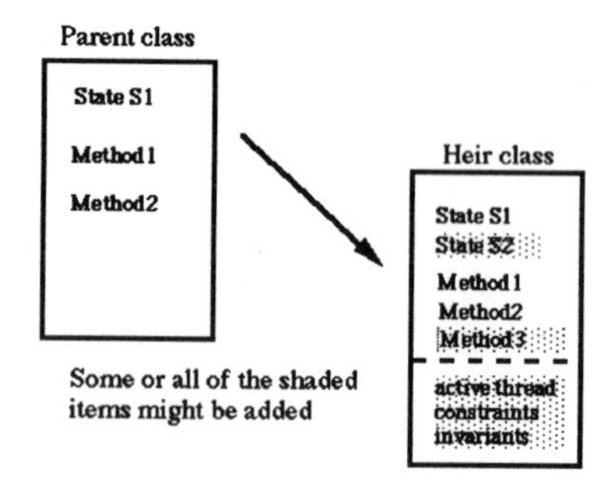

Figure 1.5: A subclass extends its parent's properties

to belong to a subclass of students, namely female students. Here the student properties have been restricted by adding an invariant condition, to be satisfied by all members of the subclass. This relationship is shown as part of the tree in figure 1.4. The subclass relationship is further discussed in chapters 13 and 12.

1.4.2 Multiple Inheritance

An object may be a subclass of more than one parent class. In this case it defines a class of objects which possess the properties of both (all) the parents. On the Venn diagram, therefore, it consists of those objects in the intersection set between the classes which are its parents. In figure 1.6 we illustrate how a class of flying objects and a class of animals can have an intersection which is a class of flying animals. In this case we have added a class of mammals, leading also to a set of flying mammals.

Notice that in this case, bats are seen to be members of a subclass both of flying animals and of mammals. Each of these is, however, a subclass of animals. So bats inherit animal properties from either of two routes. They do not, however, inherit them twice. This kind of repeated inheritance is quite common, and is discussed by Meyer[Meyer, 1988]. It may lead to an empty class, if there is no intersection between the subclasses. (There might, for

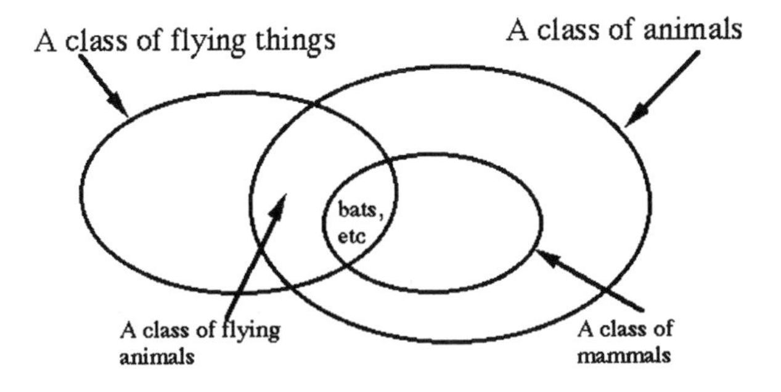

Figure 1.6: Multiple inheritance as class intersection

example, have been no flying mammals).

1.5 Clientship

As we have seen, objects can be considered to provide services for other objects, which are often called their clients. This is achieved by including in the state information of each object identifiers referring to any objects which they use, or which are otherwise dependent on them. For example, a model of a bookshelf might include among its state variables a collection of identifiers representing the set of books which are stored in it at any time.

Such ownership may be shared. The same object may be named as a server by more than one client object. In figure 1.7 the same server is shown named by two clients or, if the second name is omitted, by only one. This difference is sometimes crucial. In modeling persons, their cars or their houses may be jointly owned, but their internal organs are not capable of being shared. Present languages do not distinguish these cases well.

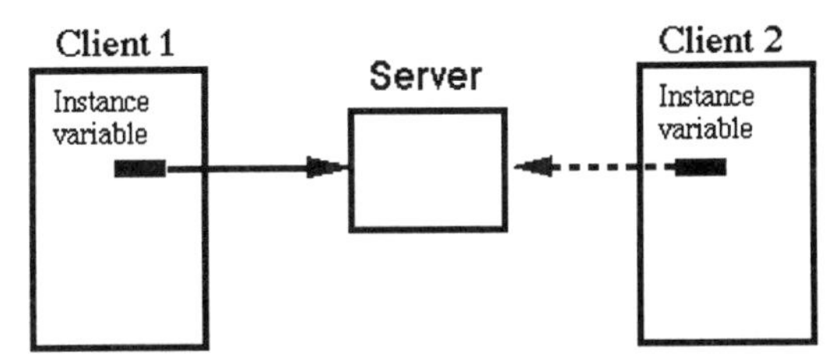

- **without the dotted link the server has a sole client.**
- **with the link its services are shared.**

Figure 1.7: The client/server relationship

As emphasised in the previous paragraph, each object has a unique identity, but the variables used to reference the server in a class description are names for objects of the class; these names do not necessarily refer to the same object at all times. In describing how I go to work, I may prescribe the use of a car, and have variables 'my-car' and 'my-destination which enables me to say that I climb into my-car and drive it to my-destination. When I change my car, or use my second car, the name my-car refers to a new instance of the class Car, and if I change my employment, my-destination may refer to a different instance of the class Destination.

The actions of the server objects are invoked by calls from the client to methods which they provide in their interfaces. Following a tradition established by SmallTalk, this is sometimes viewed as sending a message to the server object requesting it to perform its action.

1.5.1 Polymorphic Substitutability

Since each object of the child class has all the properties of the parent, a client may freely use an object of a subclass to provide the properties of the parent. If, for example, I need the basic properties of a table, say to climb on to reach the ceiling, I could use a rather simple table, having at least three legs and a top. It need have no other properties. But I would be just as well served by a desk, or a dining room table or a carpenter's workbench. In general an object from any of the descendent classes will serve to provide the features of the parent class, and I may substitute one for the other without changing the system properties. This is called polymorphism. In figure 1.8 we show how a person who needs a friend may seek the services of any other person; but of course a student, which belongs to a subclass of persons, can serve the purpose admirably. It is common to use the relation "is-a" to express this relationship

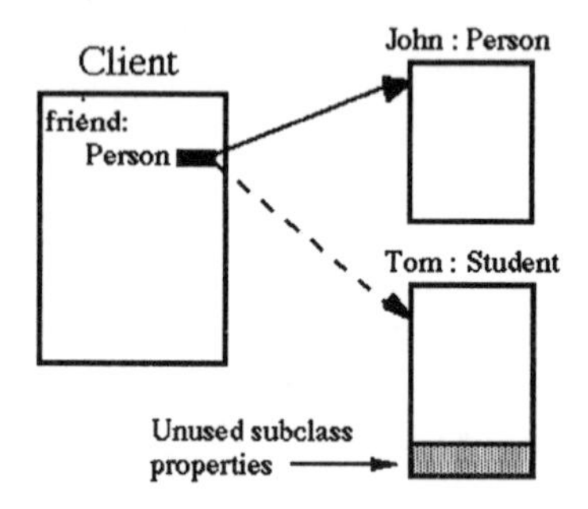

Figure 1.8: Polymorphism: a student may serve as a person

of a descendant to an ancestor. The desk "is-a" table (and so is a workbench); a student "is-a" person. This emphasises that the heir class has *all* the features of the parent, and can provide the services of a parent for a client which needs its properties.

Unfortunately, as we shall see in section 1.5.2 below, programmers and programming languages often weaken the inheritance concept, and allow the child to change some of the inherited features, yet still consider the objects to be substitutable. This leaves much uncertainty over the possible meanings of the "is-a" relation[2].

1.5.2 Abstract Classes and Dynamic Binding

It commonly happens that a class is developed by a system designer to capture some abstract concept, involving an operation which every object of the class should be able to perform. However, the subclasses are defined in such a way that this abstract operation must be implemented in a different way for each of the subclasses. A class of Shapes intended for display on a computer screen, for example, may have subclasses Polygon and Curved_shape. An operation to move the image on the screen will clearly be very different in the two subclasses, although they both must capture the more abstract concept of moving as it applies in their particular case.

In defining the parent class, it is impossible to give a detailed definition of such an operation; it must be given, differently, in the two subclasses. Most languages supporting inheritance introduce a concept of an abstract class, one or more of whose operations are defined only in respect of their calling signatures. The body of the method is specified in some way like "is abstract" or "is deferred" or "is subclass responsibility". Now, in each of the subclasses, the full definition of the method is given for the relevant data.

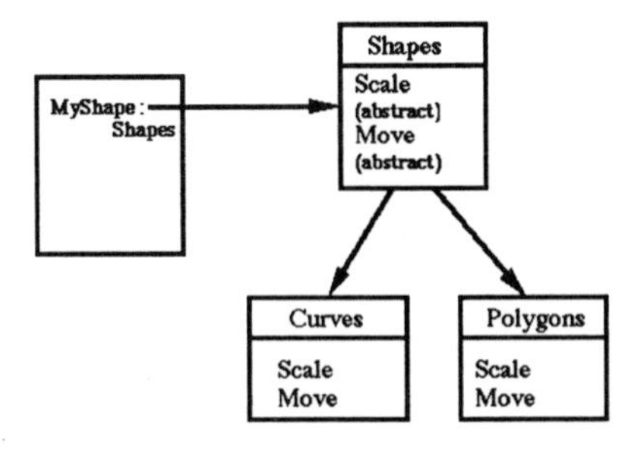

Figure 1.9: An abstract class is defined by either of its subclasses

Such an abstract class cannot be used to create instances. In figure 1.9 there are *no* objects of class Shapes. Polymorphic substitution is essential in these cases: the client having an instance variable *my_shape*, naming an object of class *Shapes*, cannot be used to refer to an object of that class, since there are none. It may, however, be used to refer to an actual object of either of its subclasses. Moreover, during the execution of the program the actual shape referred to by the state variable might be changed by asssignment from a value referring to one of the subclasses to one referencing the other. Now a call to the *move* operation of *my_shape* must invoke a different operation according to which is the subclass of the object presently referenced by the name *my_shape*.

[2] An interesting discussion of the "is-a" relation can be found in an article by Wegner and Zdonik[Wegner and Zdonik, 1988].

This selection is done dynamically (at run time) and is usually called dynamic binding, meaning binding performed at run time.

It should be noted here that this technique provides a resolution of the difference between the class as a type and as a template. In the example used in section 1.4, the class *Vector* would be an abstract parent, while the different implementations derived from templates would be alternative subclasses, each having all the (abstract) properties of the parent. Here the added properties characterising a subclass are those arising from the selection of a concrete representation.

1.5.3 Abstract Interfaces

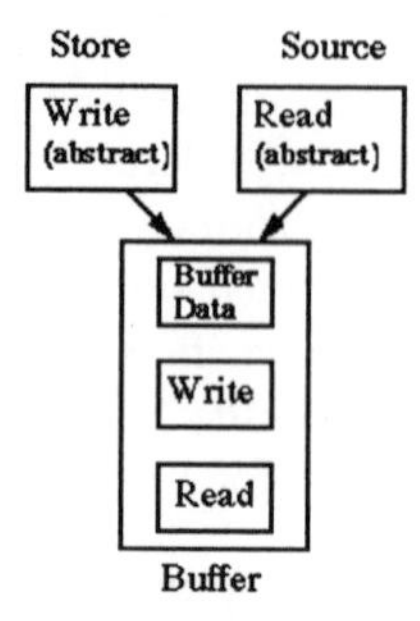

Figure 1.10: Source and Store are abstract classes accessing the store of the buffer

Coupled with multiple inheritance, abstract classes may be useful in another way: they can serve to partition the interface of a class, so that different clients have access to different subsets of the methods. This is illustrated in figure 1.10, in which a class exporting read and write operations does so in two separate partial interfaces to be used by the readers and writers respectively.

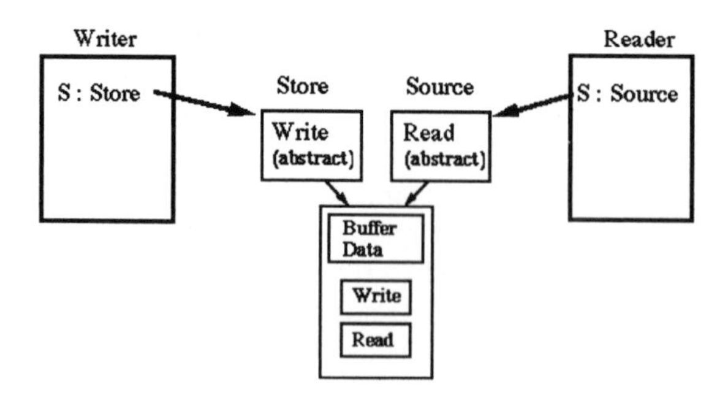

Figure 1.11: Producer and consumer system

In figure 1.11 this technique is used to provide access to the buffer for two objects of class Producer and Consumer, respectively. In a concurrent world these will be concurrently active, and the Buffer class would require constraints to synchronise their actions, ensuring that reading and writing does not occur simultaneously.

Interfacing structures like this are given a formal basis in category theory by Fiadeiro and Maibaum. Their work is described in chapter 9. In chapter 6 a family of such abstract interfaces is used to model the seats at the table in a treatment of the dining philosophers problem.

1.5.4 Aggregation

An object belonging to a class formed by multiple inheritance of two or more parent classes is able to serve as an instance of either of its parents: it combines the properties of the two classes, forming a type which is the union of the parent types.

An object may alternatively access the properties of others by clientship: it may be a client of two or more other objects, having instance variables naming each.

Schemes for system analysis for Object Oriented systems such as *Object Modelling Technique* (OMT) [Rumbaugh *et al.*, 1991] usually recognise *three* kinds of class relationship. These are inheritance, association and aggregation. The first two are implemented respectively by inheritance and clientship. However, programming languages do not provide support for aggregation, which falls in a grey area between the other two. A class is an aggregation of two others if it includes *and uses* both of its parents. For example, a telephone combines the properties of a microphone and a headphone - should it be considered to inherit both or to be a client of both? Most writers (see for example Booch [Booch, 1991] page 116) conclude the latter, but it is not clear cut. Inheriting two or more parent classes produces a class whose state is the aggregate of the states of the separate parents, and methods can be added which use the total state as if it was a single entity. In Eiffel (see [Meyer, 1988] page) and in VDM++ [Durr and Plat, 1994] a class can be formed by inheriting the same parent two or more times. A car can, for example, inherit the properties of a wheel four times. This is surely aggregation, implemented as inheritance. Perhaps one can still consider that a car "is-a" wheel (amongst other things) but it seems better to introduce a different conceptual relationship, such as "uses" or "is composed of".

1.5.5 Reuse

An important aspect of the application of Object Oriented techniques in software production has been in connection with the desire to foster techniques permitting reuse of software components. Class defininitions stored in a library form a rich source of components which can be adapted by inheritance and extension to meet the requirements of a new application. To make this flexible, language designers have permitted programmers constructing a subclass to change the definition of one or more of the inherited methods, or to omit some altogether. These possibilities destroy the logical basis of inheritance and make the subclass no longer able to guarantee to provide the properties of the parent class. The meaning of the "is-a" relation ceases to be clearly defined.

1.6 Conclusion

When objects are central to every aspect of human affairs and even of the inanimate world, it is of little surprise that they play also an essential role in the structure of software which models aspects of the real world, as illustrated in figure 1.1. This seems to apply to every problem area; object structuring transcends differences in the details of the services the objects provide for their clients in different domains.

World objects have state variables and laws of behaviour. These laws may be laws of physics in a problem of real time control, or may be laws of banking in the financial world. In software, state is represented by program variables while the laws are given by algorithms defining operations on the variables. Objects in software encapsulate the state and operations so that the state cannot be changed except by the valid operations.

Moreover there are many practical reasons why Object structuring has become popular; its powerful support for abstraction and for reuse are perhaps the most significant for the programmer.

This chapter has outlined briefly the features usually provided in a programming language intended to provide for Object Oriented programming. The rest of the book will contain views about the way in which these concepts can be put on a firm mathematical basis.

Chapter 2

Formality in Object Technology

This chapter will review the use of formal approaches in securing better quality software.

It will overview the main concepts of object technology and formal methods, and conduct a brief survey of research in the three strands described in the Preface. At the same time, the contributions which make up the remainder of the book will be set in context within their respective strands.

2.1 Formal Methods Concepts

Formal methods have been researched and applied for over 25 years, having their origin in the work of Dijkstra and Hoare on program verification, and Scott, Stratchey and others on program semantics. A major step was the development of mathematical languages for program specification, of which the most established are the Z [Spivey, 1992] and VDM [Jones, 1990] notations. The Z language was initially created by J. R. Abrial (his paper "Data Semantics" from 1974 could be considered the birth of this language), and expanded and industrially applied by many others in the 1980s, with the Programming Research Group at Oxford being the focus of much key work. The VDM language was initially developed by the IBM research laboratories in Vienna, and the University of Manchester group led by Cliff Jones became the focus of research and tool development efforts, leading to the Mural toolkit [Ritchie, 1993], and ISO standardisation [Andrews, 1993].

Both Z and VDM are "model-based" formal specification languages, that is, they use set theory and logic to build abstract models of required systems using sets, sequences, functions, and so forth. Thus a Z specification of a stack would involve selection of a particular mathematical model (such as sequences) and explicit modelling of stack operations in terms of transformations on sequences.

In contrast "algebraic" methods such as OBJ, PLUSS [Ehrig and Mahr,

1990], LARCH [Guttag, 1980] and FOOPS [Goguen *et al.*, 1978] use equational logic to provide more implicit and abstract descriptions of systems. Algebraic descriptions are more easily executable because of the restricted language used, but conversely may be somewhat algorithmic and akin to functional programs in style. Thus while an algebraic specification of a stack is highly abstract and free from implementation details, consisting exactly of the laws such as

```
top(push(x,s)) = x

pop(push(x,s)) = s
```

which characterise the stack concept, an algebraic specification of a queue is much less direct or intuitive [Jones, 1990].

Process algebras are an algebraic approach to concurrency. The main formalisms are CSP [Hoare, 1985], CCS [Milner, 1991], and the π-calculus [Milner, 1992]. The subject of each of these is an abstract concept of a "process", defined as a component which may react to external events, generate events, and carry out internal processing. Each language provides means to specify processes, combine processes, and reason about their properties (although reasoning is usually carried out in an auxiliary proof language, such as a modal logic). Substantial applications of CSP include the development of the INMOS T800 floating point unit [May, 1990], and the OCCAM language.

CCS and the π-calculus have mainly served as vehicles for extensive theoretical work. Ideas from "pure" process algebras have also contributed to hybrid languages such as LOTOS [Brinksma (ed), 1988], OBJSA [Goguen and Diaconescu, 1994] and RSL [George *et al.*, 1992], in which a process algebra component is combined with a data specification language (algebraic in the case of LOTOS and OBJSA, both algebraic and model-based in the case of RSL). LOTOS has been widely used in the telecommunications field, and RSL has been the subject of two large-scale industrialisation projects funded by the European Union.

Model-based formalisms have been used extensively in conjunction with object-oriented techniques, via languages such as Object-Z [Carrington *et al.*, 1990], VDM^{++} [Durr and van Katwijk, 1992], and methods such as Syntropy [Cook and Daniels, 1994] (which uses Z notation) and Fusion [Coleman *et al.*, 1994] (related to VDM). Whilst these formalisms are effective at modelling data structures, as sets and relations between sets, they are not ideal for capturing more sophisticated object-oriented mechanisms, such as dynamic binding and polymorphism.

In particular, the use of inheritance to specify methods in an incremental manner is difficult to reconcile with the standard Z convention that explicit equalities $\mathbf{x}' = \mathbf{x}$ must be written in operations that do not change an attribute $\mathbf{x}$. This convention can lead to the actual effect of an operation being concealed beneath extensive frame statements.

Consider the case where a method **m** is defined in a class **C**, and **C** is inherited by a class **D**. The text of method **m** refers only to the attributes of **C**, so, if read as a method of **D** under the conventional Z interpretation, it would non-deterministically assign the values of the attributes of **D** that are new in this class. For instance, if we define a class that models a dictionary (using VDM^{++} syntax):

```
class Dictionary
instance variables
  contents : set of WORD
methods
  add_word(wd : WORD) ==
        contents := contents ∪ { wd };

  ...

end Dictionary
```

and then inherit this in a class which also provides a means of assigning definitions to words:

```
class DictionaryWithDefs
  is subclass of Dictionary
instance variables
  definitions : map WORD to (set of WORD)
methods
  add_definition(wd, def : WORD) ==
        ...

end DictionaryWithDefs
```

we do not expect that the inherited **add_word** method will arbitrarily reset (or change at all) the **definitions** map.

On the other hand, a "nothing else changes" interpretation will prevent a declarative use of the method together with other methods. For instance in a Z-like language, we would expect to be able to conjoin **add_word** with other method definitions, eg:

```
  add_definition(wd, def : WORD) ==
      Dictionary‘add_word(wd) ∧
      definitions(wd) := definitions(wd) ∪ { def }
```

where **class‘feature** denotes the version of **feature** defined in **class**. Such a combination would produce a contradiction if $\mathbf{definitions}' = \mathbf{definitions}$ was an implicit consequence of **Dictionary‘add_word**

In addition, if we have a complex structure of objects, such as a sequence **dictseq** : **seq of @Dictionary** of **Dictionary** instances we expect (using the normal intuitions of object-oriented programming) that a method call **dictseq**(1)!**add_word(word)** will only affect the object **dictseq**(1) and not any other object in this sequence. We do not want to be forced to explicitly write that no other element of the sequence is modified by an operation.

To counter this problem, various approaches have been taken, such as the "rest stays unchanged" interpretation of Z schemas given by Schuman and Pitt, the Δ lists of Object-Z, and the procedural/declarative distinction in Z^{++}.

Inevitably these mechanisms add complexity to the specification language, but can provide more concise and comprehensible method descriptions than the fully explicit style of conventional Z.

Algebraic formalisms have been used less often with object-oriented techniques. The prime example of such a combination is the FOOPS language and the OOZE formalism [Alencar and Goguen, 1991], in which many-sorted, order-sorted equational logic is combined with a module structuring mechanism. Dynamic aspects of object systems, such as the set of existing objects of a class, are directly handled via the concept of a "metaclass" for each class, and methods of the metaclass. Polymorphism is treated via the hierarchy of sorts.

An example of an algebraic specification of an object class is the following specification of a buffer:

```
omod BUFFER[E :: TRIV, M :: MAX] is
 pr LIST[E]

class Buffer

at elems_ : Buffer -> List [hidden]
at empty?_ : Buffer -> Bool
at full?_ : Buffer -> Bool

me reset : Buffer -> Buffer
me get : Buffer -> Elt
me put : Buffer Elt -> Buffer
me del : Buffer -> Buffer [hidden]

var B : Buffer
var E : Elt
ax empty? B = (elems B) == nil
ax full? B = #(elems B) == max
ax elems(reset(B)) = nil
cx elems(put(B,E)) = (elems B) . E if not(full? B)
cx elems(del(B)) = tail(elems B) if not (empty? B)

 ...
```

This uses FOOPS notation. In FOOPS modules are distinguished from classes, and the effect of a method is defined on each attribute individually (here, the effect on `empty?_` and `full?_` can be derived from the effect on `elems`). The paper by Piessens and Steegmans shows how object-oriented data models can be modelled in an algebraic formalism.

Process algebras can also be used in an object-oriented manner, as was already identified in [Hoare, 1985]. More recently, the π-calculus has been used to give a semantic basis for an object-oriented extension of VDM [Jones, 1992]. Process algebras are successful in capturing some dynamic aspects of object systems, but the treatment of complex data requires either an additional specification language for this data, or a highly complex coding of data as processes, as in the π-calculus. In this book the papers by Moreira and Clark and by Clark and McCabe take the first approach, using the data specification languages of LOTOS and `April` respectively to specify object state, and the process languages to specify object dynamics. Object-oriented mechanisms such as subtyping are explicitly modelled using delegation in the `April` approach, which also allows extensions of conventional object-oriented semantics, such as transferring the **`return`** of a method to an object other than the caller.

Table 2.1 gives examples of how various object-oriented mechanisms are expressed in each of the above formalisms.

Concept	**Algebraic**	**Model Based**	**Process Algebra**
Methods	term construction	pre/post specifications	actions
Inheritance	module extension and subsorting	state and interface extension	delegation
Subtyping	theory morphisms	data refinement	simulation
Encapsulation	hidden sorts	interface constraints	internal actions

Table 2.1: Object-Oriented Concepts in Formal Methods

2.2 Formal Methods in Object Technology

Both practitioners of formal methods, and experts in object technology have investigated how formal specification can supplement object-oriented development, or how it may help to clarify the semantics of object-oriented notations and concepts.

Examples of such work include formalisation of the OMG's core object model using Z [Houston, 1994], formalisation of the OOA notation using Z, and the Fusion [Coleman *et al.*, 1994] and Syntropy [Cook and Daniels, 1994] methods. The work on OOA uncovered apparent weaknesses in its semantics, such as the lack of any relationship between the set of object identities of instances of a subtype and those of its supertypes.

The integration of formal techniques into object-oriented methods is part of a general area of work known as methods integration [Semmens *et al.*, 1992]. Industrial usage of formal methods has often been successfully achieved via such an integration: in the survey of [Austin and Parkin, 1993] it was reported that 31% of those companies using formal methods were using them in conjunction with diagrammatic semi-formal methods. This approach has definite advantages in terms of minimising disruption to existing software development practice.

Two main approaches to integration have been followed:

- use of mathematical notation to enhance diagrammatic notations, with development being carried out primarily using diagrammatic notations;
- use of a translation process from diagrammatic analysis models into formal notations, such as an object-oriented specification language. Development is then carried out mainly using the formal notations.

The first is more suitable for a "rigorous" rather than fully formal development, as the diagrammatic notations do not support proof or verification. The second is suitable when extremely high-integrity systems are being developed.

Examples of the first approach include Fusion and Syntropy, described below. The paper by Lano et al in this volume is an example of the second approach, using B AMN as the formal language, and the paper by Moreira and Clark is also an example of this approach, using LOTOS as the formal language.

Fusion grew out of work by Coleman, Dollin and others at Hewlett Packard Labs on formal enhancement of object-oriented notations such as statecharts [Coleman *et al.*, 1992]. It provides semi-formal notations (structured English) for operation pre-conditions, post-conditions and invariants, and uses the OMT object-model notation and Booch [Booch, 1991] object interaction diagram notation, together with other notations, to provide a rigorous method for sequential systems. It therefore makes use of the basic concepts of formal specification (abstraction, refinement by model exclusion, operation specification using pre and post conditions), without adopting mathematical notation.

It is now used quite extensively within Hewlett Packard, across 15–20 divisions and in areas such as printer technology, network management software and test software. Other companies have taken up Fusion in the USA and Europe. The work on Fusion identified that the main practical problems for software developers using object technology concerned the complexity of dynamic object interactions. To tackle these problems, Fusion provides techniques for specifying required operations at a system level, without forcing the developer to prematurely locate operations as methods in classes. The major omission from Fusion is concurrent specification mechanisms; this is an area of current research.

Syntropy is a more recent method in the same direction. It combines the statechart and object model notations of OMT [Rumbaugh *et al.*, 1991] with object interaction diagrams, and allows the use of Z notations on these diagrams to specify pre and post conditions, invariants and constraints. An important feature of Syntropy is its emphasis on *events* for the modelling of system requirements, rather than message passing. An event may involve a number of different objects (ie, it may trigger a transition on the statecharts of several classes or objects). This (undirected) synchronisation between objects may later be implemented by a directed message passing (method call). It is analogous to synchronisation via actions in the Object Calculus, described in the paper by Fiadeiro and Maibaum. A partial formal semantics for the notation is also provided. Unlike Fusion, a treatment of concurrency is given. Syntropy grew out of many years of industrial experience in consultancy, and has already been commissioned for use in financially critical applications in the UK.

Both Fusion and Syntropy are part of a trend for greater abstraction in object specification, away from the concepts of methods and message passing. Instead, both approaches use concepts (system operations in Fusion, events in Syntropy) at the analysis stage which permit a flexible specification of the essential elements of a system, without premature commitment to a particular architecture of objects, or of method definitions within particular classes.

The Venus toolkit for OMT and VDM^{++} provides support for the combined use of formal and diagrammatic object-oriented techniques, in a highly integrated manner [LeBlanc, 1995]. It is designed as an extension of a widely-used case tool for OMT, so enabling existing users of object-oriented techniques to increase the formality of their developments in an incremental manner. It is possible to produce an OMT specification of the set of classes, attributes and associations in a system, generate outline VDM^{++} classes from this specification, enhance these classes, and then back-translate the enhanced specification into an OMT view. The advantage of this approach is that the most appropriate editing and analysis tools may be used on the different views (diagrammatic versus formal) of a system. In particular, it is usually more effective to create and modify a class hierarchy using a diagram editor than via a textually-based tool.

Translation from diagrammatic notations to formal notations such as that performed by Venus provide a particular interpretation and semantics for the diagrammatic notations. Thus the somewhat ill-defined concept of "aggregation" in object-orientation can be given precise meanings via translation to a formal language such as LOTOS (in the paper of Moreira and Clark), VDM^{++} or Z^{++}.

In VDM^{++} the interpretation is via the concept of "indexed inheritance". If we have an aggregate of the form of Figure 2.1, then class **Engine** would

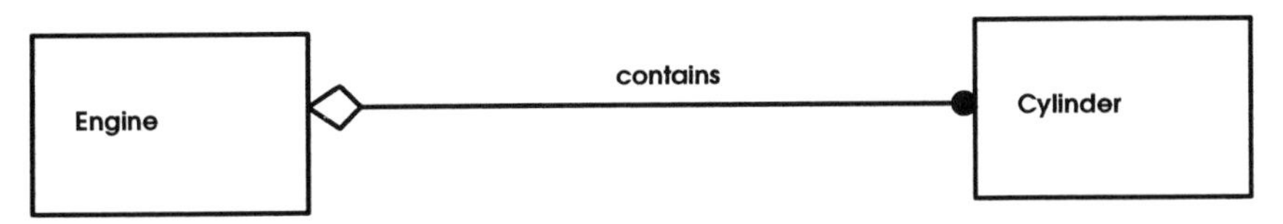

Figure 2.1: Engine as an aggregate of Cylinders

have a declaration:

```
class Engine
  is subclass of Cylinder[1,...,n]
```

for a suitable (fixed) value of **n**. This means that there are **n** distinct copies of the features of **Cylinder** in **Engine**. Features **feat** of the **i**-th cylinder are referred to as **Cylinder[i]'feat** in **Engine**.

As a result, components of an aggregation have a lifetime which coincides with the aggregation, and they cannot be shared.

In contrast, in Z^{++}, components of an aggregation are interpreted as constant object-valued attributes. The interpretation here would be:

CLASS **Engine**
FUNCTIONS
| **contains** : $\mathbb{F}(\overline{\textbf{Cylinder}})$

(using Z axiomatic definition notation). **contains** is a constant whose value is a finite set of (existing) object references. Thus the lifetime of the aggregation is contained in that of each of its components, but these components can, in principle, be shared, and continue to exist after their aggregate ceases to exist.

Their participation in the aggregate is however coincident with its lifetime. If it were additionally required that components do not outlive their aggregate, then shared components would force the destruction of their containers to be synchronised, which seems counter-intuitive.

Similarly, ambiguities of the statechart notation can be resolved by a formal interpretation, such as that given by Barroca in RTL [Barroca *et al.*, 1995]. For instance, the Booch [Booch and Bryan, 1994] and OMT methods take different interpretations regarding whether events are queued or ignored if they arrive at a time when the target object is not in a state in which they can be accepted.

2.3 Object Technology in Formal Methods

The limitations of Z and VDM for specifying large systems in a modular manner led to various investigations aimed at adding structuring mechanisms to these languages. For example, [Sampaio and Meria, 1990] proposed adding a "chapter" mechanism to decompose Z specifications into modules, and [Iachini and Giovanni, 1990] identified a similar (object-based but not object-oriented) approach using a combination of HOOD and Z. Abrial and others at BP Research also developed an object-based extension of Z, the B Abstract Machine Notation [Haughton and Lano, 1995], which has had major applications in safety-critical transport systems [Bowen and Stavridou, 1993]. The SmallVDM language [Lano and Haughton, 1993, Chapter 10] combined an object-based structuring style derived from Smalltalk with VDM notation. Other work providing minimal extensions of Z which allowed an object-oriented style of specification included [Hall, 1994], and that of Schuman and Pitt.

Researchers and practitioners were led in the direction of object-oriented mechanisms because these seemed to naturally complement model-based specification languages [Cusack, 1991, Lano and Haughton, 1993]:

- Object-orientation encourages the creation of abstractions, and formal techniques provide the means to precisely describe such abstractions;
- Object-orientation provides the structuring mechanisms and development disciplines needed to scale up formal techniques to large systems;
- Formal techniques allow a precise meaning to be given to complex object-oriented mechanisms such as aggregation or statecharts.

The first fully object-oriented extension of Z was the Object-Z language [Carrington *et al.*, 1990]. This featured the key aspects of object-oriented structuring:

- encapsulation of data and operations on that data into named modules which also define types;
- the possibility of creating subclasses of classes via inheritance, and the ability to use operations polymorphically between a subclass and its superclasses which contain a particular operation;
- the ability to use instances of a class within another class – the concept of class *composition*.

It however initially restricted composition to be acyclic (class **A** could not contain an instance of class **B** if **B** contained an instance of **A**, or other cyclic situations). Object-orientation was found to provide advantages in terms of ease of education [Swatman, 1993] and in terms of convergence with domain concepts.

Other object-oriented extensions of Z were also developed in 1989–91, such as Z^{++} [Lano, 1991], MooZ [Meira and Cavalcanti, 1991], OOZE [Alencar and Goguen, 1991] and ZEST [Cusack, 1991]. Of these, OOZE is distinctive as the only such language to be based on an algebraic specification approach, using OBJ and FOOPS as its foundation but with a Z-like syntax. In the VDM world, the Fresco language was developed as a means of providing formal specification and verification support for Smalltalk development [Wills, 1991].

A number of applications of these languages were carried out, particularly in the telecommunications and process control fields. The work of BT using ZEST is of particular note in this respect, and was sufficiently successful that ZEST is now being used in preference to more traditional telecommunications specification languages such as LOTOS and SDL [Moller-Pedersen *et al.*, 1987]. The scope and power of the languages were extended as more challenging application areas were investigated. Thus concepts of object identity and cyclic composition structures were introduced in Z^{++} and Object-Z [Lano, 1994]. The VDM^{++} language represents a significantly more ambitious approach, based on ideas from VDM, Smalltalk and the DRAGOON Ada extension [Atkinson *et al.*, 1991], it includes Ada-style concurrency and synchronisation features, traces (as in CSP), and real-time aspects taken from the ideas of Hayes [Hayes and Mahony, 1992]. As with Fusion and ZEST, the involvement of a major company (in this case CAP Gemini) has provided direct routes to significant industrial application [Durr and Dusink, 1993].

Real-time and concurrency aspects were incorporated into Z^{++} in a more declarative manner via the use of real-time logic [Lano, 1995]. Investigation of similar extensions for Object-Z are continuing. Outstanding issues in the treatment of real-time and concurrency are the transformation from continuous to discrete models of a system (covered in the paper by Durr and Goldsack) and the definition of refinement and subtyping in the presence of concurrency and real-time behaviour.

Z is not ideal as a basis for an object-oriented specification language because of its semantic complexity, particularly in the schema calculus. Attempts to merge the schema calculus and method calling have led to excessive syntactic complexity, as in Z^{++} and Object-Z. Both languages have substantially changed the syntax and interpretation of Z schemas in order to achieve an adequate formalism.

In some respects the problems of combining object-orientation and formal specification are due to the programming language origin of many of the concepts of object-orientation. Method invocation in particular is interpreted in a procedural manner, and the task of finding a suitable abstraction of method calls, or of using them in a declarative manner (as predicates on the supplier pre and post states) is still not entirely solved. Because VDM contains a procedural sublanguage, it is simpler to provide a combination of formal specification and object-orientation at this level. The modularity mechanisms of VDM are even more primitive than those of Z however, and have been entirely replaced by object-oriented mechanisms in VDM^{++}. The use of non-classical logic may provide a more natural treatment of "non-denoting object references" than the classical logic of Z-based formalisms, although it may also make learning the languages more difficult.

A significant drawback with the use of object-oriented structuring is the degree to which this complicates reasoning about specifications:

- polymorphism and dynamic binding, because these make it difficult to control which version of a method is actually to be executed at a particular invocation [Ponder and Bush, 1992];

- inheritance, because this can be used to fragment the definition of a method across many classes, making its meaning difficult to determine [Wilde and Huitt, 1991].

In addition, some object-oriented languages encourage the extensive use of aliasing and interconnected networks of objects [Hogg, 1991].

Until recently, there were no formal semantics for an object-oriented specification language. This has now been partly remedied, with an axiomatic semantics and reasoning system being provided for Object-Z [Smith, 1995] and Z^{++} [Lano, 1995], and a denotational semantics for MooZ [Lin, 1994].

2.4 Formal Underpinnings

Several areas of mathematics are being used to provide formal underpinnings to object technology: logic, type theory, category theory [Goguen, 1991] and process algebra [Hennessy, 1988]. Logic provides a way of describing and reasoning about the behaviour of objects and classes at a level which is reasonably transparent and understandable. Category theory provides a general mathematical framework in which various kinds of structure may naturally be described. By developing structures of logical theories within this framework it has proved possible to provide a concise characterisation of various object and class structures used in object technology.

Process algebra provides an alternative form of characterisation, by regarding objects and classes as processes and using the algebra to combine these in various ways. The π-calculus is particularly relevant for expressing the semantics of object-oriented systems, because its use of name passing between processes is analogous to dynamic reconfiguration in object systems.

The logic approach is exemplified in the paper of Fiadeiro and Maibaum. First order temporal logic specifications define the properties of objects, including the effects, permission constraints and liveness requirements of methods. Methods are interpreted as action symbols in the logic, whilst attributes are interpreted as attribute symbols (which are distinct from variables in the logical sense, as they cannot be quantified over, and have time-dependent values).

Category theory, via the concept of a co-limit (a generalisation of the concept of disjoint union in set theory) is used to construct systems from interconnected objects or subsystems. Connections between object theories are logical interpretations of the source theory in the target: that is, syntactic mappings σ of the symbols of the source theory $\mathbf{S}$ to the symbols of the target theory $\mathbf{T}$ such that if $\Gamma \vdash \varphi$ is a valid inference in $\mathbf{S}$, then $\sigma(\Gamma) \vdash \sigma(\varphi)$ is a valid inference in $\mathbf{T}$. Attribute symbols are mapped to attribute symbols, action symbols to action symbols, etc. Such connections allow synchronisation between components to be expressed, and an abstraction of method calling (action $\mathbf{m}$ "calls" action $\mathbf{n}$ if $\mathbf{m}$ being active implies that $\mathbf{n}$ is active – more complex relationships can also be expressed).

The paper by Malcolm illustrates a purely category-theoretic approach, whereby the concept of a *monoid* is used to express method application and history traces of processes, and diagrams in suitable categories express object structuring. Objects are objects of a category, and the object resulting from a diagram of objects is given by a limit or co-limit of this diagram. Dynamic modification of object structures are not covered by category-theory interpretations, or by the temporal logic approach of Fiadeiro and Maibaum.

Set-theoretic models of object-orientation, as given in the paper by Maung, do provide a means to discuss object identity and issues such as *subtype migration* [Wieringa *et al.*, 1993], whereby an object may be considered to be an element of different subtypes of a class at different times. Classes are modelled as sets of object identities. A clear definition of subtyping and refinement does not necessarily emerge from such models however, because of problems involving recursively defined classes.

Type theory has been extensively studied as a basis for the semantics of object-orientation: the papers by Sekerinski and by Drossopoulou and Karathanos give examples of the type-theoretic approach. See also [Cook and Palsberg, 1989, Hense, 1993]. This approach interprets classes as types (such as polymorphic and recursive types, or as existentially quantified types), and does provide definite answers to issues of subtyping, although these may be counter to standard intuitions. Work in this area has included the object calculus of [Abadi and Cardelli, 1995], in which, by analogy with the use of the

λ-calculus to define the semantics of functional languages, the semantics of object-orientation is expressed in a calculus based on the concepts of method, object and type as primitive. This formalism can be used to give a meaning to extensions of the object-oriented paradigm, such as "methods as objects".

A logical approach, whereby classes are interpreted as theories, provides an apparently natural definition of subtyping and refinement as theory extension. Thus class **D** is a subclass of class **C** if it satisfies all the properties (requirements) of **C**, via some syntactic translation. This definition satisfies the informal subtyping principles of [Liskov, 1988]. However for practical purposes, theory extension cannot be used on its own, instead, a set of more easily generated and managed proof obligations are usually used, which are sufficient but not necessary for the subtyping or refinement to hold. These obligations are decomposed on the basis of class features, and are similar to those of [Liskov and Wing, 1993].

Distinctions between subtyping and refinement have only recently been clarified. In some modelling approaches, these concepts have been combined, whilst in others, such as in the LOTOS process algebra, it is often not clear which definition of "process simulation" should be used as the basis for verification [Thomas and Kirkwood, 1995]. Similarly, determining the correct concept of bisimulation for the π-calculus is still an active area of research [Lin, 1995].

Refinement can be seen as a form of subtyping (ie, theory extension), but with additional constraints. In particular, it is not permitted to introduce new external methods in the refinement class, although new internal methods can be defined. The concept of *adequacy* from VDM (ie, that for every abstract state there is a corresponding concrete state under the interpretation of class data and attributes [Jones, 1990]) is also required for refinement but not for subtyping.

2.5 Concurrency

It is widely appreciated that objects, because of their independence and the isolation of the internal state from interference by other objects, are also appropriate units of concurrent execution. Examples of attempts to integrate concepts of concurrency and object-orientation are the POOL notation [America, 1987], DRAGOON [Atkinson, 1991] and VDM^{++}.

The match between objects and processes is not exact however. In particular, objects may be internally concurrent, if they represent a system (such as a radar track former) which must respond more rapidly to input data than can be achieved via sequential execution. In addition, several objects may be executing on the same processor, and thus can be regarded as part of the same process. In a specification language, object classes represent both conceptual and design entities, whilst processes are more a feature of the implementation of a system.

2.6 Conclusions

The combination of formal techniques and object-orientation has achieved significant advances in both fields, and has led to a greater dissemination of work in formal methods than would otherwise have occurred. For example, presentations on the object calculus have taken place at OOPSLA, and the TOOLS and ECOOP conferences frequently feature tutorials and presentations involving formal methods.

Future challenges for formal techniques in object technology include a formal treatment of *patterns* (ie., design idioms, often involving characteristic interaction between particular forms of classes, as with the MVC triad of Smalltalk [Goldberg and Robson, 1983]) and *frameworks* (domain-specified architectures for object systems in that domain). In addition, tool support for formal object-oriented methods is an area of considerable significance for further industrial take-up of these methods.

Part II

Formal Methods in Object Technology

Chapter 3

LOTOS in the Object-Oriented Analysis Process

3.1 Introduction

Object-oriented analysis (OOA) and design methods are used by the software engineering community, while formal description techniques (FDTs) are mainly used in a research environment. Our aim is to produce a rigorous object-oriented analysis method which combines OOA with FDTs to produce a practical method to be used by software engineers. It is desirable for such a method to use standard approaches. As object-oriented FDTs are still under development, and have not yet been standardized, our solution is to choose a standard FDT which can model object-oriented concepts.

We have been investigating how LOTOS [Brinksma (ed), 1988] can be integrated with object-oriented analysis. LOTOS incorporates a process algebra and abstract data types, but, as it was designed before the object-oriented approach became widely accepted, it does not directly incorporate object-oriented constructs. However, the language is suitable for writing specifications in an object-oriented style [Rudkin, 1992]. It directly supports encapsulation, abstraction and information hiding.

We have developed the Rigorous Object-Oriented Analysis (ROOA) method which is concerned with how an initial formal requirements specification can be systematically created from a set of informal requirements [Moreira and Clark, 1993, Moreira and Clark, 1994]. The resulting specification can then be the starting point of a formal development trajectory. The first task in ROOA is the production of an object model using one of the standard OOA methods, such as [Coad and Yourdon, 1991, Jacobson, 1992, Rumbaugh *et al.*, 1991]. The second task is to refine the object model to ensure that it incorporates interface objects, attributes, services, static relationships and message connections, and to identify subsystems. The third task is to build the formal model, i.e. a LOTOS requirements specification.

ROOA uses a stepwise refinement approach for the development and for validation of the specification against the requirements. The development process is iterative, allowing parts of the method to be re-applied to subsystems. Different objects can be represented at different levels of abstraction and the model can be refined incrementally.

Section 3.2 discusses the reasons for choosing LOTOS. Section 3.3 presents the ROOA method by means of an example. An important part of our method is to give a formal interpretation in LOTOS of object-oriented analysis constructs. As an example, Section 3.4 shows how aggregation and behavioural inheritance can be modelled.

This chapter assumes that the reader is already familiar with LOTOS. For those who are not, a tutorial introduction is given in [Bolognesi and Brinksma, 1987].

3.2 Reasons for Choosing LOTOS

The reasons for using formal methods are well known and need not to be repeated here. However, recent surveys have shown that there is considerable resistance to their adoption in industry [Austin and Parkin, 1993a, Craigen *et al.*, 1993]. We believe that integrating FDTs with existing OOA methods will make them more acceptable and we have been investigating how this can be done.

As our goal is to produce an object-oriented analysis method to be used by software engineers, the chosen FDT has to satisfy the following conditions:

- It is an ISO Standard.
- It is able to produce specifications in an object-oriented style.
- Support tools are available.
- The specifications are executable so that prototyping can be used.
- Concurrency is supported.

LOTOS satisfies these conditions.

Prototyping is an important tool in validating a specification against a set of requirements and enables us to find and correct omissions, contradictions and ambiguities early in the development process. The need for concurrency is because requirements analysis is concerned with modelling the real world. As entities in the real world exist in parallel, an analysis model which represents the requirements as a set of communicating concurrent objects would seem ideal. In LOTOS, concurrent objects are modelled as process instances, composed by using the parallel operators, and message passing is modelled as two processes synchronizing on an event. Inheritance is more difficult, but can be modelled [Rudkin, 1992].

An extensive set of tools, such as syntax checkers, semantic checkers and simulators, is available with LOTOS. The SMILE simulator [Eertink and Wolz, 1993] supports nondeterminism and value generation. This allows symbolic execution of a specification where a set of possible values is used rather than particular values. Many more behaviours can then be examined with each simulation than is possible when all data values have to be instantiated. SMILE uses a narrowing algorithm to determine when a combination of conditions can never be true. As we are in the analysis phase, nondeterminism can be used to model behaviour so that premature design decisions are not made.

3.3 The ROOA Method

ROOA takes a set of informal requirements and an object model and produces a formal object-oriented analysis model, i.e. a model which is expressed in a language with a formal semantics. This model acts as a requirements specification and integrates the static, dynamic and functional properties of a system in contrast to existing OOA methods which are informal and produce three separate models that are difficult to integrate and keep consistent. The ROOA model is primarily a dynamic model, but it keeps the structure of the object model.

The ROOA method provides a systematic development process, by proposing a set of rules to be followed during the analysis phase. During the application of these rules, auxiliary structures are created to help in tracing the requirements through to the final formal model.

In the following sections we present the ROOA method, by means of a running example. The example we have chosen is a version of the warehouse management problem presented by Jacobson [Jacobson, 1992]. A brief outline of the problem is:

> A company has a set of warehouses, distributed throughout the country. Each warehouse has a set of warehouse places where items may be stored. Clients may deliver items to one warehouse and, at a later date, collect them from the same or another warehouse. When clients wish to enter or remove an item, they contact the office which sends requests to the system to ensure that the item is expected or is available for collection at the correct place and time. Items may be moved from one warehouse to another if a warehouse is becoming full.
>
> After receiving requests, a computer planning system controls when the requests should be carried out by sending appropriate orders to truck drivers (for inter-warehouse movement) or to forklift drivers (to move items between the loading bay and a warehouse place). A local planning system within each warehouse allocates items to warehouse places. The system should be as decentralized as possible.

Task 1: Build an object model

The construction of the object model is performed by applying any of the existing object-oriented analysis methods, such as [Coad and Yourdon, 1991, Jacobson, 1992, Rumbaugh *et al.*, 1991].

Task 2: Refine the object model

OOA methods differ in the information they hold in the object model. Here, we add, when necessary, interface objects, static relationships, attributes, services, message connections and define interface scenarios.

An interface object models behaviour and information that is part of the interface with the system's environment. An interface scenario shows a series of services (requests and responses) that are required from the system. Its effect within the system is described by an Event Trace Diagram (ETD). An ETD showing the initiation of the redistribution of an item from one warehouse to another is given in Figure 3.1.

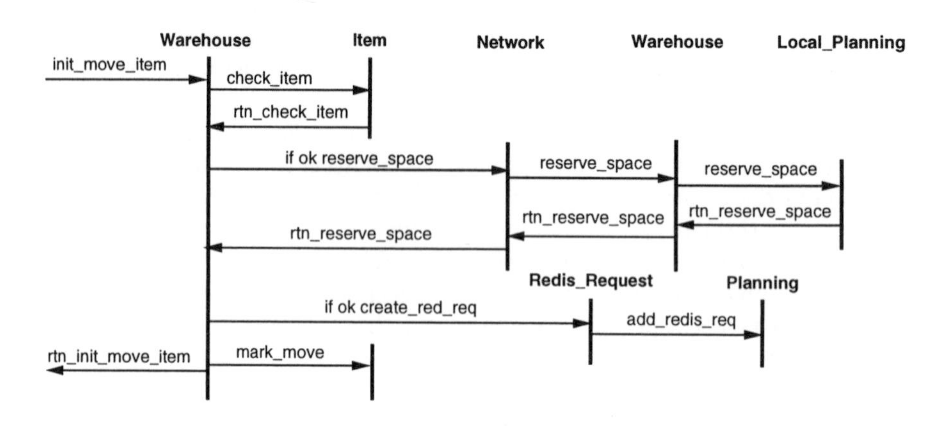

Figure 3.1: Event trace diagram for `Initiate_Redistribution`

Message connections and services are identified from the ETDs and the information is collected in the Object Communication Table (OCT). This table consists of five columns, but at this stage we only build the first four. Column one lists the class templates that form the object model; column two lists the services offered by the objects of each class template; column three lists, for each service offered in column two, the services that it requires; and column four lists, for each offered service, the clients which require that service. We use the notation <*Class_template.service*> to indicate that the *service* defined in *Class_template* is required. Part of an OCT is shown in Table 3.1. (In the table, `Warehouse` is abbreviated to `Whs`.)

Finally, we also identify obvious groupings of class templates and mark them as subjects in the object model and as composite objects in the OCT. An example is the aggregate `Warehouse` which has the components `Place` and

Class Template	Service Offered	Services Required	Clients	Gate
Whs [Whs + Place + Local_Plan]	init_move_item ...	Item.check_item Network.reserve_space Request.create_red_req Item.mark_move	Interface_ Scenario	usr1
Request [Ins_R + Red_R + Rem_R]	create_ins_req		Office	req2
	create_red_req	Planning.add_redis_req	Office, Whs	req2
	create_rem_req	Planning.add_rem_req	Office	req2
	check_expected		Whs	req2
	choose_rem	Whs.rm_fetch_items	Planning	p1
	choose_red	Truck.request_eta ...	Planning	p1

Table 3.1: Part of the OCT

`Local_Plan`. The final object model produced by applying the Coad and Yourdon method [Coad and Yourdon, 1991] is depicted in Figure 3.2. This model was created using ObjecTool[1].

Task 3: Build the LOTOS formal model

Building the formal model involves specifying the object model in LOTOS and adding a specification of:

- the dynamic behaviour of each class template,
- the dynamic behaviour of the overall system,
- the local encapsulated state of each class template,
- the information passed when objects communicate.

Formalizing the object model is relatively straightforward, but adding the dynamic and functional properties is more difficult. Much of the required information only exists in the informal requirements, although some of it is available in the ETDs and the OCT.

We start by building an object communication diagram which shows the system as a set of communicating objects. We then model each class template in the object model in LOTOS, add its behaviour, compose the objects (instances of the class templates) using parallel operators, prototype the specification and, finally, refine the specification.

[1] ObjecTool is a trademark of Object International, Inc.

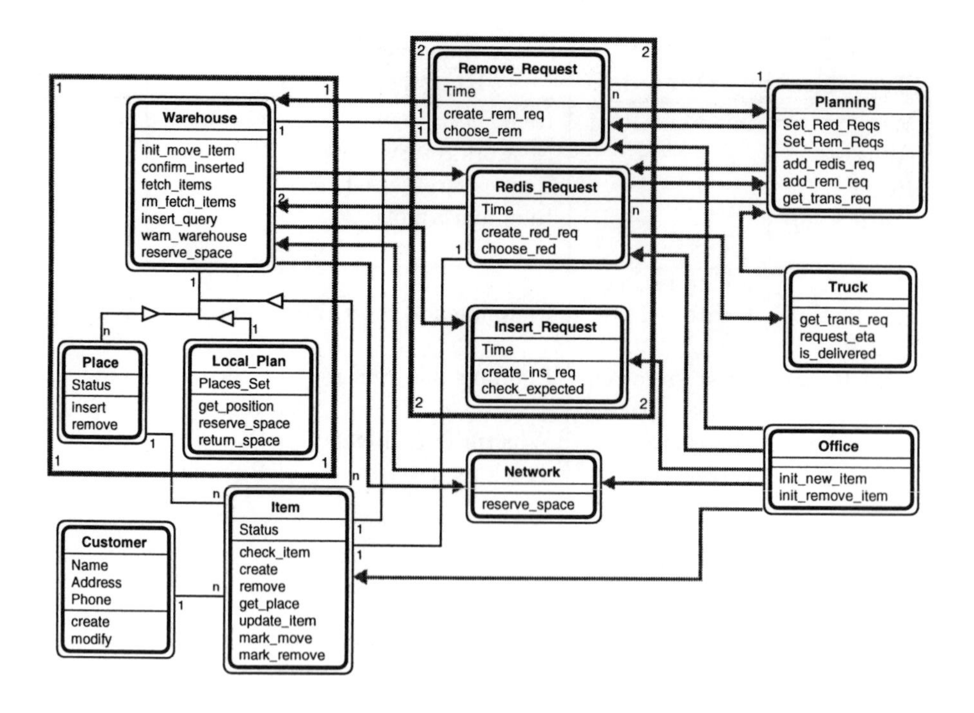

Figure 3.2: Final object model

Task 3.1: Create an object communication diagram

An Object Communication Diagram (OCD) is a graph where, in the first iteration, each node represents an object and each arc connecting two objects represents a gate of communication between them. In later iterations, the diagram is generalized to deal with multiple instances of the same class template. In the beginning, some of the objects may not be connected by arcs to the rest of the diagram. As the method is applied, these objects will either disappear from the diagram or be connected to the others. New groupings may also appear, refining the diagram. The OCD is an intermediate model between the object model and the LOTOS specification and it reflects the exact structure of the LOTOS formal model.

We apply rules [Moreira and Clark, 1993] to allocate LOTOS gates to message connections. This information is entered in column five of the OCT and is used to label the arcs of the OCD. An OCD for the warehouse management problem is shown in Figure 3.3.

Task 3.2: Specify class templates

Each object, represented as a node in the OCD, is specified by a LOTOS process definition and its state information by one or more ADTs given as parameters of the process. Message passing is represented by two processes synchronizing

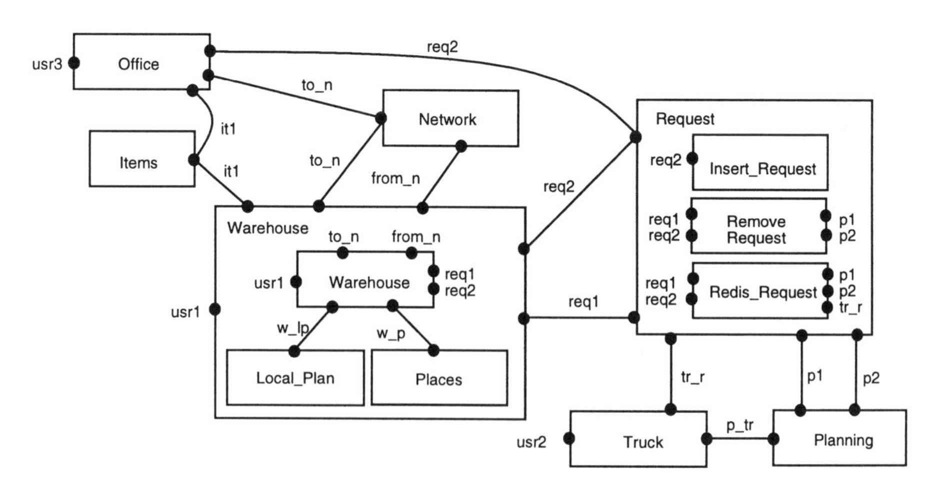

Figure 3.3: Object communication diagram

on a LOTOS event which has the structure:
<gate name> <message name> <object identifier> <optional parameters>
As an example, the class template `Item` can be defined by the following LOTOS process:

```
process Item[it1](it: Item_State) : noexit :=
    it1 !check_item !Get_It_Id(it) !Can_Move(it) !Get_Ware(it);
    Item[it1](it)
  []
    it1 !update_item !Get_It_Id(it) ?w: Warehouse_Id ?ip: Item_Place_Id;
    Item[it1](Move(w, ip, it))
  []
    ...
endproc (* Item *)
```

The process `Item` is defined recursively and uses gate `it1` for synchronization with other processes. The operator `[]` is the nondeterministic choice operator. Each service offered by the class template in the object model is, in this case, an alternative of the choice expression. The `Item`'s state information is defined by the ADT where the `Item_State` sort is defined.

Task 3.3: Compose the objects into a behaviour expression

After specifying each class template in the object model by a process definition, we follow the structure of the OCD and compose process instances into a behaviour expression, using the LOTOS parallel operators.

Task 3.4: Prototype the specification

We use interface scenarios to help prototype the specification and ensure that the execution path matches the corresponding EDTs. If we find any errors, omissions or ambiguities we must update the object model, the ETDs, the OCT, the OCD and the specification.

Below, we show part of the path generated by SMILE when the specification is executed with an interface scenario which expects successful execution of the ETD given in Figure 3.1.

```
I(usr1) !init_move_item !id2 !Id_22 !This_Time ?dest_41:Warehouse_Id
  (dest_41 ne id2)
I(it1) !check_item !Id_22 !true !id2
I(to_n) !reserve_space !id6 !dest_41
  not((Succ(0) eq h(dest_41)))
I(from_n) !reserve_space !dest_41 !id6
  id1 = dest_41
I(w_lp) !reserve_space !id14 ?ok_7:Bool
I(from_n) !rtn_reserve_space !id1 !id6 !ok_7
  dest_41 = id1
I(to_n) !rtn_reserve_space !id6 !dest_41 !ok_7
I(req2) !create_red_req ?Id_1:Red_Req_Id !Id_22 !id2 !dest_41 !This_Time
  ok_7
I(p2) !add_redis_req !id1 !Id_1
I(it1) !mark_move !Id_22
I(usr1) !rtn_init_move_item !id2 !true
```

The first synchronization occurs when `Interface_Scenario` initiates the sequence by offering the event:

```
usr1 !init_move_item !id2 of Warehouse_Id !it_id !This_Time
     ?w2: Warehouse_Id [w2 ne id2];
```

which synchronizes with:

```
usr1 !init_move_item !wid ?it: Item_Id ?t:Time ?dest: Warehouse_Id;
```

offered by `Warehouse`. Value generation is used to produce the identity of the destination warehouse, with the only externally imposed condition being that `w2` is not equal to `id2`. SMILE allocates the identifier `dest_41` and determines the restrictions on the values that it may take. The identifier `Id_22` was allocated in a similar way during an earlier synchronization. The other synchronizations are generated in the same way.

Task 3.5: Refine the specification

To refine the specification we may re-apply the whole or part of the process. During successive refinements we model static relationships, introduce object generators, identify new higher level objects, demote an object to be specified only as an ADT, promote an object from an ADT to a process and an ADT, and refine processes and ADTs by introducing more detail.

Object generators are needed when multiple instances of a class template are required. For example, as we need multiple instances of `Item`, we define the object generator `Items`:

```
process Items[it1](ids: Item_Id_Set) : noexit :=
    it1 !create ?id: Item_Id ?w: Warehouse_Id ?ip: Item_Place_Id
                 [(id notin ids)];
    ( Item[it1](Make_Item(id, w, ip, true, true))
      |||
      Items[it1](Insert(id, ids))
    )
  []
    it1 !check_item ?id: Item_Id !false ?w: Warehouse_Id
                 [(id notin ids)];
    Items[it1](ids)
where
  process Item ... endproc (* Item *)
endproc (* Items *)
```

Sending a `create` message to `Items` causes a new `Item` to be created. `Items` holds the set of item identifiers already allocated and the selection predicate `[(id notin ids)]` guarantees that each `id` generated is different from the existing ones. The event:

```
it1 !check_item ?id: Item_Id !false ?w: Warehouse_Id [(id notin ids)];
```

catches the situations where an attempt is made to synchronize with a non-existent item.

3.4 Structuring Specifications

Developing software systems requires the use of abstraction, encapsulation and information hiding. This is achieved in object-oriented methods by using the concepts of object and class while aggregation and inheritance give us mechanisms for structuring large systems. When two, or more, class templates have identical properties, we can abstract from those properties, form a superclass and then let the original class templates inherit from the superclass template. Inheritance can also be used to extend the properties already defined in a class template. Aggregation is used to form complex objects from the combination of simpler objects.

3.4.1 Aggregation

Aggregation is a relationship between several objects which allows a more complex object, the *aggregate*, to be built from simpler objects, the *components*. It is important to distinguish between an aggregate and a grouping of unrelated objects. An aggregate must be a representation of an entity from the real world. Aggregates and their components are objects and so we can talk about aggregate classes and component classes.

We can identify two different situations: aggregation with hidden components and aggregation with shared components. In either case, the number of components can be static or dynamic.

Aggregation with hiding means that the object components are not visible to other objects in the system. In LOTOS, we model this by defining a process for the aggregate class which embeds a process for each component together with an interface process. The aggregate and the interface processes are given the same name. The interface object acts as the manager of the aggregate, controlling the communication between the aggregate and the rest of the model. The components may only interact with each other or with the interface. All interactions of components with the rest of the model must be routed through the interface object. An aggregate object can have attributes which describe it as a whole, for example the number of components, and attributes which describe its individual components.

The components only exist while the aggregate exists. If the number of instances of a component is dynamic, an object generator is required. Because the components are hidden, the object generator is embedded in the process that defines the aggregate as there is no problem if components within different aggregates have the same object identifiers.

In the warehouse management problem, `Place` and `Local_Plan` are hidden components of the aggregate `Warehouse`, while `Item` is a shared component which is also accessed by `Office` (see Figure 3.2). Following the OCD in Figure 3.3, the situation for hidden components is modelled as:

```
process Warehouse[usr1, req1, req2, it1, to_n, from_n]
                    (wid: Warehouse_id) : noexit :=
  hide wp, w_lp in
    Warehouse[usr1, req1, req2, it1, wp, w_lp, to_n, from_n](wid)
    |[wp, w_lp]|
    ( Places[wp]
      |||
      Local_Plan[w_lp](id14 of Local_Plan_Id, Insert(id7,
               Insert(id8, Insert(id9, {} of Item_Place_Id_Set))))
    )
where
  process Warehouse ...  endproc (* Warehouse *)

  process Places[wp] : noexit :=
    wp !insert ?ip: Item_Place_Id ?it: Item_Id;
    ( Place[wp](Make_Place(ip, it))
      |||
      Places[wp]
    )
  where
    process Place ... endproc (* Place *)
  endproc (* Places *)

  process Local_Plan ... endproc (* Local_Plan *)
endproc (* Warehouse *)
```

We have a dynamic number of places, and so an object generator `Places` is required, but we only have a single local plan. `Local_Plan` is initialized with the identifiers of the places available in the warehouse.

Aggregation with hiding is useful in providing information hiding and supports a top-down development of the system. Many authors [de Champeaux *et al.*, 1993, Rumbaugh *et al.*, 1991, Smith and Smith, 1977] consider aggregation with hiding the only interesting situation.

Aggregation with sharing means that the object components can be accessed by other objects in the system. Rumbaugh [Rumbaugh *et al.*, 1991] argues that this should be treated as an ordinary conceptual relationship. However, we believe that there are advantages in showing such a relationship as an aggregation as it can improve our understanding of a system and may give directions for reusability. While a conceptual relationship is more likely to change when it is put in a different context, an aggregation may not change. Because of this, we propose modelling aggregation within the object state, instead of modelling it as an extra argument in the process template, as we do with conceptual relationships.

Although a shared component, such as `Item`, can have a relationship with a hidden component, such as `Place`, it cannot use this information to communicate directly with the hidden component. Access must be through the interface object. The deletion of a shared component may only occur when all objects that have access to it have been removed.

3.4.2 Behavioural Inheritance

Inheritance is used for two separate purposes: for the inheritance of behaviour and for the inheritance of implementation. The inheritance of behaviour is concerned with the subtype-supertype relationship and with *inclusion polymorphism*, that is with the possibility of using an object of a subtype anywhere that an object of the supertype is expected. *Incremental* or *implementation inheritance* is concerned with the creation of a *derived class template* by the modification of a *base class template*, that is with reusability and code sharing.

In many object-oriented languages, subtyping is implemented by means of a restricted form of implementation inheritance. The restrictions are given in terms of the *signatures* of the offered services and should ensure that the *contravariance rule* [Cardelli, 1984] is respected. These restrictions mean that the use of inclusion polymorphism does not introduce run-time errors, but they do not necessarily provide behavioural inheritance. We want an object of a subtype to have all the properties of an object of the supertype and for it to be usable as an alternative for the supertype object. We also require that behavioural inheritance supports our ability to reason about the effect of complex software [Bar-David, 1992]. If we know about the effect of a superclass, i.e. its specification, then we should be able to *inherit* that knowledge when we are reasoning about the effect of one of its subclasses. We do not get this when subtyping is only defined in terms of signatures.

A definition of the subtype-supertype relationship that satisfies behavioural inheritance has been given by Liskov, [Liskov, 1988]:

> S is a subtype of T if, for each object o_1 of type S, there is an object o_2 of type T such that, for all programs P defined in terms of T, the behaviour of P is unchanged when o_1 is substituted for o_2.

Although behavioural inheritance is difficult to enforce in a programming language, it can be enforced in a formal specification language such as LOTOS.

Suppose that we have a class `Sub` which inherits from, and modifies, a superclass `Super`. The properties of class `Super` can be defined in a LOTOS process `PSuper` as described in [Moreira and Clark, 1993]:

```
process PSuper[g](state: State_Sort, id: Id_Sort)
                         : exit(State_Sort, Id_Sort) :=
    g !selector_1 !id ... ; ... exit(state, id)
  []
    g !modifier_1 !id !val1 ... ; ... exit(F1(state), id)
  []
    ...
endproc (* PSuper *)
```

The attributes of objects of class `Super` are components of sort `State_Sort` and their object identifiers are of sort `Id_Sort`. The simplest behaviour, shown here, is for a process to offer its services as the alternatives of a LOTOS choice expression. Two kinds of service are offered: *selectors* and *modifiers*. A *selector* will return the value of an attribute to a client while a *modifier* will update an object's attributes. The ADT operation `F1` returns a new `State_Sort` value.

The class template `Super`, from which objects of class `Super` can be instantiated, is defined as:

```
process Super[g](state: State_Sort, id: Id_Sort) : noexit :=
   PSuper[g](state, id)
     >> accept upd_state: State_Sort, id: Id_Sort
        in Super[g](upd_state, id)
endproc (* Super *)
```

It is possible for processes `Super` and `PSuper` to be amalgamated, but if class `Super` is to be a superclass then a process such as `PSuper` with exit functionality is required [Rudkin, 1992].

Let us now consider the case where a class template `Sub` inherits from `Super`, but modifies the service `modifier_1`, adds service `modifier_2` and adds attributes defined in `Ext_State_Sort`. For incremental inheritance to provide behavioural inheritance, a modified service must return the same values and must have the same effect on the attributes defined in `State_Sort` as the service it is modifying. The only allowed change is in the values of the added attributes [Bar-David, 1992]. (It is not therefore possible to modify a pure selector.) We first define a *modifier process* `M`:

```
process M[g](state: State_Sort, ext_state: Ext_State_Sort, id: Id_Sort)
        : exit(State_Sort, Ext_State_Sort, Id_Sort) :=
```

```
    g !selector_1 !id ... ; ... exit(any State_Sort, ext_state, id)
  []
    g !modifier_1 !id ?x: Val1_Sort ... ; ...
    exit(any State_Sort, Nf1(ext_state), id)
  []
    ...
endproc (* M *)
```

As no service may be deleted from the superclass Super, all event offers on gate g in PSuper must also occur in M. Although the modifier process repeats the external interface of the services inherited from the superclass, it does not repeat the internal actions.

To ensure that the same values are returned, where a value !vi appears in a structured event in PSuper, the corresponding parameter in M must be ?x: Vi_Sort. As changes to the attributes defined in State_Sort are carried out in process PSuper, process M offers the value any to achieve proper synchronization on exit. Only changes to the attributes defined in Ext_State_Sort can be defined in M and this is shown, within the modified service modifier_1, by the operation Nf1. The properties of the subclass Sub can now be defined as:

```
process Sub[g](state: State_Sort, ext_state: Ext_State_Sort,
               id: Id_Sort) : noexit :=
  (  ((PSuper[g](state, id) >> accept upd_state: State_Sort, id: Id_Sort
                              in exit(upd_state, any Ext_State_Sort, id)
     )
     |[g]|
     M[g](state, ext_state, id)
    )
   []
    g !modifier_2 !id ... ; ...  exit(F2(state), Nf2(ext_state), id)
  ) >> accept upd_state: State_Sort, upd_ext_state: Ext_State_Sort,
             id: Id_Sort in Sub[g](upd_state, upd_ext_state, id)
endproc (* Sub *)
```

Operations F2 and Nf2 change the attributes defined in State_Sort and Ext_State_Sort, respectively. As process M synchronizes with process PSuper on gate g, we have multiway synchronization between M, PSuper and the process used to define the client object.

It is possible, by examination of the LOTOS text, to ensure that a modifier process does not affect the attributes of the base class or change the values returned. Static checking can therefore be used to ensure that incremental inheritance is only used to implement behavioural inheritance.

A feature of our approach is that all the services inherited from class Super must also occur in M. If the external interface of Super is changed then M must also be changed so that it remains compatible. It is sometimes stated that an advantage of the object-oriented approach is that a change in a superclass is automatically propagated to all its subclasses and requires no change in the subclass. It can be argued that this is not a good idea. If the external behaviour of a superclass is changed then this should be known by users of the subclass.

3.5 Conclusions

We have been investigating how OOA methods and FDTs can be combined in the software development process. Because our goal is the creation of an analysis method to be used by software developers, we have chosen an FDT which has an ISO Standard, is suitable for writing specifications in an object-oriented style, has support tools, produces executable specifications and supports concurrency. An FDT that satisfies all these requirements is LOTOS.

In developing the method, we took the existing OOA methods as a starting point. We have developed ROOA. ROOA builds a formal object-oriented analysis model that acts as a requirements specification of a system. Once we have a formal requirements specification, we can use it in a formal development strategy where an implementation is obtained from successive mathematical transformations of the specification.

ROOA builds on the object model produced by any of the existing OOA methods. However, instead of producing three separate models, it uses a set of rules to systematically create a single integrated object-oriented analysis specification from the static properties captured in the object model and the dynamic and functional properties given in the informal requirements. The ROOA model is primarily a dynamic model, but it keeps the structure of the object model. As the resulting specification is executable, prototyping can be used to find and correct omissions, contradictions and ambiguities early in the development process.

An important part of ROOA is to give a LOTOS interpretation of object-oriented analysis constructs. In this chapter we have shown how class templates, message passing, aggregation and behavioural inheritance can be modelled in LOTOS.

Chapter 4

The Impact of Inheritance on Software Structure

4.1 Introduction

In most object-oriented languages an inheritance hierarchy is assumed to represent a type inclusion hierarchy. However, inheritance, viewed from an architectural level of abstraction, can be seen as a simple extension of the module/package inclusion mechanisms (referred to here as *import/export mechanisms*) found in object-based languages such as Modula-2 [Wirth, 1988] and Ada [Booch, 1987]. One way to express the notion that inheritance is an extension of import/export mechanisms is to say that it allows the client-server relation between an exporting module and an importing module to be transitive, so that a client can transparently pass on services which it imports (i.e. inherits from parent classes) to its own clients. However, this extension, while useful, is far short of the expressive power needed to guarantee that valid type inclusion relations are properly captured in a class hierarchy. This chapter demonstrates these observations by presenting formal models of inheritance hierarchies, and of module hierarchies with import/exports, and defining a translation function from the former to the latter.

Object-oriented programming is beginning to mature. Abandoning the use of the term 'object-oriented' as a byword for 'good' (a practice noted by Meyer in the preface to [Meyer, 1988], researchers and practitioners now feel the need to promote the technology less keenly than the need to diagnose and correct its weaknesses. Formal models are useful in this diagnosis.

Previous formal models of inheritance have concentrated on its semantics as a programming language construct, and have used denotational methods [Wolczko, 1989], [Kamin, 1988] [Cook and Palsberg, 1989], [Hense, 1993]. Most notably, Cook & Palsberg use the notion wrappers to describe the semantic modifications made to a superclass through the act of inheriting it (chiefly the rebinding of the variable self to objects of the child class) [Cook and Palsberg,

1989]. They demonstrate that their semantics can be proved equivalent to the operational semantics of object-oriented languages. This model has even been proposed as a basis for a new object-oriented language which has an explicit wrapper construct, which serves to increase the amount of code reuse possible [Hense, 1993].

Cook & Palsberg conclude on the basis of their model that inheritance provides "expressive power not available in other languages". In our experience, this conclusion is often taken to mean that inheritance provides code reuse without compromising software quality. In fact, this is very far from the case. There have been numerous warnings from the object-oriented community that inheritance mechanisms are potentially harmful [Halbert and O'Brien, 1987], [Sakkinen, 1989], [Armstrong, 1991], [Magnusson, 1991], [Rumbaugh, 1993]. Indeed, Magnusson has compared inheritance to the infamous goto construct [Magnusson, 1991]. Our concern in this chapter therefore is to de-emphasise the code reuse power of inheritance and analyse instead its impact upon software structure. Structuring is a more important issue than reuse; it is better that a program be well-structured, but larger, than smaller, but baroque because intricate code reuse techniques have been employed.

Our models start from a much higher level of abstraction than the denotational approach. Our concern is not with the implementation details of classes, but with their role as software components organised into a hierarchy, and whether inheritance helps us to understand the relations between components and the concepts they represent. We do not need to model instance variables, state, or values, only the methods classes provide for client classes (e.g. we assume that clients access an object's state via *get* and *set* methods).

We demonstrate that inheritance with classes is comparable to import/export relations between modules, and to with/use relations between packages, as found in the object-based languages Modula-2 and Ada 86 respectively. We present a formal model of multiple inheritance hierarchies, a formal model of module hierarchies based on the notion of abstract data type, and a translation function from the former to the latter. We do so by exploiting Meyer's dictum that *class = module = type* [Meyer, 1990]. However, we choose this rule only because it is the most practical and commonly-followed design heuristic in our experience. The fact that classes and modules are useful for implementing types does not mean the three ideas should be confused. Our models generalise those in [Armstrong *et al.*, 1992], which dealt with mapping single inheritance into a restricted form of module hierarchy.

The chapter is organised as follows. Section 4.2 gives an informal analysis of inheritance hierarchies which serves to explain the terminology underlying our model. Section 4.3 is a similar discussion of the principles behind module/package hierarchies (as found in Modula-2 and Ada). We restrict our attention to type-module hierarchies in which each module/package implements an underlying abstract data type. Section 4.4 presents the inheritance model and section 4.5 the type-module hierarchy model. Section 4.6 presents the translation function. This function is *encapsulation preserving*, that is, we translate each class into a module with exactly the same external client interface, whilst ignoring the internal details of both. The translation shows that as regards the interfaces of the classes in a hierarchy, inheritance provides an automatic way of importing inherited methods into a class and then re-exporting some or all of them to clients. This automatic process can easily be mimicked by an import/export mechanism.

We are not suggesting that nothing would be lost in the translation from inheritance to module hierarchies if the translation function were to be applied mechanically to programs. For example, inheritance is attractive to programmers because object-oriented languages interpret it as a type inclusion relation between a superclass and its subclasses. In languages which implement module hierarchies based on import/export, there are no such facilities. In the translation from inheritance hierarchies to module hierarchies, type inclusion abstractions would be lost.

However, our models do reveal the fundamentally flawed nature of multiple inheritance as a type inclusion mechanism: *a inherits b* is no more a basis for accepting class a as a subtype of class b than the fact that module a imports operations from module b, and makes them externally visible to its clients. Inheritance mechanisms necessarily capture only implementation relations between classes. The dynamic substitutability of subclasses for superclasses, which is so essential to reliable polymorphic programs, is not more strongly guaranteed by the use of inheritance than by the use of less elaborate module inclusion constructs.

4.2 Multiple Inheritance Hierarchies

Inheritance hierarchies form directed graphs where the nodes are classes and the edges are inheritance paths which induce a partial ordering on them. Figure 4.1 shows the topology of a typical hierarchy. Each class c can be viewed as a module defining a (sub)type and a set y of operations (methods) over that type. A class c may be either a subclass or a root class. Root classes have no superclass, and constitute complete type definitions on their own.

A subclass has $1 \ldots n$ superclasses. We number each superclass from left to right. The set of n superclasses is referred to as the *parents* of class c. A superclass may itself have a superclass, and so forth, until a root class is reached. The set of classes which are transitively related to c by inheritance are known as its *ancestors*.

A subclass defines only a partial set of operations over a type. This set is made whole only when the inheritance mechanism "unites" these operations with those of the ancestor classes. We call the resulting set of methods the *protocol* of c. A slight complication is that c may over-ride a method defined in an ancestor, in which case the ancestor method is replaced in c's protocol. The protocol is the complete set of methods that clients of c can use. Within the protocol we can distinguish the set y of methods defined in c, from its inheritance, that is, the set of methods it inherits from ancestors, without redefining them.

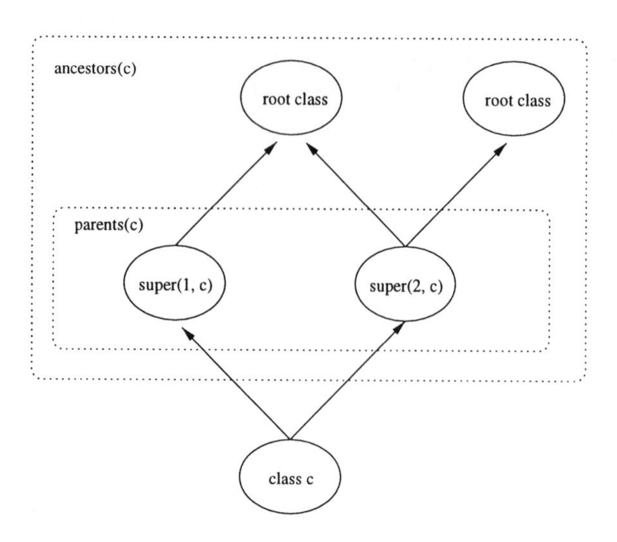

Figure 4.1: Inheritance class hierarchy

4.3 Type-Module Hierarchies

Module hierarchies form directed graphs where the nodes are modules, and the edges are import/export relations. In principle, a module need only contain a collection of operations, but we will restrict our attention to type-module hierarchies, where each module implements an underlying abstract data type. We also assume that it is not possible for module a to import from module b if b imports from a. This will help simplify our formal model, as well as reflecting good programming practice. It also ensures that the hierarchies are partially-ordered by import/export relations. Figure 4.2 shows the general form of a type-module hierarchy and the relations between modules.

A type-module exports a type and a set of operations over that type, which other type-modules may import.

A type-module m may be a basic module. In this case, it is constructed only from language-defined simple and composite types, and does not import from any other type- module.

Otherwise, m has $1 \ldots n$ subcomponents, where a subcomponent is defined as a type-module whose type and operations m imports. We assume that the subcomponents can be ordered in some trivial way, for example, by the order of their importation in m's code.

Each subcomponent may be a basic type-module, or may itself have subcomponents from which it imports. The complete set of type-modules needed to implement a module m are referred to as its components.

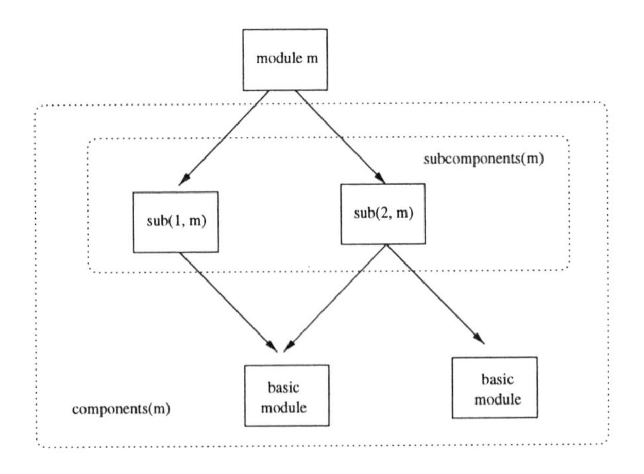

Figure 4.2: Type module hierarchy

4.4 A Model Of Multiple Inheritance Hierarchies

4.4.1 Basic Definitions

A multiple inheritance hierarchy (**MIH**) can be defined as follows:

MIH $=< C, Y, roots, super, parents, ancestors, methods, protocol, inheritance >$

where C is a set of names of classes and Y is defined as follows:

> **Methods**. Each $y \in Y$ is a set of methods possessed by a class. Let Id be a set of identifiers. Each individual method is modelled as a maplet from its name to its body, that is, a specification of the functionality of the method:
>
> $$y : Id \twoheadrightarrow Body$$
>
> Because methods are modelled as maplets, one method may be overwritten by another where both share the same name, as follows:
>
> $$\{..., id \mapsto s1, ...\} \oplus \{id \mapsto s2\} = \{..., id \mapsto s2, ...\}$$
>
> where $\oplus$ (read as "overwritten by") is defined as follows:
>
> $$(g \oplus f)(x) = \begin{cases} f(x) \text{ if } x \in \textbf{dom } f \\ g(x) \text{ if } x \in \textbf{dom } g \text{ - } \textbf{dom } f \end{cases}$$

We define the other parts of MIH as follows:

$roots \subseteq C$

where each $r \in roots$ is a root class of the hierarchy.

$super : C \times \mathbf{N}^+ \nrightarrow C$

maps a class to its n'th superclass (if it has one) where

$$\mathbf{dom}\ super = (C - roots) \times \mathbf{N}^+$$

and $\mathbf{N}^+$ is the non-zero naturals.

$parents : C \rightarrow \mathbf{P}\ (C)$

$parents(c) = \{x \in C | (c, n, x) \in super\}$ and $\mathbf{P}$ is the powerset symbol

$ancestors : C \rightarrow \mathbf{P}\ (C)$

where $ancestors(c)$ is simply $parents^+$ the transitive closure of $parents$. We can also define it recursively:

$$ancestors(c) = \bigcup_{x \in parents(c)} ancestors(x)$$

The defining conditions for a multiple inheritance hierarchy are as follows:

$$\forall c \in C, (c, c) \notin ancestors(c) \text{ and if } c \notin roots \text{ then:}$$
$$\exists c' \in roots, (c, c') \in ancestors(c)$$

In other words, no class is its own ancestor, and each class must either be a root class or have at least one root class as an ancestor.

$class\_methods : C \rightarrow Y$

$class\_methods(c)$ is the set of methods defined over the class c.

$instance\_methods : C \rightarrow Y$

$instance\_methods(c)$ is the set of methods over instances of c that are defined in class c.

$methods : C \rightarrow Y$

where $methods(c) = instance\_methods(c) \cup class\_methods(c)$ (note that the domains of $class\_methods(c)$ and $instance\_methods(c)$ are disjoint).

The distinction between class methods and instance methods is made to reflect the intuitions of object-oriented programmers, but is not crucial to the model.

Based on the above, we can now define the protocol a class presents to clients:

$$protocol : C \rightarrow Y$$

where $protocol(c)$ = if $c \in roots$ then $methods(c)$ otherwise:

$$\bigcup_{x \,\in\, parents(c)} protocol(x) \oplus methods(c)$$

The effect of method redefinition (formally, $\oplus$) is important. At run-time clients will not be able to invoke all the methods of classes in $ancestors(c)$ on instances of c. Consider, for example, the case where an inherited method m is redefined as m' by a class c. In this model m' is considered to be a different method from m, since it maps $id(m)$ to a different body. However, from the client's point of view m' will seem to have replaced m. In other words, if a method of c, say m', and a method m belonging to some class in $ancestors(c)$, are such that $id(m') = id(m)$, then m' replaces m in $protocol(c)$.

We can also give a simple definition of a class c's *inheritance*, that is, the methods which are in its protocol, but which are implemented in ancestor classes:

$$inheritance : C \rightarrow Y$$

where $inheritance(c) = protocol(c) - methods(c)$

4.4.2 Discussion: Assumptions in the Model

Our model makes some simplifying assumptions about the nature of multiple inheritance hierarchies.

Firstly, methods cannot be inherited multiple times along different inheritance paths. We assume that where the repeated inheritance is required, the method will have been renamed for each path. The principle of renaming is described in [Meyer, 1988]. Secondly, different identifiers may map to the same body, but since each is a different maplet, each constitutes a different method.

In the absence of renaming, the properties of sets and the union operator are sufficient to ensure that duplicate inherited methods are included only once in the protocol of a class.

4.5 A Model of Type-Module Hierarchies

4.5.1 Basic Definitions

A type-module hierarchy (**TMH**) can be defined as follows:

TMH $=< M, Q, basics, nth\_sub\_comp, subcomponents, components, exported\_ops >$

where M is a set of names of module definitions, and Q is defined as follows:

> **Operations** Each $q \in Q$ is a set of operations. Like a method (defined in section 4.4.1), an operation is modelled as a maplet from its name to the body defining it, as follows:
>
> $$q : Id \mapsto Body$$
>
> At this level of abstraction, the body of an operation is the same as the body of a method (i.e. a specification of functionality), and hence $Q = Y$.

We define the other parts of **TMH** as follows:

$basics \subseteq M$

where for each module$m \in basics, subcomponents(m) = \emptyset$, i.e. m is at the bottom of the hierarchy.

$nth\_sub\_comp : M \times \mathbf{N}^+ \rightarrow M$

$nth\_sub\_comp(m)$ is the n'th sub-component module of m, where

$$\mathbf{dom}\ nth\_sub\_comp = (M - basics) \times \mathbf{N}^+$$

$subcomponents : M \rightarrow \mathbf{P}(M)$

$$subcomponents(m) = \{x \in M | (m, n, x) \in nth\_sub\_comp\}$$

$components : M \rightarrow \mathbf{P}(\mathrm{M})$

where $components(m)$ is simply $subcomponents^+$ the transitive closure of $subcomponents$. It may also be defined recursively as the set of the modules needed to define m:

$$components(m) = \bigcup_{x \in subcomponents(m)} components(x)$$

$exported\_ops : M \rightarrow Q$

where $exported\_ops(m)$ is the set of operations m exports.

The defining conditions for a module hierarchy are as follows:

$$\forall m \in M, (m, m) \notin components(m) \text{ and if } m \notin basics \text{ then:}$$
$$\exists m' \in basics, (m, m') \in components(m)$$

In other words, no module is a subcomponent of itself, and each module is either a basic module or is built from at least one basic module.

4.5.2 Discussion: From Inheritance to Module Hierarchies

In the above definitions, the function $exported\_ops(m)$ may seem to provide a way to distinguish between operations that a module defines anew, and those it imports from others. However, consider the set of all operations that m is potentially built from, i.e. the exported operations of all the modules in $components(m)$. Call this set $usable\_ops(m)$. It is distributed union of the operations exported by the modules in $components(m)$:

$$usable\_ops(m) = \bigcup_{x \in components(m)} exported\_ops(x)$$

Normally, one would expect these two sets of operations to be disjoint. The members of $exported\_ops(m)$ will be implemented by calls to a subset of the operations in $usable\_ops(m)$, imported from members of $subcomponents(m)$. These will be implemented as calls to other operations in $usable\_ops(m)$, and so forth. However, consider the implications if the import/export relationship were allowed to be transitive. Operations in $usable\_ops(m)$ could then be imported into m and simply re-exported, so that:

$$exported\_ops(m) \cap usable\_ops(m) \neq \emptyset$$

This is comparable to the inheritance of operations in $usable\_ops(m)$. In moving from multiple inheritance to a module hierarchy, we simply have to ensure that:

$$exported\_ops(m) \cap usable\_ops(m) = inheritance(c)$$

This transitivity property will enable import/export patterns to mimic inheritance in our translation model. Import/re-export relations allow an operation to be 'passed through' a series of modules to clients. In simple terms, a translation function need only flip the inheritance hierarchy upside down, and replace inheritance relations with import/re-export relations (see Figure 4.3).

Operationally speaking, a multiple inheritance hierarchy can be turned into a module hierarchy like so:

Translation Procedure:

1. For each class c with no children, declare a module m.
2. If $c \in roots$ then go to step 3. Otherwise, apply the following steps for each $n \in 1 \ldots |parents(c)|$:

   a) Translate $super(c, n)$ into a module exporting translations of $methods(super(c, n))$. Do this by applying the translation procedure recursively.

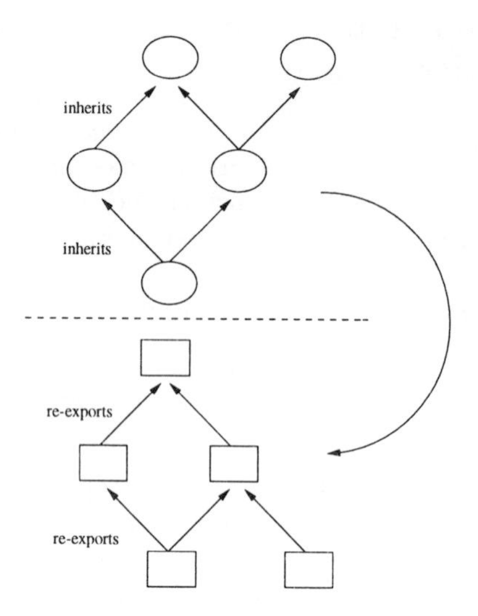

Figure 4.3: Flipping an inheritance hierarchy

b) Make this module $nth\_sub\_comp(m, n)$ by importing its operations into m.

c) Re-export from m the subset of these operations corresponding to those methods c inherited, without redefinition, from $super(c, n)$.

3. Translate $methods(c)$ into operations implemented in and exported by m.

4.5.3 Discussion: Language Issues

Import/export is not transitive in object-based languages, so imports cannot be re- exported. However, transitivity can easily be mimicked. For example, one need only define an operation op_2, which does nothing but call an imported operation op_1, and then export op_2. An overloading mechanism such as Ada's means that one can even preserve the name of op_1 in op_2 [Booch, 1987].

Of course, the useful thing about inheritance, is that if c inherits from x, c is treated as type compatible with x, with attendant reuse benefits. For example, storage types for x instances can handle instances of c as well. Given current object-based languages, if one were to translate an inheritance hierarchy into a type-module hierarchy the type inclusion relations would be lost. Thus it would not really be practical to translate a program that depended on type inclusion relations from Eiffel into, say, Ada.

However, the purpose of the models is to suggest a novel perspective on inheritance. Initially, we can view it as an import declaration which means 'import all the methods of class x into this class c, and re-export them to its clients'. The programmer can then decide to discard unwanted 'imported' implementations by redefining them in class c. In other words, the essential purpose of multiple inheritance is to reuse implementations. More meaningful relations between classes, such as IS-A subtyping will only be captured incidentally, if at all.

4.6 Relating Multiple Inheritance and Module Hierarchies

We now define a translation function, Ω, that takes an inheritance hierarchy:

MIH $=< C, Y, roots, super, parents, ancestors, methods, protocol,$ $inheritance >$

and yields a module hierarchy:

Ω (**MIH**) $=< M, Q, roots, nth\_sub\_comp, subcomponents, components,$ $exported\_ops >$

where $Q = Y$ and $M = C$ (see above) and we make the following definitions:

$$basics = roots$$

$$\forall c \in C, n \in \mathbf{N}^+ \text{ if } c \notin roots, \text{ then } nth\_sub\_comp(c,n) = super(c,n)$$

$$exported\_ops(c) \cap usable\_ops(c) = inheritance(c)$$

From these definitions it follows that:

$$subcomponents(c) = parents(c) \text{ and } exported\_ops(c) = protocol(c)$$

Theorem: For every inheritance hierarchy **MIH** there is an equivalent module hierarchy Ω(**MIH**) which preserves encapsulation.

Proof: We show that Ω(**MIH**) satisfies the defining conditions of a module hierarchy, given in section 4.5.1, using the fact that **MIH** is an inheritance hierarchy and so satisfies the conditions stated in section 4.4.1.

For example, since $subcomponent = parents$, it follows that:

$$subcomponents^+ = parents^+$$

Hence, since $M = C$ and $basics = roots$:

$$\forall m \in M, (m,m) \notin subcomponents^+ \text{ and if } m \neq base \text{ then: }$$
$$\exists m' \in basics, (m,m') \in subcomponents^+$$

Since *components* is defined as $subcomponents^+$ (section 4.5.1) we have the defining condition for a module hierarchy, as required.

Furthermore, since, $exported\_ops(c) = protocol(c)$ for every class c, it follows immediately that Ω is encapsulation-preserving.

4.7 Summary & Conclusions

This chapter has presented formal models of multiple inheritance and type-module hierarchies. A mapping from the former to the latter was described. The models illustrate that inheritance does not in principle provide the modelling power necessary for constructing valid type inclusion hierarchies, even though, as denotational models have shown, it provides extra expressive power for code reuse. This expressive power may affect the code reuse, but it has no impact upon the interfaces that classes present to clients. Inheritance was viewed as a form of import/export mechanism, in which operations from ancestor classes are automatically imported, and can be then be either re-exported, blocked, or replaced. Since this is an ad hoc process, the assumption that the type implemented by a subclass is included in the type implemented by a superclass is fundamentally unsound.

The basic problem is, of course, that programming languages type systems are based on a syntactic notion of type. These systems are used to represent a more semantically powerful notion of type, in which types are defined in terms of their mathematical properties. In languages with monomorphic type systems, the question of whether one mathematical type includes another never arises, so it is safe to represent them as programming language types. In functional languages, the same question is resolved using powerful type inference mechanisms with a basis in mathematical logic. In object-oriented languages, it is resolved simply on the basis of inheritance. This is not sufficient. Furthermore, the application of covariant or contravariant typing rules [Shaffert *et al.*, 1986] can do little to ensure that the functional properties of a supertype are preserved by a subtype. For instance, a method redefinition may preserve the syntactic type signature of an inherited method, but implement a different behaviour.

The authors feel that despite the problems associated with inheritance, type inclusion hierarchies can and should be built into programs. However, unless one avoids the redefinition of inherited behaviour altogether, one cannot convincingly demonstrate that subtypes are universally substitutable for supertypes without the use of formal techniques. After all, this demands the proof of a universal quantification (that the subtype behaviour implies the supertype behaviour in all contexts), which cannot be demonstrated by testing.

Fortunately, suitable notations do exist for specification and proof of subtype hierarchies, as do design paradigms within which to apply them; for example, refinement techniques [Morgan, 1990], and Meyer's programming-by-contract [Meyer, 1988], [Meyer, 1990]. In [Armstrong, 1991] a specification method is proposed which enables both IS-A subtyping and reuse-oriented techniques to be specified. The two-tiered specification language Larch is used as a basis. The top tier is used to specify an IS-A subtype hierarchy, whilst the bottom tier records code reuse decisions.

It is both puzzling and worrying that the object-oriented community has been slow to consider the role formal methods can play in controlling the fundamental weakness of inheritance mechanisms. This may be because object-oriented programmers see inheritance as "an implementation issue, not a design issue", as Cox puts it [Cox, 1986]. Unfortunately, if inheritance hierarchies are treated as type inclusion hierarchies, and client programs depend on this fact, inheritance becomes a design issue.

Part III

Object Technology in Formal Methods

Chapter 5

D_Parlog^{++}: Object-Oriented Logic Programming with Distributed Active Classes

5.1 Introduction

D_Parlog^{++} is an OO extension of a distributed version of the concurrent logic programming (CLP) language Parlog. A major difference between this language and other OO extensions of CLP languages is that the classes in D_Parlog^{++} are active, they have there own methods and, like the class instances, are implemented as Parlog processes. Classes are also the mechanism for distributing the computation. We just load different class objects onto different machines of a local area network. All the active instances of a class will reside on the same machine as the class, but an inheriting class, and hence all its instances, can reside on another machine.

All object methods, whether for class or instance objects, are invoked by sending a message to the object denoted by its unique object identity. A class object's identity is its public name, such as `student`, given in the program. An instance identity can either be program assigned, or system generated. The fact that instance objects can have public names is another distinguishing feature of the language.

The language is highly concurrent. All objects can process their stream of incoming messages concurrently. That is, as soon as a message is accepted, and the method for the message is activated, the object is able to accept the next message. In addition, each the actions of a method can be executed concurrently. So D_Parlog^{++} objects can exhibit the high degree of internal concurrency of actors [Agha and Hewitt, 1987]. On the other hand, the actions of a method can be sequentialised if need be, and much more easily than in the actor paradigm. It is simply a matter of defining the method as a sequential, rather than parallel, conjunction of actions.

This chapter assumes some familiarity with concepts of concurrent object oriented programming and logic programming, ideally concurrent logic programming, but familiarity with Prolog will probably suffice.

5.2 Informal Computation Model of D_Parlog++

Shapiro and Takeuchi[Shapiro and Takeuchi, 1983] first demonstrated the potential of combining CLP and OOP to form a new programming paradigm. They investigated the possibility of implementing active objects, with encapsulated state, as processes in a CLP language. In their model, an object is implemented as a tail-recursive process that passes the updated state components of the object as arguments in the recursive call. The identity of the object, is the identity of an input stream argument of the process into which all message for the object are merged. An object processes the message at the head of its input stream, and performs actions which may compute new values for some of the state components, according to its current state and the form of the received messages. It then recurses with the new state components and the tail of its input message stream.

Following their model, several language proposals such as Polka[Davison, 1989], Vulcan [Kahn *et al.*, 1987] and A'UM[Yoshida and Chikayama, 1988] have since emerged. Each is essentially syntactic sugar for defining the CLP process that will implement each instance object.

Most of these languages use a class declaration (or an equivalent) just as the vehicle for defining the methods of the instances of the class. The class declaration is compiled into a procedure definition in the CLP language. Creating an instance of a class is then just syntactic sugar for a direct invocation of this procedure as a recursive process. In this approach, no global information about these created instances is kept. Thus, no managing operations on these instances can be conveniently performed. For example, there is no state component of any object holding the identities of all the instances of a particular class, so one cannot easily broadcast a message to all the members of the class. In addition, objects cannot be given public names, so in order for one object O to send a message to another object O', O must be passed the identity of O' in a message, or be given the identity as the value of one of its state components when it is created. Finally, inheritance is generally implemented as message delegation, and for each created instance of a subclass there will be a string of processes forked, one for each level in the class hierarchy. This is perhaps acceptable in a non-distributed system, but is not accepted if the different processes might reside on different machines.

A different semantics is adopted in D_Parlog++ for interpreting the class structure, for object naming and for inheritance Both the instances and the class are mapped into tail recursive processes, so the class is also an active object. An instance process in D_Parlog++ is created by sending a message to its class, not by a direct procedure call. The class will invoke the process representing the active instance, but it will also add the identity of the created instance to one of the class state components that holds the identities of all the instances. To broadcast a message to all the instances of a class we just ask the class for its list of current instances.

A class has a public name, which is a name such as `course` given to it in the program. The message that requests that a new instance be created can also give the instance a public name. So we can create a course instance with the public name, such as `cs225`. Other objects can send messages to this instance using this public name. The public naming of objects is possible because the underlying Parlog system [Crammond *et al.*, 1993] allows for the creation of named streams that become the input message streams of the processes that represent the objects. By adding a name server, we can allow these named streams to be on different machines of a local area network. Thereby, we can distribute the classes and their instances.

Inheritance of instance methods and state is not via message delegation. Instead, as in SmallTalk [Goldberg and Robson, 1983], the methods and state declarations of the super classes are copied down to give the instance definition of an inheriting class. Inheritance of class methods is still via via message delegation. This is all that we shall say about implementation. For further details we refer the reader to [Wang, 1995].

The informal semantics of the Object Oriented Programming model of D_Parlog++ can be summarized as follows:

1. Encapsulation is by class structure which, in its instance declaration section, specifies the common attributes(state) and capabilities(methods) of the individuals of the **class**, and, in its class declaration section, captures the attributes and capabilities of the **class** itself.

2. A class structure has a program specified identity, the class name, and is activated and instantiated by the system as a class process(**class**) when it is loaded. Different classes can be loaded on different machines. It will run actively until it is terminated. During its life time, it creates instances on request, and manages and provides global information to all its instances. A class can be asked for the list of identities of its current instances. The instance creation and deletion methods can be program defined. If not, default methods are added to the class.

3. An instance belongs to a class from which it is created. It has a program specified or a system generated identity and becomes an active process as soon as it is created on the machine on which its class resides. Instance processes also respond to messages from the outside world by taking actions defined in the associated methods. Although instances of a class embody the same set of instance methods defined by the class, they may

have different functionalities owing to the different instantiations of their states. An instance can be asked what class it belongs to.

4. Objects communicate with each other by sending and receiving messages using object identities and a message send of the form `ms => O`. The messages are added to the back of the queue, on time order of arrival. Two mechanisms for making a response to a received message are used, one is by binding variables in the message, the other is by explicitly sending a reply message to the enquirer identified by keyword **sender**. The former exploits the distributed unification of the underlying D_Parlog system [Wang, 1995]

5. Classes can be linked via single inheritance. Inheritance of instance methods and state is via the copy semantics. An inheriting class has direct access to all the instance state components of its super classes and the inherited methods are appended to its methods. Inheritance of class level methods is via message delegation. There is no automatic access to the super class state information. Explicit inquiry messages must be sent to access a state component of a super class.

6. Methods in an inheriting class, for the same message patterns, will override those in the inherited classes. An express message send to **super**, of the form `ms ==> super`, can be used to invoke an overridden method, an message send to **self**, express or normal, will always access an overriding definition, if it exists. Express messages, which can only be only be sent to **self** or **super**, go to the front of the objects message queue. They are always the next messages to be processed.

5.3 An Introductory Example

Our first example is a stack class definition given in Program 5.3.1. This does not have any program defined class level state components or methods. The 'at laotzu' is an optional specification of the machine on the local area network on which the compiled class definition must be loaded. This is then the machine on which the **stack** class object and all its instance objects will reside.

The program will actually be expanded by the D_Parlog++ compiler into the class definition of Program 5.3.2 which has a class state component, `Members`, that will hold the identities of all the created instances. It also has the default instance create and destroy methods which make use of built in procedures `fork_instance` and `terminate_instance` that actually do the forking and killing of the active instance objects, and it has methods for accessing and manipulating the `Members` list. Of these, only the `members/1` method should be used in application programs. The `add_a_member/1` and `delete_a_member/1` methods are only used if another class inherits from this one. More on this later. Notice that the instance definition also has added state components holding the instance identity and the class identity of the instance. It also has a method for accessing the class identity. This provides a runtime type test for instances.

```
class stack at laotzu with
{
 instance_definition
   states  Size, StackPointer:=0, Buffer:=[].

   methods
     push(ELEMENT) : StackPointer < Size ->
        Buffer := [ELEMENT | Buffer],
        StackPointer := StackPointer + 1.

     pop(ELEMENT)  : StackPointer > 0 ->
        Buffer = [ELEMENT | RestBuffer],
        Buffer := RestBuffer,
        StackPointer := StackPointer - 1 .

     top(ELEMENT) : StackPointer == 0 ->
          ELEMENT = -1 .
     top(ELEMENT) : StackPointer > 0  ->
          Buffer = [ELEMENT | RestList].
}.
```

Program 5.3.1 Declaration of a stack class structure

```
class stack at laotzu with
{
 class_definition
   states  Members := [].
   methods
     members(R) ->
        R = Members.

     add_a_member(InstanceID) ->
        Members := [InstanceID | Members].

     delete_a_member(Id) ->
        remove(Id,Members,RMembers),
        Members := RMembers.

     create(StackId, SIZE, STACKPOINTER, BUFFER) ->
        fork_instance(stack,StackId,SIZE,STACKPOINTER,BUFFER),
        Members := [StackId | Members].

     destroy(StId) ->
        remove_from_list(StId, Members, RestMembers),
        terminate_instance(StId),
        Members := RestMembers.

 instance_definition
   states  _class, _self, Size, StackPointer:=0, Buffer:=[].
   methods
     push(ELEMENT) : StackPointer < Size ->
        Buffer := [ELEMENT | Buffer],
        StackPointer := StackPointer + 1.

     pop(ELEMENT)  : StackPointer > 0 ->
        Buffer = [ELEMENT | RestBuffer],
        Buffer := RestBuffer,
        StackPointer := StackPointer - 1 .

     top(ELEMENT) : StackPointer == 0 ->
          ELEMENT = -1 .
     top(ELEMENT) : StackPointer > 0  ->
          Buffer = [ELEMENT | RestList].

     class(C) -> C=_class.
}.
```

Program 5.3.2 System expanded version of program 5.3.1

5.3.1 State Declaration and State Variables

The State declaration is a sequence of variable names/ variable assignments following the keyword **state**. An assignment uses the := operator, and has the syntax as:

Variable := *Default Value*

In Program 5.3.2, states are declared both for class and instance sections as:

```
Members := [].
Size, StackPointer := 0, Buffer:=[].
```

Default values are used to instantiate state variables of objects when they are activated or created. In the above, the class state variable `Members` has an empty list as its default value, and the last two of the instance state variables have default values of 0 and [] respectively; the third instance state variable, `Size` must get its initial value from a create message because no default value is given for it. The `_self` state variable can have its value supplied in the create message, for a program assigned identity for the instance, or have it assigned by the `fork_instance` primitive, for a system assigned identity. The `_class` state variable gets its value from the `fork_instance` call. Note that the language follows the Prolog convention that variables have alphanumeric names beginning with an upper case letter or _.

The scope of the class state variables is confined to the class methods. The instance state variables have scope all the instance methods of this class *and* all the instance methods of any inheriting or inherited classes. So an instance method in an inheriting class can directly access and update an instance state variable of any of its super classes, and, conversely, a method in a superclass can access or update a state variable of a subclass. This is *not* the case for the class state variables because of the different inheritance method used - message delegation - for class methods. However, part or all of an object's state can be made accessible to other objects if corresponding inquiry methods are declared. For example, in program 5.3.2 the class state variable `Members` is made accessible by the system provided `members/1` method. The message invoking this method must be of the form `members(L)`, where L is an unbound *reply* variable. The match of this message against the selector `members(R)` will bind the `R` variable of the method to `L`, The unification action `R=Members` of the method will become `L=Members`, which will bind the reply variable `L` to the current value of `Members`.

Variables such as `R` are local to the method in which they appear. This is true of all variables used in methods that are not state variables, they are always local to the method, and have scope the entire method.

The class state is initialized by the system according to the default values given for its state variables when the class definition is loaded and the class object is created. Initial values for an instance's state variables are generally given in its create message. These initial values can override default values given in the class definition.

Methods can be declared to change the values of some or all of the state variables. The new values for the state variables are indicated by use of explicit assignments in the method. In Program 5.3.2 one can find at the end of the `push/1` method the following assignment:

```
StackPointer := StackPointer + 1.
```

This assigns to the state variable `StackPointer` a new value which is the old value plus one. This new value will be seen by the next method invocation of the object.

5.3.2 Method Declaration

A method declaration establishes the connection between a message pattern and the associated actions an object would take when such a message is received. All the methods have the format of:

Selector **:** *Guard* **->** *Actions Terminator*

The Selector The *Selector* is the message pattern which will trigger the method, and is an atomic formula with the form:

mess(t_1,....,t_k)

where t_1,....,t_k are terms. Such a pattern is said to have name: **mess** and arity: k. The pair *mess/k* is called the skeleton of the method selector. The set of different message skeletons appearing in the methods of an object definition (including any inherited methods) are called the *acceptable skeletons* of the object. Methods with the same message skeleton are called **polymorphic methods**. Upon receiving a message, an object will first check to see whether or not the message has an acceptable skeleton, if it has, the message is then used to match against the selectors of all the polymorphic methods for that skeleton. (These could be inherited methods if they have not been overridden by locally defined methods.) If the new message does not have an acceptable skeleton, the message is regarded as unrecognizable and will be discarded after an warning message is sent back to an error handling process(object) in the sender's machine. A message which does match a method's *Selector* will trigger the evaluation of the *Guard* part, if there is one, hoping to commit to the method. If there is not a guard part, or the guard succeeds, the execution is said to *commit* to the method and the object is said to have *accepted* the message.

The Guard A method is called a **guarded method** if the *Guard* is included in the definition. In general, a guard is a conjunction of conditions which determine whether or not the *Actions* of a method, whose selector matches the received message, should be executed at this juncture. State variables can have their values accessed, but not updated, in the guard. When a guard is evaluated to a **true** value, the *Actions* initiated. If the guard fails (is **false**), the message will be matched against the selectors of the polymorphic methods that appear textually later in the sequence of method definitions. If none of the polymorphic methods has a selector that matches the message, or if any those that do have a guard that fails, the message is skipped over and the next message in the message queue for the object is tried. A skipped message will be tested again against each of its polymorphic methods as soon as some message for the object is accepted. If the message queue for an object is empty, or all the pending messages are skipped over, the object suspends until a new message arrives. Certain messages may not be accepted, even though the object has methods for their skeletons, because the state of the object is such that the actions of the methods are not valid. Such messages are skipped over and retried after any message is accepted in the hope that the state of the object will have been changed by that message.

Since D_Parlog^{++} inherits the committed choice properties of Parlog[Clark and Gregory, 1986], a method, once committed to, must be such that *Actions* taken will always succeed. Otherwise, the failure is regarded as a runtime error of a program and may cause the error termination of the whole application.

The Actions The *Actions* are a Parlog conjunction of individual actions, conjoined by the Parlog ',' and '&' conjunction operators. Actions conjoined by ',' are executed concurrently, those conjoined with '&' are executed sequentially. One of the actions can also be conditional action expressed using the Prolog style conditionals `Cond -> Actions` or `Cond -> Actions1 ; Actions2`. Here the `->` is read as *then* the ; as *else*. Nested conditionals are not allowed. The `member/2` class method has a conditional action.

Variables shared between the actions act as communication channels, particularly for concurrent calls. Suspension or resumption of these calls follows the Parlog operational semantics for concurrent conjunction. Essentially calls suspend if they need a value of variable `X` that is being computed by some other call, until that other call generates a sufficiently instantiated binding for `X`. The suspension and resumption is automatic, and is provided by the underlying Parlog execution mechanism. The action calls can be calls to programs defined in a code_definition section that is private to the class, or be calls to built-in predicates of D_Parlog^{++}.

Actions can also be explicit state update assignments. These assignments take effect *after* all the other actions have started execution, so these other actions see the values of the state components that the object had when the message was received. The state updates determine the state for the *next* message. The updates are always executed concurrently. In addition, message send operations are allowed within the *Actions*, which makes it possible for direct communication between objects. A message sent to `sender` can be used to give a response to a message. As we have seen, a unification that binds a variable in the received message can also be used to give a response. Finally, messages can be sent to `self` or `super`. More on this later.

In Program 5.3.2, two methods are declared for the instance as:

```
push(ELEMENT) : StackPointer < Size ->
                Buffer := [ELEMENT | Buffer],
                StackPointer := StackPointer + 1.

pop(ELEMENT) : StackPointer > 0 ->
               Buffer = [ELEMENT|Rest],
               Buffer := Rest,
               StackPointer := StackPointer - 1.
```

The first of the above two methods determines what a stack instance should do upon receiving a `push(ELEMENT)` message, a message with skeleton `push/1`.

In its guard, the stack pointer is checked against the size of the stack to see if there is at least a space for a new element, if so, the element is put into the buffer and the stack pointer is increased accordingly. If not, the message will be skipped over. The second method will invoke a contra-action. Note the crucial difference between the use of the assignment `Buffer:=[ELEMENT|Buffer]` in the `push` method and the use of the unification `Buffer=[ELEMENT|Rest]` in the `pop/1` method. The former updates the value of the state variable, the latter unifies its current value against the pattern `[ELEMENT|Rest]` thereby binding the variables `ELEMENT` and `Rest`. `ELEMENT` is the reply variable received in the `pop/1` message. Suppose a sequence of messages like:

pop(E1), pop(E2), push(10), ...

is received by an instance of the stack class created with a size of 3 and the stack pointer at 0, the first two `pop/1` messages will be skipped and the object will suspend until the first `push/1` message is received and accepted. This `push/1` message puts a number 10 in to the buffer and increases the stack pointer to 1. After it is successfully resolved, the first skipped `pop/1` message is accepted, because the guard of its method will now succeed. Then, the successful actions will bind reply variable `E1` of the message to 10, and set the stack pointer to 0 again. The second `pop/2` message will be retried but will again be skipped.

The Terminator Two kinds of *Terminator*s can be used to conclude the definition of a method. They actually control when an object can start processing the next message after committing to the method. The **concurrent** '.' terminating a method allows an object to receive and process the next message right after the commitment to the method. This will maximize the parallelism by allowing many messages to be processed concurrently. One important fact to remember is that if a state component is changed within the current actions, the actions triggered by the next received message will not see its current value. Instead, the new actions will see either an unbound new state component, or the value generated by the current actions. Actions of a method will automatically suspend if they need to access a state component that is changed by any previous invoked method, and for which the new value has not yet been computed.

In some situations, especially when messages should be dealt with sequentially, forcing an object to finish current actions before it can proceed to the next message is vital. For example, suppose in Program 5.3.2, in addition to putting popped terms in the messages as replies, the method has to write the terms into a file stream strictly according to the order in which terms are popped. If the `pop/1` method is allowed to proceed to the next message immediately after the commitment, successive pop messages may not be written to the file stream in the same order as they have been accepted because of the concurrently executed *write/2* calls. In this situation, the **sequential** terminator '&.' can be used to terminate a method `M`. This delays the commencement of processing of the next message until the all the actions of the method have terminated.

5.4 Communication Between Objects

Communication between objects in D_Parlog^{++} is based on the asynchronous message passing paradigm. Asynchronous messages are the only media for interactions between objects. However, apart from carrying data, a message sent to another object can carry response variables. Using a response variable that is bound immediately a message is accepted (immediately an action for the message is committed to), we can implement synchronised communication. The sender simply waits until the response variable is bound. This provides a synchronous mechanism in an asynchronous message system. The form of the interaction is given below.

```
At sender's side:      send(Arguments) : guard ->
                                         preceding-actions,
                                         message(Data, Signal) => receiver,
                                         wait(Signal) &
                                         subsequent-actions &.

At receiver's side:    message(Arguments, Signal) : guard ->
                                                    Signal = accepted,
                                                    other-actions.

or                     message(Arguments, Signal) : guard ->
                                                    other-actions &
                                                    Signal = copied.
```

`wait/1` is a Parlog primitive that suspends until its argument is a non-variable. The sequential conjunction operator which follows the call to the *wait/1* primitive, and the '&.' sequential terminator of the method, means that the future activities on the sender side will be suspended until the `Signal` variable is bound.

On the sender side that are two forms of message receive for the signal message. One binds the `Signal` variable immediately the actions for the method commence, the other only binds it when all these actions have terminated. This is because the unification that binds the variable is preceded by the sequential '&'. In the latter case the sender will resume only when the action it requested has terminated. This makes the invocation of the method behave more like a remote procedure call.

5.4.1 Atomic Transactions

At the sender side, the `other-actions` can contain other message sends to the same receiver. The fact that the sender waits until the first message is acknowledged before sending another message means that the second message cannot overtake the first. In contrast, two message sends:

```
ms => O, ms' => O
```

executed in parallel might well result in message `ms'` arriving at `O` before message `ms`.

So using response signals is the way we can guarantee order of arrival of messages at another object. However, this technique will not prevent the sequence of messages sent from an object O having messages sent from other objects interleaved. These will be messages that have been sent while the object O was waiting for a response. (Remember that all objects are active and are executed concurrently.) To solve the interleaving problem, we can send a sequence of messages as in:

```
(Msg1,Msg2,...,MsgK) => O
```

This guarantees that the sequence will be preserved on arrival at the receiver object, i.e. that the messages will not get out of order (`Msg1` will arrive before `Msg2` etc), and no other messages sent from elsewhere will get interleaved with the sequence. This is useful for keeping the atomicity of a sequence of access and update operations performed on an object.

5.4.2 Self Communications

An essential feature in object-oriented programming is **self** communication. In D_Parlog++, the keyword **self** always refers to the identity of the object which is currently executing the method in which the **self** appears. During compilation, a **self** appearing as a argument in a call or in a message send will be replaced by the extra state variable `_self` added to every instance object. This variable will be bound to the object identity when the object is activated. Use of **self** in class methods is replaced by the name of the class.

Ordinary Self Communication In an ordinary **self** communication, the message is sent back to the object using the normal message send operator `=>`. It will go onto the back of the current input message queue for the object, just like a message sent to the object from outside the object. The situation is just like a person putting non-urgent work, which should be done by themselves, at the bottom of the 'in' tray labeled 'to be done'.

Privileged Self Communication The special message send operator ==>, instead of =>, can be used to send a privileged or express **self** communication. In a privileged **self** communication, the message (or sequence of messages) will be put at the front of the current message queue of the object, to become the next message to be acted on. Only one express message can appear in any method.

Example of self communication Program 5.4.1 shows a fragment of the instance definition section in a class called **library_counter** which is responsible for issuing books. The program serves as a demonstration for the two kinds of self communication, rather than efficient implementation.

In the program, the `renewing/3` polymorphic methods are used to renew a book borrowed by an user. If the guard tests of the first two methods fail, the third method simply sends two messages in an atomic sequence to the object itself. The purpose of the atomic communication is to return the book and then borrow it *immediately*. This prevents the book from being reserved by other users after it has been registered as returned but *before* it has been re-borrowed. The timing of the two changes to the records database is not so critical as long as the changes are performed atomically. Their net effect is to record a new return date for the book, with the same borrower.

The express self communication of a `borrowed/2` message in the second method for `borrowing/3` is in order to ensure that the `borrowed/4` message is immediately sent to the Library_records object linked with the counter object. Note that the invoked `borrowed/2` method also waits until an acknowledgment of receipt of the message comes back from library records before terminating. Since the method rule is also syntactically terminated with an &., no other message will be accepted by the library counter object until this acknowledgment is received. Hence the book will not appear to be available to any other user once a borrowing request for the book has been accepted by the library counter object but before that request has been recorded in the `Library_records` object.

```
class library_counter with
{
 ................................
 instance_definition
   states
     DueDate, Library_records
   methods
     borrowing(Borrower,Book,Ans):
             over_due(Borrower,Reply)=>Library_records,
             Reply==yes ->
        Ans = (no, 'you have an overdue book!').
     borrowing(Borrower, Book) ->
        borrowed(Book, Borrower) ==> self,
        Ans = yes.
     borrowed(Book, Borrower) ->
        borrowed(Book, Borrower, DueDate,Ack) => Library_records
        wait(Ack) &.
     returned(Book)  ->
        returned(Book) => Library_records.
     renewing(Borrower,Book,Ans) :
             over_due(Borrower,Reply)=>Library_records,
             Reply==yes ->
        Ans = (no,'you have an overdue book!').
     renewing(Borrower,Book,Ans) :
             reserved(Book,Reply) => Library_records,
             Reply=yes ->
        Ans = (no,'book is reserved!)'
     renewing(Borrower,Book,Ans) ->
        ( returned(Book), borrowed(Book, Borrower) ) => self,
        Ans = yes.
     ................................
}.
```

Program 5.4.1 A fragment of a class for a library counter.

Another example As a second example of the use of express self communication, Program 5.3.1 is extended to Program 5.4.2 with a new method `evaluate/1` which sends an atomic sequence by an express self communication in order to pop the current top two elements in a stack, then to push the result. The result is evaluated by a call to a procedure, `evaluate/4`, that is private to the class.

This is a very interesting example since it *requires* concurrency of evaluation of the messages sent to the object. Remember that when the '.' terminator of a method is used that the object will be able to start processing the next message immediately after it commences evaluating the method. This is essential. The call to `evaluate/3` will be suspended waiting for the values of the reply variables `First` and `Second` of the two express `pop/1` messages send back to the object. But the object will be able to process these messages even though the call is suspended. Note also that it does not matter that the `push/1` message may be received and acted upon before its argument `Result` has been computed. If `Result` is unbound when the message is received, the variable `Result` will be pushed on the stack. When `Result` gets bound, its value will automatically replace the variable on the stack. This is because values are passed between action calls and between objects using unification. If any other object should pop the `Result` variable before it was bound, it would simply get a reference to the variable as the value. Any computation it needed to do on the value would just suspend until it was bound. When it was bound, *distributed* unification would ensure that it saw the value, even if it was on another machine.

```
class stack with
{
 instance_definition
   states
     Size, StackPointer:=0, Buffer:=[].
   methods
     %... Other method definitions .....

     evaluate(OP) : StackPointer > 1 ->
        evaluator(OP, First, Second, Result),
        (pop(First), pop(Second), push(Result)) ==> self.

 code_definition
     mode evaluator(OP?, First?, Second?, Result^).
     evaluator(+, First, Second, Result) <-
        Result is First + Second.
     evaluator(-, First, Second, Result) <-
        Result is First - Second.
     evaluator(*, First, Second, Result) <-
        Result is First * Second.
     evaluator(/, First, Second, Result) <-
        Result is First / Second.
}.
```

Program 5.4.2 Extensions to program5.3.2.

5.5 Inheritance and Delegation of Classes

The single inheritance mechanism is adopted in D_Parlog^{++}. Though simple in both syntax and semantics, it provides great possibility for code reuse and flexibility on class organization. (Smalltalk, the original OO language, only has single inheritance). A class inherits from another by having [**isa** *SuperClass*] added to its title line. What is inherited from a superclass include all the state components and the methods of the instance definition section. A superclass may inherit from yet another class and so on. For efficiency, a copy policy is adopted for instance methods. All the instance state components and the methods of the superclasses are added as extra state components and methods of the instance at the bottom of the hierarchy. This is also the inheritance method of Smalltalk. The class state components and the class methods are not copied, inheritance at the class level is achieved via message delegation.

To examine the single inheritance mechanism of D_Parlog^{++} in more detail, consider Figure 5.1 in which **current** *inherits* from **account**. This single inheritance link is implemented by two class declarations in Program 5.5.1.

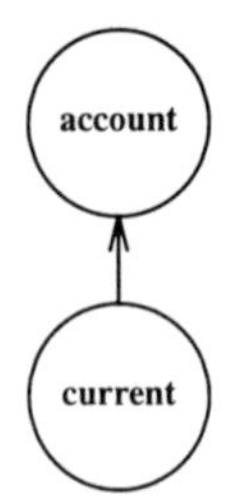

Figure 5.1: Inheritance graph for 'current account'

We do not give the create methods etc that will be automatically added to each class definition, but suffice it to say that the methods will be similar to those that were added to Program 5.3.1. The difference is that the create method for the inheriting class, **current** will have a message send:

```
add_a_member(InstanceID) =>  account
```

where `InstanceID` is the identity of the newly created instance. This is to ensure that any new current object has its identity added to the membership list for the **account** class as well as for the **current** class. Correspondingly, when a current object is destroyed, a `delete_a_member/1` message will be sent the **account** class. Finally, because it does inherit, the `add_a_member` and `delete_a_member` methods for **current** are slightly different. They must forward any received message to the **account** class. Below we give `add_a_member` method for **current**.

```
add_a_member(InstanceID) ->
   Members := [InstanceID | Members],
   add_a_member(InstanceID) => account.
```

The Superclass - account The first class, called **account**, defines the basic information and some simple operations for a bank account.

The instance state component includes variables for the name of the account holder, the balance of the account and the accumulated interest over a certain period. Two methods are shown in the instance definition. The first is a simple inquiry method which returns the present balance of the account, and the second is a method to calculate, on request, the interest earned by the account. Notice that the second instance method *calculate_interest/0* uses a state component `InterestRate` which is not defined yet. This is possible in

D_Parlog^{++} because of the copy policy. In this example, there is no particular advantage of defining the state component `InterestRate` only in a subclass of **account** instead of in the class. It is done just to demonstrate the flexibility of using inheritance for organizing classes. It does assume, however, that no instance of the account class will be directly created, only instances of subclass in which `InterestRate` is a state variable will be created. However, more advantages can be obtained if the same idea is used for methods, and realized by using **self** communication. That is, a **self** method can be invoked that is not yet defined, and will only be defined in a subclass.

```
class account with
{
 instance_definition
    states  Holder, Balance := 0, AccInterest := 0.
    methods
      balance(BAL) -> BAL = Balance.

      calculate_interest ->
         AccInterest := AccInterest + InterestRate * Balance.
}.

class current isa account with
{
 instance_definition
    states  InterestRate, CreditLimit.
    methods
      credit(AMOUNT) ->
        Balance := Balance + AMOUNT.

      debit(AMOUNT, ANSWER): Balance >= AMOUNT ->
        ANSWER = AMOUNT,
        Balance := Balance - AMOUNT.
      debit(AMOUNT, ANSWER)  ->
        ( ( VitualBalance is Balance + CreditLimit &
            VitualBalanc >= AMOUNT ) ->
          ( ANSWER = AMOUNT,
            Balance := Balance - AMOUNT );
          ( ANSWER is Balance + CreditLimit,
            Balance := - CreditLimit)
        ).

     collect_interest ->
       Balance := Balance + AccInterest,
       AccInterest := 0.
}.
```

Program 5.5.1 Current account class through inheritance

The Subclass - current Next, the program defines another class **current**, which inherits the **account** class.

As we discussed above, when creating an instance, the subclass will put the identity of the instance into its own member list, `Members`, as well as sending an **add_a_member/1** message to its superclass.

In **current** class, two instance state variables are defined, one is the expected `InterestRate`, and the other is the `CreditLimit`, which will hold the maximum amount an account can be overdrawn. Thus, for creating an instance of the **current** class, one might use a create message as:

```
create(AccountID, 0.02, 400, 'Debbie', 2000, _) => current
```

Notice that initial values for the bottom object are given first, but also that all state components, including the inherited ones, need to be mentioned in the create. The value of the last state variable, `AccInterest`, is given as '_'. This indicates that the default value given in the class definition should be used. The 2000 given for the initial value of `Balance` overrides its default value.

In addition to the two methods inherited from its superclass, instances of the **current** class accepts three more message skeletons. The `credit/1` method is very simple, and just adds the `AMOUNT` into the balance of the account, while the `debit/2` polymorphic methods are more complicated. To debit an account with an `AMOUNT`, the balance is first compared with the amount. If there is enough credit, the withdrawal is granted by assigning the `AMOUNT` to the `ANSWER` variable and the balance is deduced accordingly. On the other hand, if the balance is insufficient, but the shortage can be covered by the credit limit, the withdrawal is also granted with the balance going into a debt status. The sum of the balance and the credit limit is, however, the maximum amount for a withdrawal if the credit limit can not cover the shortage. Finally there is a method `collect_interest/0` for transferring the accumulated interest into the balance and clearing the accumulated interest. As can be seen in these methods, instance state variables declared in the superclass can be accessed directly in the instance methods of the subclass.

Overriding Another class **high_rate**, which is also a subclass of **account**, is declared in Program 5.5.2 in order to explain method overriding. The **high_rate** class omits the class definition section, and has only one instance state component `InterestRate`. As for instance methods, the `credit/1` method is similar to that of the **current** class, while the `debit/2` method is simpler because of the disposal of the state component `CreditLimit`.

An extra method `calculate_interest/0` is defined in the same class, which uses the same message pattern as the one inherited from the superclass **account**. The method, however, contains a different action. When locally defined polymorphic methods have the same skeleton as some inherited method, the locally defined methods will override the inherited one. This means that if the `calculate_interest` message is sent to an instance of the **high_rate** class, the local method instead of the one inherited from the superclass will be invoked. An overridden method will not be tried even when all the overriding ones fail to accept a message. The message will be skipped. Overridden methods must be explicitly invoked using a *super* communication.

```
class high_rate isa account with
{
 instance_definition
    states
      InterestRate..
    methods
      credit(AMOUNT) ->
         Balance := Balance + AMOUNT.

      debit(AMOUNT, ANS) ->
         ( Balance >= AMOUNT ->
           ( ANS = AMOUNT,
             Balance := Balance - AMOUNT );
           ( ANS = Balance,
             Balance := 0 )
         ).

      calculate_interest ->
         AccInterest := (AccInterest + Balance) * (InterestRate + 1).

      %....other methods......................
}.
```

Program 5.5.2 High_rate account class through inheritance

5.5.1 Super Communication

Instance methods which are not overridden when inherited from a superclass are used in the normal way because they are **inherited**, and are owned as a part of the inheritor. Overridden methods, though still inherited, will not be tried even when all the overriding methods fail to resolve the triggering message, they are obstructed by the overriding methods. However, an object can still invoke overridden methods if **super** communication is used.

```
class high_rate isa account with
{
 instance_definition
    states InterestRate, Threshold := 500..
    methods
      %....Other methods......................

      calculate_interest : AccInterest >= Threshold ->
         AccInterest := (AccInterest + Balance) * (InterestRate + 1).
      calculate_interest ->
         calculate_interest ==> super.

      %....Other methods......................
}.
```

Program 5.5.3 The high_rate account using super communication

Program 5.5.3 redefines the `calculate_interest/0` method in Program 5.5.2, and contains two polymorphic methods overriding the `calculate_interest` method of **account**. The first method will apply when the accumulated interest(`AccInterest`) has exceeded a certain amount(`Threshold`). The second method is committed to only if the guard of the first method fails. It will issue a super communication to force the inherited and overridden method defined in the **account** class into action. In another class, which in turn inherits the class `high_rate`, other conditions may be placed on the definitions of `calculate_interest/0` methods. In this way, the interest can be calculated circumstantially at different levels in the inheritance hierarchy.

5.5.2 Self Communication in an Inheritance Hierarchy

In D_Parlog^{++}, a **self** communication can be issued from either a class or an instance method. A **self** communication message sent from an instance method of a class goes to the instance that executed the method, which may be an instance of same subclass which inherited the method. However, a **self** communication in a class method does go to the class in which the method appears. The difference is due to the different implementation policies for inheritance of class methods and instance methods.

¿From an instance object's point of view, when a self-communicated message is received, the instance will try to resolve the message first by using the methods declared in the bottom class of the hierarchy of which it is an instance.

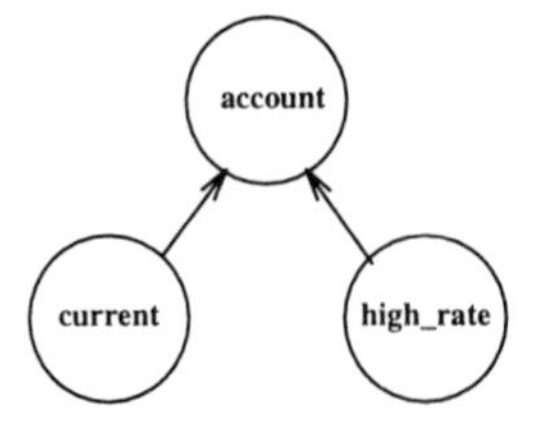

Figure 5.2: An inheritance tree

```
class account with
{
 class_definition
    %...class states and methods declaration................
 instance_definition
    states  Holder, Balance := 0, AccInterest := 0, TaxRate.
    methods
      %....Other methods ..................................

      net_interest(NET_INTEREST) ->
         gross_interest(G_INTEREST) ==> self,
         NET_INTEREST = G_INTEREST * (1 - TaxRate).

      gross_interest(G_INTEREST) ->
         G_INTEREST = InterestRate * Balance.
}.
class capital isa account with
{
 class_definition
    %...class states and methods declaration................
 instance_definition
    states  InterestRate, MiniBalance.
    methods
      %....Other methods ..................................

      gross_interest(G_INTEREST) ->
         ( Balance >= MiniBalance ->
              G_INTEREST = InterestRate * Balance ;
              G_INTEREST = 0
         ).
}.
class high_rate isa account with
{
 instance_definition
    states  InterestRate, Threshold := 500..
    methods
      %....Other methods......................

      gross_interest(G_INTEREST) : AccInterest >= Threshold ->
         G_INTEREST = (AccInterest + Balance) * InterestRate.
      gross_interest(G_INTEREST) ->
         gross_interest(G_INTEREST) => super.

      %....Other methods......................
}.
```

Program 5.5.4 Self communication in an inheritance hierarchy

To explain the idea more clearly, the inheritance tree in Figure 5.2 is implemented by the class declarations in Program 5.5.4. In the **account** class, there are two methods for calculating net and gross interests of an account. To calculate the net interest, the `net_interest/1` instance method first uses a **self** communication to invoke a `gross_interest/1` method for calculating gross interest, and then deduces the amount liable to the tax from the gross interest carried back by the self-communicated message.

For an instance of the **account** class, the `gross_interest/1` **self** communicated message will be handled by the `gross_interest/1` method defined in that class. However, for an instance of the **capital** class, the message will be handled by the overriding method definition for `gross_interest/1` given in the **capital** class, and similarly for an instance of the **high_rate** class. Thus, different formulas are be used to calculate the gross interests for **capital** and **high_rate** accounts even though this is calculated as part of a method, `net_interest`, that they both inherit from their shared parent.

The super communication technique mentioned before can also be used in association with self communications to bring the overridden methods into effect if, in a subclass, the imposed local conditions of the overriding methods fail to select a suitable method for resolving a self-communicated message. For example, in the **high_rate** class of Program 5.5.4, the second of the polymorphic methods `gross_interest/1` uses a super communication to call the overridden method in the superclass if the guard part of the first method fails.

5.6 Concluding Remarks

We have introduced most of the key features of D_Parlog^{++}. One feature we have not mentioned is modelled on the 'become' of actors [Agha and Hewitt, 1987]. Instead of a state update, as a last action an instance object method can metamorphise the object into an instance of another class *without* changing the object's identity. This can be used to restrict, or totally change the methods of an instance object, when it enters a certain state.

The language is fully implemented and running on a network of machines at Imperial College. We believe that it has a unique blend of features and is a very powerful, and quite declarative, distributed object oriented programming language. Major influences on its design were the previous OO extensions of CLP mentioned in the introduction, actors, and ABCL/1 [Shibayama and Yonezawa, 1987]. ABCL/1 is an actor derived language which also has the concept of express messages. It is actually more general, for any object can send an express message to any other object in ABCL/1. However, the language does not have inheritance, and methods are sequential.

One weakness of D_Parlog^{++} is that methods cannot be allowed to fail, because there can be no backtracking on the commitment to use a method. To overcome this drawback, we have also developed a much more elaborate language called DK_Parlog^{++} [Clark and Wang, 1994]. The 'K' stands for 'Knowledge'. It reflects the fact that as well as the concurrently executable methods of D_Parlog^{++}, an object can also have *knowledge* methods expressed as sequentially executable, but backtrackable, Prolog rules. There is a penalty. Calls on knowledge methods only return answers when the method terminates, and the method execution is entirely sequential. So, there is far less concurrency when knowledge rules are invoked. Both languages, and their implementations, are further described in [Wang, 1995].

Going in the opposite direction, we are also considering restricting the use of reply variables in messages so that D_Parlog^{++} can be implemented without the need for an underlying distributed unification scheme. The key idea here is to declare which occurrences of variables in both a sent message and a message receive pattern are reply variables, and to insist that each reply variable of a method appears in exactly one action of the method, an action which assigns a value to the variable using a special reply variable assignment operation.

Now, when a message is sent, we know what its reply variables are and we can create temporary mailboxes on the sender's machine to hold the reply bindings for these variables *if* the target object is on another machine. For each such reply variable X we also fork a process that reads the mailbox value and assigns it to X. This process suspends until the mailbox gets a value, and kills the mailbox when the value is read. The mailbox identities replace the variables in the sent message, which will also be delayed if it still contains unbound variables. The assumption is, that since these variables were not declared as reply variables, their values are being computed on the sender side, so we should delay transmission until their values are computed.

When a receiving method wants to 'assign' to one of its reply variables, and this assignment action finds that the variable already has a value which is a mailbox identity, it instead sends the reply value to the mailbox. Again, this transmission will be delayed if this value still contains unbound variables, until their variable free values are computed.

Even with this restriction on the reply variable mechanism, all the example programs of this chapter can be easily rewritten and will still execute as described.

We have not given any formal semantics to D_Parlog^{++}. We leave this as a challenge to the OO semantics community.

Chapter 6

Concurrency and Real-time in VDM^{++}

This chapter discusses problems arising from the requirements of concurrency in Object Oriented systems, and with particular reference to the solutions adopted in the VDM^{++} specification language[1].

6.1 Concurrency and Object Structuring

6.1.1 Why Concurrency

There are two main reasons why we may wish to construct systems showing concurrent behaviour.

1. Concurrency may be an important aspect of the world being modeled. The problem domain consists of co-existing elements, so a model of the domain recognises concurrently active components; thus concurrency is an essential part of the problem abstraction.

2. Alternatively, the purpose may be to speed up a large computation. For example, a finite element computation, can be subdivided into parts which can be computed concurrently, so gaining time on a multi-processor ma-

[1] VDM^{++}[Dürr and N.PLat, 1994] is an object structured specification language developed as part of the AFRODITE project with support from the CEC under ESPRIT III project number 6500. As its name implies, it is an extension of the well established language VDM [Jones, 1990]. Most of the features of VDM used in VDM^{++} are drawn from the standardised version VDM-SL [Andrews, 1993][Dawes, 1991].

chine. This can lead to highly parallel systems, with hundreds, or even thousands of executions of some basic algorithm executing simultaneously.

Language features most appropriate for supporting concurrent execution may be different in these two problem domains. Though its use for the second kind of problem is quite possible, the selection of features in VDM^{++} has been oriented towards the first type of system.

6.2 The World Model

In the real world concurrency is the norm. The world consists of collections of entities, people, machines, etc. which *coexist*. Each pursues its life and objectives, modifying its own state and the states of other objects in its environment, concurrently with others.

As described in chapter 1 and illustrated by figure 1.1, each world object is characteristically modeled in a program by an *object*. Such world entities may be passive or active.

The state of a world entity is represented in the model by its data variables; its laws of behaviour are captured in the algorithms of the code. If the world objects are active then they must be concurrently active.

If an object in the real world changes its state spontaneously, by its own actions, then in the computer model its algorithms should execute concurrently with those of other world objects. They will be concurrent *processes*.

In a system composed of concurrently active entities there will be contention for the use of shared resources and services which require synchronisation of actions. Hoare introduced the *monitor* [Hoare, 1974] as a safe way of constructing correct programs sharing a resource. A monitor consists of a resource encapsulated in a module which exports operations to use it. The monitor itself, rather than its users, is responsible for its integrity. It is natural to consider a second monitor-like category of object, which manage resources, and are protected from inappropriate concurrent executions by their own internal organisation.

In VDM^{++}active classes are recognised by the presence of a *thread* part. All other classes used in a concurrent system are, *by default*, monitors which allow no internal concurrency, protecting their resource by imposing the rule that methods execute at most one at a time, in *mutual exclusion* with all executions of other methods of the class and *self exclusion* with concurrent executions of itself. This restricted form of control can be overridden by the presence of a synchronisation section introduced by the keyword *sync* in the class definition.

6.3 Synchronisation and Threads in VDM++

Objects in VDM++are structured in much the way described in section 1.3 of chapter 1. Each object encapsulates its instance variables and methods, while exporting the method names and signatures as the interface accessible to client objects.

To allow the definition of the thread activities of an active class and the specification of the synchronisation rules for a passive class, a class definition may have sections labelled respectively, thread and sync, as appropriate for the problem. Thus:

class *SomeName*

instance variables etc ....

methods etc ...

thread
— defines algorithms of spontaneous action.
— absent in passive class

sync — defines synchronisation of the methods

end *SomeName*

6.4 Describing the Behaviour of Active Objects

Active objects are characterised by the presence of a so-called *thread*, which specifies the behaviour of the object once it has been started.

The thread section has the form:

thread < *thread* >

VDM++ allows two kinds of < *thread* >:

1. *Declarative threads*, which specify threads in an *implicit* manner.
2. *Procedural threads*, which cater for the specification of threads in an *explicit* manner.

Classes with threads may be inherited. To avoid sharing data between concurrent program executions, care must be taken if inheritance leads to more than one thread part.

6.4.1 Declarative Threads

A declarative thread specification can be regarded as an implicit way of describing the activities to be carried out by the object.

The most abstract form of thread specification is an invariant which is to be maintained by the on-going thread activities of the object. In the modeling of physical systems such invariants may appear as differential equations, as we shall see in section 6.7.

Periodic Obligation

In many problems an object must obey periodically some algorithm. It is like a loop with a delay so that cycles are repeated with a fixed periodicity. A periodically recurring obligation for the invocation of some method of the object has the following form:

```
thread
  periodic (< time indication >)( < method >)
```

where < *time indication* > is a natural number, and < *method name* > is the name of a method defined for the class. Such a periodic obligation expresses that the method should be invoked at the beginning of each time interval of length < *time indication* >. There are two uses for this construct. First, the requirements of the world may dictate it; for example a lighthouse is to emit a flash of light every twenty seconds, or a clock to chime every hour.

```
class Clock
  ...
  ....

  methods
    chime ()
    is not yet specified

  thread
    periodic (3600)(chime)

end Clock
```

Secondly, as we shall see in section 6.12 this description can also be used for the periodic sampling of the variables in simulating continuous real time control, or the integration of differential equations in physics.

6.4.2 Procedural Threads

A procedural thread provides for the explicit specification of the behaviour of an object in terms of a *sequence of statements*. Statements are separated by a semi-colon, which is interpreted as a sequencing operator; the object always performs the actions defined by those statements in strict sequential order. The statements allowed are the same as those allowed in VDM-SL operation definitions, [Andrews, 1993] augmented with the *delay* statement and with the *select statement* (as in POOL [America, 1987] and Ada[US , 1983]) which provides the possibility of specifying conditions in which an object may, in the course of its sequence of operations, execute a method in response to an external call.

6.4.3 A New Look at the Dining Philosophers

The Dining Philosophers problem is *the* classic example of a collection of active objects competing for the use of a shared resource. It has been used as an illustration in many works on concurrency and has therefore the advantage of requiring little introduction to describe it to the reader. In this chapter it is given in a very simple form. The philosophers are taken as typical examples of active objects. Each philosopher belongs to a class Philosopher whose members have a thread which loops forever: after becoming hungry, he decides he wishes to eat and eats until he has had his fill; he then thinks until he again becomes hungry. None of these changes of state is determined by any request from outside. This can be modeled roughly as follows[2]:

class *Philosopher*

instance variables
T : @*Table*;
my_seat : *Nat*

methods

Eat ()
is not yet specified — a random period of eating ;

Think ()
is not yet specified — a random period of thought

[2] The operations *eat* and *think* are given as local methods of the Philosopher class. This is for simplicity in presentation. It could well be a better design to provide them as exported operations from some other object.

```
thread
  while true
  do   (T!Place[my_seat]‘pick_up_forks;
        self!Eat;
        T!Place[my_seat]‘lay_down_forks;
        self!Think)


end Philosopher
```

Here the type @*Table* denotes a reference type for objects of class Table. Control of the contention between philosophers in the use of their shared forks will be treated after the following section, while the notation

```
Seat[my_seat]‘pick_up_forks
```

will also be explained later.

6.5 Concurrency and Contention

In a system in which there exist objects executing concurrently, there may be contention for the use of shared resources. It must be possible to specify synchronization constraints on the method executions involved in accessing the resource.

In particular, methods of passive objects may be concurrently invoked causing contention for the access to their state variables, with undetermined and possibly erroneous behaviour. It is therefore necessary to constrain the executions of methods of an object, so that an object becomes like a ‘monitor’ in the terminology of Hoare[Hoare, 1974].

In VDM^{++} the *default* behaviours are as follows:

1. From the viewpoint of the caller, method executions are atomic. This means that the caller waits until execution is complete before proceeding. (A normal procedure calling mechanism)

2. Methods are executed in a mutually exclusive manner.

3. At most one execution request for a given method is dispatched at a time.

The specifier may need to generalise this behaviour; the synchronization part can be used to describe a wide range alternative disciplines. If an object has a synchronisation part, the default no longer holds.

The synchronization part has the form:

sync <synchronization specification>

VDM^{++} allows two kinds of synchronization specification:

1. *Declarative synchronization control*, which caters for the specification of synchronization constraints in an *implicit* manner.

2. *Procedural synchronization control*, which caters for the specification of synchronization constraints in an *explicit* manner using *traces.*[3]

6.5.1 Declarative Synchronization Control (permission predicates)

A declarative style of specifying synchronization constraints can be achieved through the use of *permission predicates*[4], also called permission *guards*. A permission predicate states conditions which must hold to ensure that a method can be executed in a safe manner. It has the form:

sync
  per $< method\ name > \Rightarrow < condition >$

where <*method name*> refers to the method of that object to which the constraint applies, and <*condition*> is a guard in the form of a boolean expression specifying the condition which must hold for the valid execution of the method. The scope of this condition does not include the formal input parameters of the method. Note that this does not mean that if the condition holds the method is executed, but only if the method has been invoked (by the object itself or by other objects) then failure of the <*condition*> means that its execution must be blocked.

An arbitrary number of permission predicates can be given. If two or more are given for a particular method, then they all must be satisfied for an execution to be permitted. This allows an existing set of rules to be extended, (for example in a subclass) so strengthening the control.

If no permission predicates have been defined for a certain method, then there are no restrictions on the execution of that method.

---

[3] A third technique, enforcing synchronised behaviour by the use of answer-statements in a procedural thread specification (see section 6.4.2), is not described here.

[4] Permission conditions form part of deontic logic and were introduced into the specification technique FOREST[P. Jeremaes, 1986]. They were used to specify synchronisation conditions in the definition of the programming language DRAGOON [Atkinson, 1991].

Events

Permissions are controlled by the *current* state of the object, and transitions between states occur at times determined by *events*. An event is an occurrence in the process of execution happening, in principle, at an instant in time. In practice it may be a time interval which is small compared with the times which are considered significant in the problem. The essence of an event, however, is that it is *indivisible*. In programming terms, this means that any actions which form part of the event are executed in a way which cannot be interrupted by other actions, and any data which are changed by the event are not accessible to any other active objects for the duration of the event.

Method executions generally take time. However, there are three events which can be recognised as associated with any method execution. Consider a method m.

1. A caller requests its invocation in an event req(m). Whether or not the execution of the event is possible at that time, the request will be noted by the system software.

2. At some time, the method execution will be dispatched. Accepting the dispatch is a new event act(m).

3. Finally, the execution will be completed. The instant of completion of the execution will be recognised in the event fin(m).

It is convenient to maintain counts of the number of each event since the system was started up. In VDM^{++}these counts are represented by *history functions* using the symbols #req, #act and #fin respectively.

The meanings of these symbols may be given by formulae such as:

```
on req(m) #req(m) := #req(m) + 1
```

In VDM^{++}however, they are implicitly defined, and may be used in defining permission predicates.

There is, in fact, a further event of interest for each *object*. That is its *creation*. At that event each of the counters for all its methods is set to zero. If my_object is an object with a method m, then:

on create (my_object)
    $\#req(my\_object.m) := 0$ **and** $\#act(my\_object.m) := 0$ **and**
    $\#fin(my\_object.m) := 0$

Permission Guards

Three types of guard may exist (which may be combined using logical conjunction):

1. *History guards* describe constraints in terms of events that have occurred in the past.
2. *Object state guards*, describe constraints in terms of the current internal state of the object.
3. *Queue condition guards* describe dispatching rules in terms of the states of the queues of method calls awaiting execution.

We will describe each of these separately in the following sections.

History guards A history guard is defined in terms of the following functions, which return values of implicit counters recording the histories of past method invocations:

```
#act, #fin, #req :  method_id → Nat0
#act (m) yields the number of times m has been activated
#fin (m) yields the number of times m has been completed
#req (m) yields the number of times the activation of m
                                            has been requested
```

There are also the following useful derived functions:

```
#active (m) = #act (m) - #fin (m)
    -- the number of currently active instances of m
#waiting (m) = #req (m) - #act (m)
    -- the number of outstanding requests
    -- for activation of m
```

In implementation, such history functions are in the nature of semaphores. For example, the function #active(m) is a counting semaphore which is 'upped' at the start of each instance of m and 'downed' at its completion

The following gives some examples of common cases: Some of these may become available as "macros" or "generic synchronisation specifications" in the VDM++toolset.

```
--mutex(a,b)
    per a ⇒ #active(a) + #active(b) = 0;
    per b ⇒ #active(a) + #active(b) = 0;
-- alternation (a,b)
    per a ⇒ #fin(b) = #act(a);
```

```
    per b ⇒ #fin(a) > #act(b);
--cyclic (a,b,c)
    per a ⇒ #act (a) = #fin(c);
    per b ⇒ #fin(a) > #act(b);
    per c ⇒ #fin(b) > #act(c);
-- readers and writers (reading r, writing w)
    per w ⇒ #active(w) + #active(r) = 0;
    per r ⇒ #active(w) = 0;
```

Queue condition guards Sometimes it may be necessary to control method invocation by the state of the queues of waiting calls. The commonest situation is to define priority invocation:

```
--priority mutex(a,b)
    per a ⇒ #active(a) + #active(b) = 0;
    per b ⇒ #active(a) + #active(b) = 0
                     ∧ #waiting(a) = 0;
```

Object state guards Object state guards specify constraints in terms of instance variables of the object.

6.5.2 Example of a Passive Class

A common example of a passive object is a buffer, providing its clients with storage for some kind of items, and presenting operations to insert and remove items, often referred to as put and get operations.

We assume in the following that there is some already defined class, SystemTypes, in which at least the type item is defined.

class *Buffer* **is subclass of** *SystemTypes*

instance variables
$store : item^*$;
init $store \triangleq$ **len** $store = 0$

methods
$put\,(i : item) \triangleq$
[**ext wr** $store$
post $store = \overleftarrow{store} \frown [i]$];

$get\,()$ **value** $i : item \;\triangleq$
[**ext wr** $store$
pre len $store > 0$
post $i = \overleftarrow{store}\,(1) \wedge [i] \frown store = \overleftarrow{store}$];

$size\,()$ **value** $n : \mathbf{N} \;\triangleq$
[**ext rd** $store$
post $n =$ **len** $store$]

end *Buffer*

In a concurrent program there may be calls overlapping in time from different clients for use of the services offered by the operations. The default of *mutual exclusion* ensures that at any time when a method starts executing the buffer is in a well defined state. However, this may not always be what the system designer requires, or it may not be sufficient. For example, the put and get operations write to the store, but size only reads from the store, without changing it. The so called *readers and writers* protocol demands that at most one writer may be busy at a time, and readers and writers may not be concurrently busy, but any number of readers may execute at once.

The default condition of mutual exclusion is sufficient to ensure that actions *put* by the Producer p and actions *get* by the Consumer c can never conflict.

A Bounded Buffer

Suppose now that the producer produces items faster than the consumer can consume them; the buffer size then grows without bound. Conversely, if the consumer works faster than the producer it will eventually attempt to remove an item from an empty buffer, violating the precondition of the get method, and creating an error. To make the system more controlled, we can extend the synchronising constraints of the buffer so that the producer becomes blocked if it tries to send a new item to a full buffer. This is done by adding the further synchronisation constraints:

```
per put => #act(put) - #fin(get) < maxbuffsize;
per get => #fin(put) - #act(get) > 0;
```

The second can replace the precondition.

A bounded buffer class can be created from the existing buffer by inheritance. A subclass may be formed with the same or *stronger* synchronising properties as its parent. In the present case, the parent had mutex constraints by default. The VDM++rule makes this constraint disappear when a synchs part is included. Thus this inheritance is not valid subclassing unless the mutex constraint is added.

class *BoundedBuffer* **is subclass of** *Buffer*

sync
$mutex(put, get, size)$;
per $put \Rightarrow$ **#act** $(put) -$ **#fin** $(get) < maxbuffsize$;
per $get \Rightarrow$ **#fin** $(put) -$ **#act** $(get) > 0$

end *BoundedBuffer*

a client of the buffer with an instance variable $b : @Buffer$ may now accept the assignment of an object reference to b:

```
b := BoundedBuffer!new;
```

where advantage is taken of polymorphic substitution of the `BoundedBuffer` subclass for the parent class `Buffer`.

In an exactly similar way the readers and writers protocol, discussed in section 6.5.1 can be applied to the Buffer. Since this time the constraints are weaker than mutex, the mutex conditions do not have to be replaced in the inherited class. One should note, though, that the class created is not a true subclass of the Buffer, since it has behaviours not possessed by the parent.

class *RWBuffer* **is subclass of** *Buffer*

sync
per $put \Rightarrow$ **#active** $(put) = 0 \wedge$ **#active** $(get) = 0 \wedge$
#active $(size) = 0$;
per $get \Rightarrow$ **#active** $(put) = 0 \wedge$ **#active** $(get) = 0 \wedge$
#active $(size) = 0$;
per $size \Rightarrow$ **#active** $(put) = 0 \wedge$ **#active** $(get) = 0$

end *RWBuffer*

The *size* operation is a reading operation, and may be executed as many times as there are requests unfulfilled. The *put* and *get* operations are both writers, so need full exclusion of readers and other writers.

6.5.3 A Return to the Dining Philosophers

The forks used by the philosophers constitute shared resources. They must be protected from contention in their use by the active philosophers by synchronising the actions to pick them up. They are made into attributes of a shared table object, derived from a class Table. To provide each philosopher with a separate seat at which he can obtain his requirements we first define a class for the places at the table.

class *Seat* ;

types
forkState = IN_USE | FREE

instance variables

f : *forkstate*

methods
pick_up_forks ()
is subclass responsibility;

lay_down_forks ()
is subclass responsibility

end *Seat*

A seat object forms an abstract interface to the table, in a way described in chapter 1 section 5.3. It has two methods whose definition is postponed to the subclass.

The table is an aggregation of five instances of the seat. It therefore presents to its clients a family of five instances of each of the methods pick_up_forks and lay_down_forks. Where the multiple inheritance causes ambiguities of names, the names are dis-ambiguated by using a prefix; in the standard VDM^{++} form they are labelled Seat[1]'pick_up_forks, etc.

class *Table* **is subclass of** *Seat*[1 . . . 5]

functions
left : **N** → **N**
left (*n*) $\triangleq$
n **mod** 5 + 1

instance variables — No new instance variables

methods
Seat[*i*]'*pick_up_forks* () $\triangleq$
[**ext wr** *Seat*[*i*]'*forkstate*, *Seat*[*left*(*i*)]'*forkstate*
post *Seat*[*i*]'*forkstate* = IN_USE ∧
Seat[*left*(*i*)]'*forkstate* = IN_USE];

$Seat[i]`put\_down\_forks\,() \triangleq$
$[\ \mathbf{ext\ wr}\ Seat[i]`forkstate, Seat[left(i)]`forkstate$
$\quad \mathbf{post}\ Seat[i]`forkstate = \text{FREE} \wedge$
$\qquad Seat[left(i)]`forkstate = \text{FREE}\ ]$

sync
— default mutex must be strengthened with a state guard
— ensuring that each action of picking up forks occurs
— only when both are free

$\mathbf{per}\ Seat[i]`pick\_up\_forks$
$\quad \Rightarrow\ Seat[i]`forkstate = \text{FREE} \wedge$
$\qquad Seat[left(i)]`forkstate = \text{FREE}$

end *Table*

The overall structure of the system is illustrated in figure 6.1, using the notation of OMT.

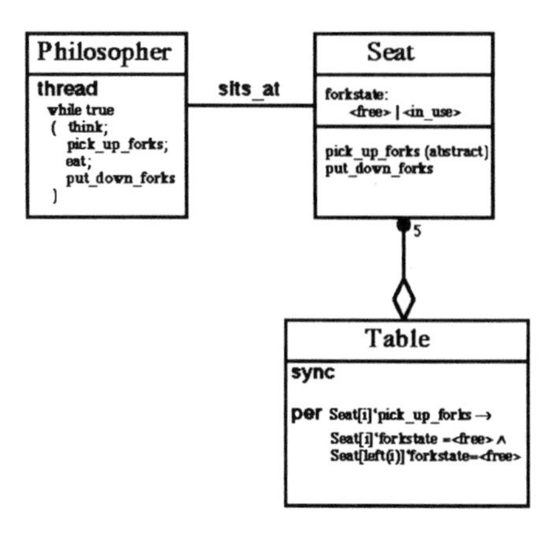

Figure 6.1: OMT diagram illustrating the structure of the Philosophers' Table

6.6 Other Ways of Expressing Synchronisation

6.6.1 Traces

Traces are an elegant way of specifying event sequences. If methods are considered atomic actions then the rules governing their sequences can be defined by defining a trace.

Consider the buffer again. Suppose we want a *unibuffer* which can only hold one item. If no items are to be lost, put and get actions must alternate. This is the behaviour of a message box. Its state alternates between full and empty.

The rule of alternation can be expressed as a trace:

```
<(put;get)*,{put,get}>
```

The first bracket defines the allowed interleavings of operations from an alphabet listed in the following brace. The sync part of the class could therefore be given in the form:

sync
  general Messagebox = ⟨(put;get)*, (put,get)⟩

One may consider the permissions as a step towards implementing the trace. However, traces are less general than permissions since they always assume the methods are atomic and mutually exclusive. There is no concept of method executions overlapping in time.

6.7 Feasibility and the Periodic Obligation

A periodic obligation requires that the method named be executed exactly once in each time interval. Ideally, this would be at the precise periodic time. However, it is necessary to allow tolerance on this to handle possible contention for the use of the processor. The specifier must show that:

- The duration of execution of the periodic method is much less than the specified periodicity.
- that the time required for all the computations initiated by invocations of methods by external clients is still available.

These are not, of course, capable of proof in a formal sense. Time requirements which cannot be met on some processor may be satisfied on a faster one. However, there are a few considerations which should be mentioned.

1. There must be no synchronisation rules which could block the periodic method. It must always be permitted to dispatch the periodic method. Formally one can write:

   ```
   per periodic-method => true
   ```

2. This means that the periodic method must not appear in any trace alphabet constraining the execution sequences for the class.

3. Care must be taken that, if the periodic method itself makes calls on some external method, that it in turn satisfies the same conditions.

It is not forbidden to inherit a class with a thread, thereby forming a class which has multiple periodic obligations. The same considerations apply to each.

If a class has been formed whose time obligations are not feasible on the available processor, it may be possible to "anneal" [Goldsack and Lano, 1996] the class, forming several objects whose combined effects are the same as those intended. Then the required time responses may be obtained using a distributed or multi-processor system.

6.8 Specifying Real-time Systems

6.8.1 Real-time Requirements Engineering

We consider a real-time system to be one in which the correctness of its result is determined not only by the value of the result but also by the moment in time that result is available. Thus, there exists a *deadline*, an instant at or before which an event must happen or by which a 'calculation' must be completed.

Sound development of real-time systems is based on:

1. Specifying a high level real-time requirement.
2. Having a "calculus" to show that after decomposition the "refined" specification still has time behaviour within the higher level requirements.

To give a simple example, the figure 6.2 shows a method a with a specified maximum duration. It is implemented as a sequence of steps, $a1$; $a2$ etc. then it is evident that the accumulated time must satisfy the overall duration requirement.

Similarly, a method is composed of statements; each statement has an upper bound on its duration which may depend on the data (both state and parameters). These must not exceed the required duration of the complete method.

This makes real-time specification orthogonal, in some sense, to the corresponding functional specification, although the decomposition criteria may be related.

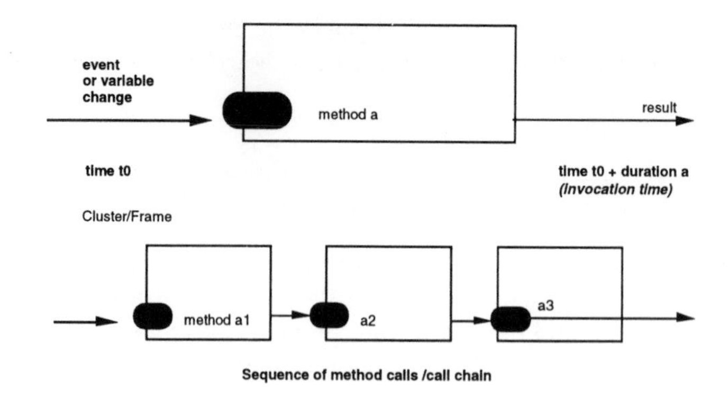

Figure 6.2: Sequence of method calls/call chain

6.9 Language Facilities

To specify a real time system we need certain new language features:

"Real-time"

1. Access to the current real time on the system clock, and expression of relations of time behaviours.
2. Concepts of *duration* and *delay* to express time behaviour of events (especially method activations and completions).

"Design"

1. The use of continuous time in the specification (representing the continuous behaviour of objects in the real world)
2. A technique of sampling to provide the mapping of continuous variables to their discrete representations in the computer.

6.9.1 The Notion of Time

When, in a static specification, a reference is made to the current time, it refers to the exact instant at which the *pseudo execution* of the specification, "evaluates" the expression or "executes" the action which references it. In VDM++ this expression has the syntax:

```
now
```

The value returned by this expression is an observation of the monotonically increasing value of the environment time variable. It will be known with a resolution implicitly defined for the entire system. In many cases specifications will use only differences between two 'moments', making the absolute value of the time less interesting.

A special interpretation is given to the meaning of the `now` expression in a postcondition. Here the value `now` means the time of the completion of the specification statement, at which instant the post-condition holds. The old value $\overleftarrow{now}$ refers to the time of the corresponding start event. For a complete method M these are the events `#fin(M)` and `#act(M)` respectively

Duration of a Method

With the meaning of `now` described above, a method specification can be given as follows:

methods
mymethod ($p1 : par1$) $\triangleq$
[**pre** *pre-condition* — held at $\overleftarrow{now}$
post *post-condition* — holds at now]

As a consequence, in a post-condition, the time interval

$$< \overleftarrow{now}, ..., now >$$

represents the time interval during which the method is active, whereas

$$now - \overleftarrow{now}$$

gives the duration of the method.

With the use of these expressions one can express also properties which require constraints over time dependent variables. However, a clear separation between functional specification of a method, by means of its post-condition, and expressions about the time involved with the execution of the specified method is desirable. For that purpose a special timed post condition is used.

Timed Post-conditions

In general the result of an execution of a method depends on:

- the state S at $\overleftarrow{now}$
- the input parameters par

The first dependency is the normal post-condition and can be characterised by a relation:

$$\textbf{post} \quad F(par, S, \overleftarrow{S})$$

A similar dependency holds for the time it takes to complete the method:

$$\textbf{duration of method} = \tau \; (\; par, \; \overleftarrow{S})$$

A timed post condition, optionally present after the normal post-condition, describes the constraints in the real-time domain, which are imposed on every implementation of this specification. It is typically a requirement specification. Its syntax is:

$$\textbf{time post} < \textit{time-expression} >$$

As an example one could add to the previously specified method *mymethod*:

$$\textbf{time post duration}(mymethod) \leq 25ms$$

expressing an acceptable upper bound for the execution duration.

6.10 Continuous Models

6.10.1 Continuous Time Variables

Most variables of interest in a real world system are continuous functions of time. The discrete nature of the variables in a computer are an approximation to reality, an implementation detail. To provide a truly abstract specification of a real time system requires, therefore, a concept of *continuous real time*.

In specifying a class to model some world concept, the "instance variables" block of a normal class is supplemented or replaced by a block defining physical variables which are continuous functions of time. For example:

time variables
  input $x : \mathbf{R}$;

$y : \mathbf{R}$

The function values x and y change continuously over time either according to rules given in the requirements as mathematical formulae, with the time as an independent variable, or if it is an *input* it may change as an observation value of a physical quantity in the environment. Just as a method may only be expected to perform correctly if some precondition is satisfied over the parameters supplied by the caller, so the continuous system will act correctly only if the input variables satisfy some requirements. For example, a central heating system may not be designed to operate to maintain the room temperature if the external temperature falls below -10°C. Such a condition is given as an *assumption* for the variable described as an *input*. Similarly, if the assumptions about the input are met, the system will be expected to perform in a way described in an *effect* clause. For a single output variable, this is analogous to a postcondition in a normal class; for the whole object it is an invariant.

Classes with such timed variables can be related to each other by inheritance. Clientship relations are also possible; however, the coupling between input and output variables has to be achieved through a topology statement at a 'higher level' (workspace).

The use of the method interface is reserved for the discrete system behaviour and provides the specification of a sampling mechanism, when used on continuous variables.

6.11 Continuous Classes

For brevity we call classes with time variables, continuous classes. A high level model may be composed entirely of continuous classes. Its topology is established through an independent workspace in which the objects are created and the connections established and in which the system invariants interpreted for the chosen topology may also be specified.

The continuous classes are idealised model components. The relationships between input and output variables are specified in one or more *effect* clauses; in many practical problems these take the form of differential equations.

Because all continuous object are idealised, no synchronisation is needed: conceptually they simply have the role of maintaining the invariants represented by the continuous differential equations at all instants of time. Evaluation of these invariants requires no time.

6.11.1 An Example of a Continuous Model.

Consider a simple electrical circuit as depicted in figure 6.3.

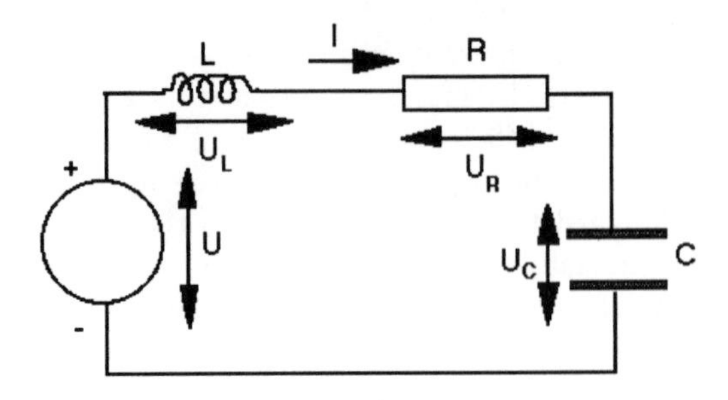

Figure 6.3: Circuit

In this circuit the main invariant at the system level is known as *Kirchhoff's Law.* relating the sum of the voltages:

$$-U + U_R + U_L + U_C = 0$$

With the current I and voltage U as continuous functions of time, differentiation yields the familiar differential equation:

$$-\frac{dU}{dt} + L\frac{d^2I}{dt^2} + R\frac{di}{dt} + \frac{1}{C}I = 0$$

To describe the classes involved in this example we first introduce a class *Component*.

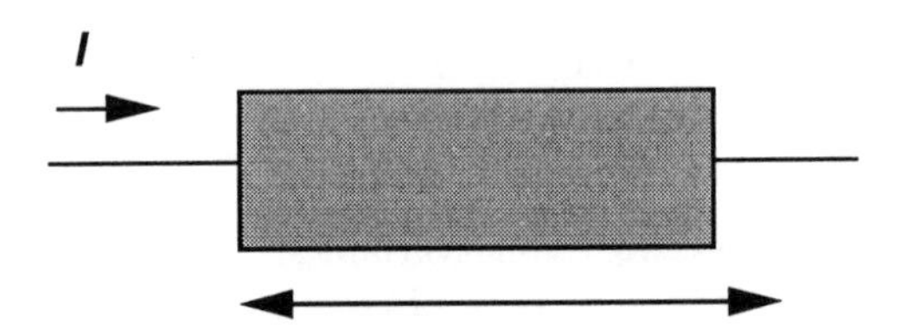

Figure 6.4: One component object

Such a component can be described by the relation between the Voltage *over* and the current *through* it. Each component has, as a general property, its connection capability, and we describe this by an instance variable to hold a reference *prev* to the previous Component object in the circuit, the one which supplies its input current.

This class serves as the superclass of all the more detailed components.

The specification of the Component class can be given by:

class *Component*
time variables
input: I_{in}: *current*
I_{out}: *current*
U : *voltage*
effect U: $U(t) = H \cdot I_{in}(t)$;
effect I_{out}[5]: $I_{out}(t) = I_{in}(t)$;

instance variables
name: **char***
value: **R**
`-- Denotes the magnitude in appropriate units of the`
`-- component (R, L or C) as the case may be.`
prev: @ *Component*
end *Component*

The voltage generated by the component is supposed related to its input current by a transfer function H.

6.11.2 Subclasses of Component

The special components of the example circuit can now be given as *subclasses* of this *Component* class. In each class, assumption clauses may be introduced concerning the input conditions to which may be safely subjected. The proper declarations for these values are supposed to be contained in one *SystemConstants* class, which can be inherited by each other Class, but which is omitted here for simplicity.

The operator H is given more concrete form in each subclass, as an expression of the physics of the kind of element it describes.

class *Resistor* **is subclass of** *Component*
time variables
`-- no new variables needed`
effect U : $U = R \times I_{in}$
assumption I_{in} : $I_{in}^2 \times R \leq maxdiss$

instance variables
maxdiss: **R** `-- constant per object`
end *Resistor*

[5] For the components we deal with in this example this effect clause is valid. If one wished to include transformers and current splitters, this effect could not be given in so simple a way.

class *Inductor* **is subclass of** *Component*
  effect U : $U = L \times \frac{dI_{in}}{dt}$
  assumption I_{in} : $\frac{dI_{in}}{dt} \leq maxRate$
end *Inductor*

class *Capacitor* **is subclass of** *Component*
  effect U : $U = \frac{1}{C} \int_0^t I_{in}\, dt$
end *Capacitor*

We can include also here a class for the source, with a definition of the applied voltage. For example:

effect $U = U_0 sin(\omega t + \phi)$

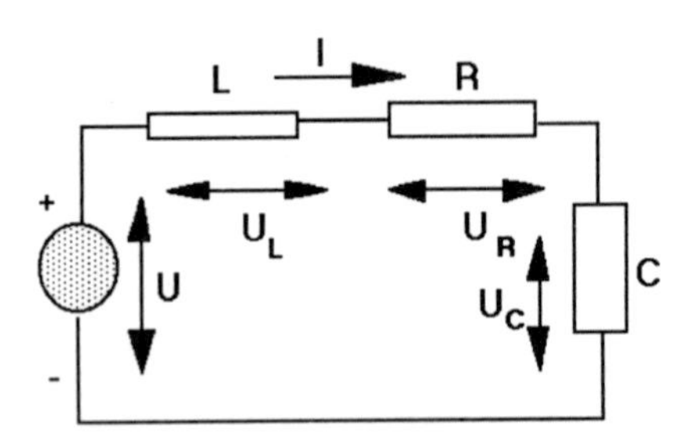

Figure 6.5: Component topology

The circuit itself is now an *aggregation* of these components. It can be constructed as a Workspace class as follows:

```
class Workspace
  instance variables
    s: @Source;    c: @Capacitor;
    r: @Resistor;  l: @Inductor;
  methods
    initial-method() ≜
    s := Source!new ; r := Resitor!new;
    l := Inductor!new ; c := Capacitor!new;
-- set the network structure
  topology [ post
        s.prev = c
      ∧ c.prev = r
      ∧ r.prev = l
      ∧ l.prev = s
    ];
-- set the input current sources
    topology [ post
```

$$
\begin{aligned}
& s.I_{out} = l.I_{in} \\
\wedge\ & l.I_{out} = r.I_{in} \\
\wedge\ & r.I_{out} = c.I_{in} \\
\wedge\ & c.I_{out} = s.I_{in}
\end{aligned}
$$

```
    ];
auxiliary reasoning
    Kirchhoff's Law:
      -s.U + l.U + r.U + c.U = 0
end Workspace
```

This "auxiliary reasoning" part is equivalent to the differential equations given earlier, when the *effect* clauses are substituted for the corresponding variables.

The specification described above, with its workspace, gives the full formal specification of our $Model_0$; all the components, all the connections and all physical relations are specified. However such a model is not directly implementable by a digital computer, though it could be used for real electronic analogue components. This specification is therefore not biased towards either a construction with hardware components or in software.

To provide a route from the analogue $Model_0$ towards a model in the discrete time domain, a domain switch is needed, which preserves all the functionalities of our $Model_0$.

We will now show in outline how the domain switch can be achieved in this example.

6.12 The Principle of Discretising

For the route towards an implementation (or a simulation) we will have to move to the *discrete time domain*. The sampling mechanism provided by the periodic thread of VDM^{++} described in section 6.4.1 provides support for the specification of this route. For each component we now introduce a *discrete-time equivalent* of the continuous-time component classes. It is convenient to make this a subclass of the continuous component, since it must still access the continuous input, in order to sample it.

The continuous time variables are replaced, as the main vehicle for state representation, by *sequences* of variable values. The ith element of each sequence will conceptually hold part of the system state at the time $i \times \Delta t$. Effects clauses from a continuous class are replaced in the discrete world by invariants and/or method post conditions. Active objects continually generate new values, extending the sequences, as the effect of a periodic thread; at each sampling time the value of the input is sampled (with a little non determinancy in the time) to become the next value in the sequence. The next value of the voltage is computed from the new input and its predecessors, using some appropriate approximation to the differentiations or integrations implied by the effects clauses. Inputs to passive objects will become, in the implementation, arguments of methods invoked by the environment which "owns" them. However, at the present stage each component has a reference to the one supplying the input current.[6] Here is a possible form for the *DiscreteComponent*. The input current is sampled by calling a method of the predecessor in the circuit.

class *DiscreteComponent* **is subclass of** *Component*

instance variables
$ID_{in} : \mathbf{R}^*$;
$UD : \mathbf{R}^*$;
currvalue : *Real*; ;
nextU : *Real*;
init objectstate $\triangleq$ **len** $ID_{in} = 0 \wedge$
len $UD = 0$
inv objectstate $\triangleq$ **len** $ID_{in} =$ **len** UD

methods
getCurrent () **value** *Idiscrete* : $\mathbf{R} \triangleq$
[**ext rd** I_{in}
— the inherited continuous variable
post $\exists t \in \overleftarrow{now} \ldots now \cdot Idiscrete = I(t)$];

makesample () $\triangleq$
(currvalue := prev!getCurrent;
— get next sample by calling the predecessor's getCurrent
nextU := Z ID_{in};
— The Z domain transfer function
[**ext wr** ID_{in}, UD
post $ID_{in} = \overleftarrow{ID_{in}} \frown [currvalue] \wedge$
$UD = \overleftarrow{UD} \frown [nextU]$]
)

thread
periodic (Δt)(makesample)

end *DiscreteComponent*

This generates a series of sampled values in the sequences,(ID_{in}, UD from the the continuous values sampled at 'exactly' the times:

$0, \Delta t, 2\Delta t, \ldots, n\Delta t$

[6] Earlier we asserted that for each component $I_{in} = I_{out}$. This makes the output current redundant, so it has been omitted in the following.

Full treatments of this transition can be found a many text books on numerical control and signal theory. See, for example, [van den Enden and Verhoeckx, 1989]. The relation between the continuous functions and the sample values can be defined formally by the *Z-transform* and the choice of a *Numerical Integration* method. The relations in the effect clauses are the Z-domain transformations of the differential equations and can be transformed into difference equations.

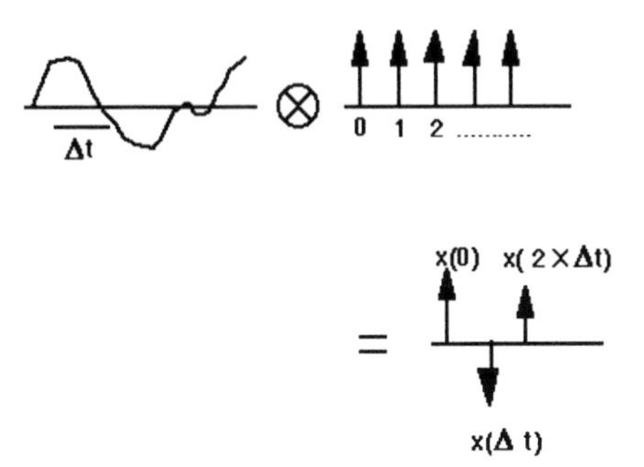

Figure 6.6: Z-transform

For example, the Inductor can now be formed by refining the Discrete component, relacing the sampling method by the following, in which we have used the clientship relation, to enable the component to access its predecessor in the circuit to obtain its newest sample value. A very simple 'trapezoidal' method is used here to evaluate the time derivative of the current, to provide the transfer function in this case.

makesample () $\triangleq$
(currvalue := prev!getCurrent;
nextU := L $\times$(*currvalue*-I_{in}(**len** I_{in}))/Δt;
— The Z domain transfer function made explicit
[**ext wr** ID_{in}, UD
post $ID_{in} = \overleftarrow{ID_{in}} \frown [currvalue] \wedge$
$UD = \overleftarrow{UD} \frown [nextU]$]
)

A further refinement notices that none of the sequence of values is actually used except the two most recent values of the current and the most recent voltage. These can be maintained by a simple device of updating the values of normal instance variables.

class *InductorSimulator* **is subclass of** *Component*

instance variables
I_{last} : **R**;
U : **R**;
currvalue : *Real*; ;

prev : @*Component*;

methods
getCurrent () **value** *Idiscrete* : **R** $\triangleq$
[**ext rd** I_{in}
— the inherited continuous variable
post $\exists t \in \overleftarrow{now} \ldots now \cdot Idiscrete = I(t)$];

makesample () $\triangleq$
(currvalue := prevComp!getCurrent;
U := L $\times(currvalue\text{-}I_{last})/\Delta t$;
— The Z domain transfer function made explicit
I_{last}: = *currvalue*;
)

thread
periodic (Δt)(makesample)

end *InductorSimulator*

6.12.1 Reasoning about Sampling

The above approach is valid only if the calculation time of the *makesample* method is much smaller then the system repeating period Δt. This extra requirement can be added to the definition of the method:

timed-post duration(*makesample*) $\ll \Delta t$

This building of the simulator equivalent using the Z-transform can be applied to each linear differential equation and is essentially the controlled development route from the differential to the difference equations.

6.13 Synchronising the Components

Each simulator object derived so far relies on its own active thread to re-establish the relation between its input current and output voltage at each system cycle.

For a system with more than one component this is not satisfactory: instead, sampling of its input by each element should be synchronised, so that the state changes of all the components occur as a single atomic event. This is best effected by invoking all the sampling methods from a single periodic thread at the level of the workspace. This is a straightforward transformation of the system which is not given in full here; details are left to the reader.

Chapter 7

Integrating Formal and Structured Methods in Object-Oriented System Development

This chapter describes systematic approaches for the formalisation and refinement of domain and analysis models, expressed in the OMT notation of Rumbaugh [Rumbaugh *et al.*, 1991], in the B Abstract Machine Notation (AMN) [Abrial, 1995 to appear] and the Z formal language [Spivey, 1992]. Whilst B provides a method and tool support for fully formal development from specifications to (3rd generation language) code, the main support for validating specifications against requirements which it provides is via animation. Similar, although more limited, support can be provided for Z.

The techniques described in this chapter are intended to improve the rigor of requirements specification, and the feasibility of formal development from these formalised requirements. The techniques define formal procedures for moving from diagrammatic analysis notations to mathematical specifications in B [Abrial, 1995 to appear].

Both the static and dynamic models of the OMT method are addressed. A comparison of the effectiveness of two alternative approaches to formalisation of dynamic models is performed, based upon the proof requirements generated by these approaches. A strategy for refinement to code and reuse of existing developments in the context of the B methodology is also described. Extracts from two case studies are used to illustrate the formalisation approaches:

1. A safety critical shipboard system;

2. A lift system.

The benefits of integrating formal and structured methods for high-integrity systems are becoming recognised by a number of groups, for example the work of Coleman at Hewlett-Packard [Coleman *et al.*, 1992], or of Hill at Rolls-Royce Associates [Hill, 1991]. The structured notations assist in the correct development of specifications (correct with respect to the user requirements), and the formal notations assist in the correct development of code (correct with respect to the specification). Both aspects are clearly essential for a safety-critical system. The IEC 65A 122 standard for safety-critical software also recommends the use of both structured and formal methods for software of the highest integrity level [IEC, 1992].

A substantial amount of work has been performed on the formalisation of structured method notations such as SSADM [Eva, 1992] and OMT [Rumbaugh *et al.*, 1991] in formal specification languages. Some Z-based approaches are described in [Polack and Whiston, 1991, Pyle and Josephs, 1991, France, 1992]. Particularly in the security domain, significant software development and formal proofs of systems have been performed using an integrated approach [Draper, 1993]. In addition, software engineering methods such as FUSION [Coleman *et al.*, 1994], developed by Hewlett-Packard, are also adopting a combination of formal and diagrammatic notations (in this case, a VDM variant and OMT). Formal languages are used to add semantics to informal notations, and the informal notations are used to provide an easily comprehensible presentation of the formal specifications.

However there needs to be some means by which the quality of the formalisation and the strength of the link between the informal and formal models can be assessed. In this chapter we provide systematic mapping techniques which can be used to give a mathematical meaning to the data and dynamic models of the OMT notations in an initial formal specification. This mapping not only helps guide the correct development of code, but assists in the practicality of proof and the possibility of reuse of user theories for resolving the proof obligations which arise in development. OMT has been chosen because the meaning of its notations is relatively well-defined and unambiguous, and because its notations are consistent with standard notations used in software engineering.

In Section 7.1 we introduce the B AMN language. In Section 7.2 we describe the representation of data models, in Sections 7.3 and 7.4 we describe two alternative representations of state models. In Sections 7.5 and 7.6 we describe experience of applying the techniques.

7.1 B Abstract Machine Notation

B Abstract Machine Notation (AMN) arose from the work of Jean-Raymond Abrial and others on the creation of a specification language and method capable of supporting the mathematical development of large systems. In 1986 BP Research at Sunbury initiated a project in its Information Technology Research Unit to look at ways of improving the cost effectiveness and quality of the software development process, both for software developments within BP, and for potential suppliers of software to BP. The Z language and method was selected as the candidate for this improvement technology, and in the period 1986 to 1989 there was a concentrated effort in developing a toolkit to support the B abstract machine notation which emerged, and the processes of specification, design (refinement) and coding. This effort also involved collaboration between BP and leading formal methods groups in a number of UK universities, and potential industrial users of the method.

By 1990 an alpha-test version of the toolkit was undergoing trials in a number of organisations, including IBM, British Aerospace, GCHQ and RSRE. Commercial courses in B technology were established by the BP group in 1991. Papers describing the work of BP and of the French groups were published, in the UK Refinement Workshop [Abrial, 1991, Lee *et al.*, 1991] and in the FORTE '91 conference [DaSilva *et al.*, 1991].

B applications at BP included the specification, design and coding of a database and an interface for storing refinery models and simulating movements of materials within a refinery. This was successfully completed in 1991. 20,000 lines of code were automatically generated from the refinement of an initial 6,000 line specification, at a considerable saving in resources over conventional development. In 1988 the use of B in the SACEM computerised signalling system was initiated. This system, which controls RER commuter trains in Paris, represents the largest application of B technology to date, and involved the formal specification and verification of 63% of the 21,000 line system. Manual verification was used after verification conditions had been automatically generated. The total validation effort was of the order of 100 person years [Bowen and Stavridou, 1993]. The initiative for this work came after the discovery of errors in previous safety-critical software systems for the French railways, and the consequent concern of the developers and procurers of these systems to improve their safety. The companies involved in this work were GEC Alsthom, MATRA Transport and RATP.

The use of B has continued in France, with two further safety critical railway applications being developed by GEC Alsthom using B [DaSilva *et al.*, 1991]. These were a subway speed control system, of 3,000 lines of code, and a further speed control system, of 16,000 lines of code. MATRA is currently involved in the redevelopment of an assembler-based metro control system using B technology, and established assessment and certification procedures for B-developed systems in the railway domain exist in France [Chapront, 1992]. The uptake of B in France represents one of the largest commitments to the use of formal methods in the safety critical domain to date.

The B Abstract Machine Notation is a formal specification language, which also contains elements which are close to conventional 3rd generation programming language constructs, particularly those of Ada. A brief introduction to the syntax of B Abstract Machines is given here. Full details of the language can be found in [Abrial, 1995 to appear].

The outline form of a B abstract machine is:

`MACHINE` $\mathbf{M}(\mathbf{params}_{\mathbf{M}})$

```
CONSTRAINTS Constraints_M
CONSTANTS constants_M
PROPERTIES Prop_M
VARIABLES variables_M
DEFINITIONS definitions_M
INVARIANT Inv_M
INITIALISATION Init_M
OPERATIONS
  z ← m(y)  ==
     PRE Pre_m,M
     THEN
       Op_m,M
     END
  ....
END
```

The machine **M** can be parameterised by a list of set-valued or scalar-valued parameters. The logical properties of these parameters are specified in the `CONSTRAINTS` of the machine. Optionally, constants, corresponding to axiomatic definitions in Z, can be declared in the `CONSTANTS` section. The definitions of these constants are given in the `PROPERTIES` section.

The variables of the machine are listed in the `VARIABLES` section. The constraints on the variables, including the typing of the variables, are specified in the `INVARIANT` of the machine. `DEFINITIONS` of mathematical abbreviations can be given in terms of the state variables and constants. The initialisation operation of the machine is specified in the `INITIALISATION` section of the machine. The methods or operations of the machine are listed in the `OPERATIONS` section. Input parameters are listed after the name of the operation in its definition, and output parameters are listed to the left of an arrow from the operation name.

It is possible to link and extend machines via the `USES`, `SEES`, `INCLUDES` and `EXTENDS` structuring mechanisms. These have the following roles:

- `SEES`: to support the read-only sharing of a component between other subsystems of a development, and the independent refinement of this component;
- `USES`: to support the definition of a shared read-only component between machines in a single subsystem of a development, whose state can be linked with the states of the machines which use this component;
- `INCLUDES`: to support the incremental addition of operations and state to a specification of a subsystem;
- `EXTENDS`: as with `INCLUDES`, but with all operations of the included machine becoming operations of the extending machine.

The term 'subsystem' means a single refinement path from a single abstract machine (possibly constructed from a set of other machines using the above constructs) through a sequence of refinements to an implementation. A B AMN development will usually consist of a linked set of such subsystems, with the implementations of some subsystems being defined in terms of the specifications of others via the IMPORTS construct.

The modularity provided by machines and the structuring mechanisms enables specifications and implementations to be decomposed into small, easy to verify, components, which may also correspond to intuitively natural domain entities (for example, **Ship** or **Compartment** in the load planning system of Section 7.5).

7.2 Representation of Data Models

Entity relationship attribute (ERA) models provide an overview of the abstract entities in a system being specified, and support partitioning of a system into conceptually coherent subparts. They present candidates for modules and (in B terms) machines in the design of the system. From a user's viewpoint, they provide a simple description of the static properties of the system.

In this chapter we will use OMT object model diagrams [Rumbaugh *et al.*, 1991]. Object models consist of named boxes for entities or object classes, in which all non-object valued attributes of the class are listed, together with their types. Relationships between entities are represented by lines. The number of participants in a relationship is indicated by means of circles or numeric annotations: a filled circle at the end of a line indicates that the adjacent entity may occur zero or more times in this relationship with one instance of the source entity. An unfilled circle indicates zero or one occurrences. A line with no circles indicates a 1:1 relationship.

Inheritance and specialisation can also be explicitly signalled on an object model diagram. An example of an object model, of a set of lines and boxes, is shown in Figure 7.1.

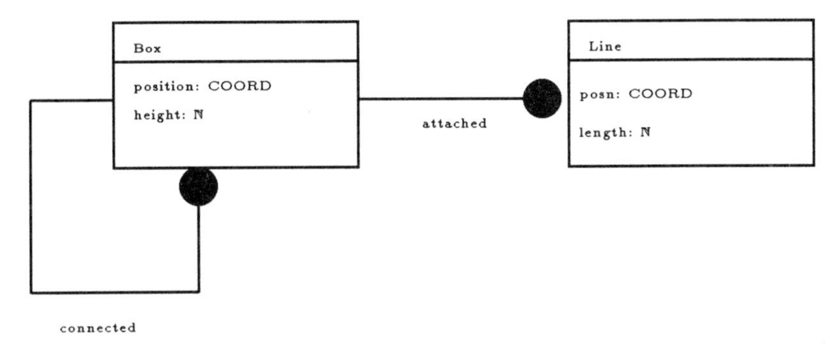

Figure 7.1: Object model of lines and boxes

In this system, each box may be **connected** to many other boxes and may be **attached** to many lines.

Approximately, entity types in a OMT object model will be expressed as B machines, encapsulating the sets of possible and existing instances of the entity type, and a set of functions representing attributes of the type. Links between entity types are represented using the B inclusion mechanisms for machines.

The following process model can be used to map analysis models expressed in OMT object model notation into systems of B machines:

1. identify the *families* of entity types in the data model – that is, the sets of entity types which are subtypes of a given type T which has itself no supertypes;

2. identify the *access paths* which are needed by operations and attributes of the types within each family to types in other families;

3. on this basis, produce a directed acyclic graph, whose nodes are the families, and whose edges are inclusion relationships USES or SEES between the nodes;

4. define machines for each family, following the procedure outlined below, and include machines in other machines using the relationships identified in the previous step.

Note that step 3 may not always be possible. Cycles $\mathbf{A} \rightarrow \mathbf{A}$ are allowed in the graph, but cycles $\mathbf{A} \rightarrow \mathbf{B}$, $\mathbf{B} \rightarrow \mathbf{A}$ or longer cycles are not allowed (they would lead to cycles in the machine inclusion relations, which are not allowed). If such cycles are required in the system (rather than being a feature of the general domain), then the entities concerned must all be placed in a single abstract machine.

The simple case of an entity without subtypes (that is, of a family containing a single type) will be considered first. In step 4 in this case, each concept Entity with attributes or links att1 : T1, att2 : T2, ..., attn : Tn will have a corresponding machine **Entity** of the form

```
MACHINE Entity
SETS
  ENTITY
VARIABLES
  entities,
  att1, att2, ..., attn
INVARIANT
  entities <: ENTITY &

  att1 : entities --> T1 &
  att2 : entities --> T2 &
  .....
  attn : entities --> Tn
  .....
END
```

This machine models a collection of Entity instances, rather than a single entity. If only one instance of the Entity was required, we would omit the declaration of **entities** and have instead a declaration of a machine **Entity**, together with variables **att**1 : **T**1, **att**2 : **T**2, etc.

The set **ENTITY** is the domain of object identities, and **entities** records the current set of Entity instances which are known to the system.

If there are relationships between entities in the object model, then some of the T1, ..., Tn will themselves involve other entities, say Entity2, Entity3 .... In this case we must SEE or USE the associated machine:

```
MACHINE Entity
SEES Entity2, Entity3, ...
...
END
```

We use SEES if we only need to use the object identity sets **ENTITY**2, **ENTITY**3, etc, in the invariant of **Entity** (ie, to provide a range type for a link of Entity), and we use USES if we need to be more specific and use the set of existing entities **entities**2, etc, as range types in the invariant. If there is a 1-1 link for example, such a stronger typing constraint would be needed. We may also use a parameter to place a bound on the maximum number of instances of a given entity which we will allow:

```
MACHINE Entity(maxEntity)
CONSTRAINTS
  maxEntity >= 1
  ....
INVARIANT
  entities <: ENTITY &
  card(entities) <= maxEntity
  ...
END
```

If we have Entity2 inheriting Entity1, then we need to place the constraint **entities**2 $\subseteq$ **entities**1 in the invariant of the machine representing the supertype (most abstract entity).

```
MACHINE Entity1
SETS
   ENTITY1
VARIABLES
  entities1, entities2
INVARIANT
  entities1 <: ENTITY1 &

  entities2 <: entities1
```

```
...

END
```

Similarly with exclusive subtypes Entity1 and Entity2, the constraint **entities1** $\cap$ **entities2** $= \varnothing$ is added to the invariant.

The examples in Section 7.5 and Section 7.6 provide a detailed illustration of the above method.

7.3 Representation of Dynamic Models

7.3.1 Process Model for Formalisation

Dynamic models for systems are often expressed in the form of *statecharts*, which define a set of states that an object can be in, and a set of transitions between states. These notations are based on those of Harel [Harel, 1987]. Transitions can depend upon the state of other instances of the type whose dynamic model is expressed in the chart, and can involve requests for transitions in other statecharts.

One technique that can be used for formal modelling of such systems is to create a machine encapsulating a set of instances of each object type, and the sets of instances of this object type that are in a given state. This representation regards such states as defining a 'transitory subtype' of the object type whose instances can be in these states. Transitions from one state to another are modelled by B AMN operations, with suitable preconditions and cases to express the different situations that can arise. One advantage of this scheme is that modularisation of the specification follows the structure of the problem domain, making it easier to trace the location of required changes to a system as a result of changes in the domain. Synchronisation between subsystems is expressed by means of operations which call operations from subsystem machines using the 'multiple generalised substitution' operator $||$. The semantics of this operator is similar to that of $\wedge$ in Z operation schemas: it specifies that the state transitions defined by its operands should both be performed, but does not specify any procedural order by which these transitions can be achieved. Naturally, the sets of variables updated by the two operands should be disjoint.

The associated process model extends that for transforming static data models into machines described in Section 7.2. Steps of recognition of families of entity types and the creation of associated machines are still performed. However, each entity type E may now have an associated dynamic behaviour description given by a state model. For each state S in this model, a new 'transitory subtype' S of E is considered to exist, and an associated variable **ss** is added to the machine representing the family of types to which the entity type E belongs.

Each transition between states in this model becomes an operation in the associated machine. Activities within states also become operations of the associated machine. Synchronisation between instances of entities from different families is expressed using || combination of transitions from the individual machines, in a machine which includes all the 'sibling' machines in the development. At this level, or at the level of systems which use the set of machines, any required ordering of operations can be imposed (such as that an activity performed in a state is executed immediately after an instance arrives in that state).

The general steps are therefore:

- create machine **Entity** for each entity type family Entity in the data model, and a variable **states** $\subseteq$ **entities** for each state in the chart for Entity;
- create operations of **Entity** for each transition and activity;
- express synchronisation between subsystems by means of operations which call operations from subsystem machines using the 'multiple generalised substitution' operator ||.

In more detail:

- **states become sets of instances**;
- state sets are disjoint;
- state sets make up the entire set of entities;
- **attributes and links become functions**;
- the domain of an attribute may be a proper subset of the instance set of the entity (it may only make sense in certain states of the entity);
- **initial states become creation operations**;
- the creation operation of the **Entity** specification initialises an instance to have the initial state (ie, it adds it to the variable representing the set of instances in the initial state);
- **transitions between states become operations moving an instance from one state set to another**;
- transition preconditions become operation preconditions;
- transition preconditions depending on the state of other instances of Entity can be expressed;
- **activities in a state become operations that do not change the state** (and which are preconditioned by membership of the state of which they are activities).

The process will be illustrated by the lift control system example of Section 7.6.

7.4 Alternative State Representation

It is clear that the above approach becomes infeasible once the number of states of an entity to be handled climbs above 8 – 10. As an alternative, we have the following process:

- Create a machine **Entity** for each entity type family Entity in the data model, and a status variable **status$_i$_flag** : **entities** $\rightarrow$ **STATUS_SET$_i$** for the **i**th statechart factor for Entity, where **STATUS_SET$_i$** is the set of state representatives for the states in the **i**th factor;
- create operations of **Entity** for each transition and activity in each statechart. Transitions which occur in more than one statechart require that two or more of the **status$_i$_flag** variables are simultaneously updated;
- synchronisation between subsystems is expressed as in the previous approach.

The overall process is thus quite similar to the previous approach, however:

- **states become sets of instances** defined by the inverse images of elements of the **STATUS_SET$_i$** sets under the **status$_i$_flag** functions;
- these state sets are therefore disjoint and exhaustive of **entities**;
- the creation operation of **Entity** initialises each **status$_i$_flag** to the initial state of each statechart factor;
- transitions between states become operations updating the **status$_i$_flag** variables of the affected factors.

If each factor is itself complex, we may attempt to separate out the definition of different factors into different machines. `DEFINITIONS` may be used to define the state sets as the inverse images of certain **STATUS_SET$_i$** sets under the **status$_i$_flag** functions. The ship loading case study of Section 7.5 and the alarmclock development shown below are examples of this approach.

7.4.1 Comparison

The state function approach has the following advantages over the set based approach:

- fewer variables are used in the machines representing entity types;
- in general, fewer proof obligations are generated.

Conversely, it has the following disadvantages:

- statements regarding membership of a state become less clear, or require the use of `DEFINITIONS` to define the state sets from the state functions. Definitions are not transmitted from one machine to another by inclusion constructs such as `INCLUDES`, so must be duplicated in each machine in which they are needed;

- the proof obligations are generally more complex and more difficult to resolve automatically or manually.

A comparative study of the two approaches was carried out on the lift system. Figure 7.2 gives (1) the number of generated proof obligations for both the state function based and the set based approaches, (2) the number of proof obligations remaining after autoproof, and (3) the number of proof obligations remaining after 30 minutes of interactive proof effort for each development.

Machine	*Generated POB's*	*Non-Autoproved*	*Non-Interproved*
Lift (set)	49	12	8
Lift (function)	23	14	12
Door (set)	23	2	0
Door (function)	9	0	0
Button (set)	26	4	0
Button (function)	12	0	0
LiftSystem (set)	7	1	0
LiftSystem (function)	7	1	1

Figure 7.2: Comparative proof difficulty for set and function approaches

7.4.2 Alarmclocks

As an extended example of the second approach to statechart formalisation, consider a simple specification of an alarmclock (with attached windows, which we ignore here). The dynamic model of the system is shown in Figure 7.3. This involves a factored representation, whereby transitions in each factor of the statechart can occur independently if they have distinct names. If a transition is used in more than one factor of a statechart then it can only occur if the state transitions described in both components happen simultaneously. The main benefit of a factored statechart is that the number of states is reduced (effectively the states of the system are formed from the cartesian product of the state sets in each factor).

Note that the transitions **on** and **off** are actually composed transitions:

on == **on_alarm/W.open_window**

off == **off_alarm/W.close_window**

where **W** is the window instance associated with the alarm clock instance acted upon by the transition.

The machine derived from this state model is:

```
MACHINE
  AlarmClocks(ALARMCLOCK)
SETS
```

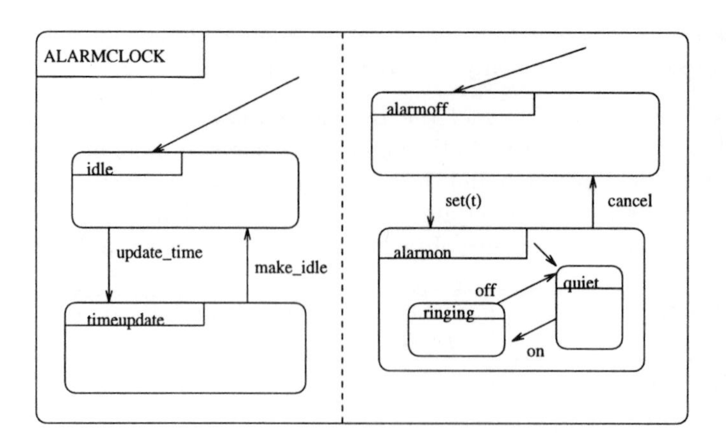

Figure 7.3: Statechart of alarmclock

```
  ALARMSTATE1 = {idle1,timeupdate};
  ALARMSTATE2 = {alarmoff,ringing,quiet}
     /* alarmon is an abstract state */
VARIABLES
  alarmclocks,
  alarmstate1,
  alarmclocktime,
  alarmstate2,
  alarmsettime
DEFINITIONS
  setalarmclocks == alarmstate2~[{ringing,quiet}];
  ringingalarmclocks == alarmstate2~[{ringing}];
  unsetalarmclocks == alarmstate2~[{alarmoff}]
INVARIANT

  alarmclocks <: ALARMCLOCK &
  alarmstate1 : alarmclocks --> ALARMSTATE1 &
  alarmclocktime : alarmclocks --> NAT &
  alarmstate2 : alarmclocks --> ALARMSTATE2 &
  alarmsettime : setalarmclocks --> NAT

INITIALISATION
  alarmclocks, alarmstate1, alarmclocktime,
     alarmstate2, alarmsettime := {}, {}, {}, {}, {}
OPERATIONS

   add_alarmclock(ac) =
           PRE  ac : ALARMCLOCK
           THEN
```

```
            ANY  nn
            WHERE
              nn : NAT
            THEN
              alarmclocks := alarmclocks \/ {ac} ||
              alarmstate1(ac) := idle1 ||
              alarmstate2(ac) := alarmoff ||
              alarmclocktime(ac) := nn
            END
          END;

update_time(ac) =
          PRE ac : alarmclocks &
              alarmstate1(ac) = idle1
          THEN
            ANY tt
            WHERE tt : NAT &
                  tt > alarmclocktime(ac)
            THEN
              alarmstate1(ac) := timeupdate ||
              alarmclocktime(ac) := tt
            END
          END;

make_idle(ac) =
          PRE ac : alarmclocks &
              alarmstate1(ac) = timeupdate
          THEN
              alarmstate1(ac) := idle1
          END;

set_alarm(ac,tt) =
    PRE ac : unsetalarmclocks &
        tt : NAT
    THEN
      alarmstate2(ac) := quiet ||
      alarmsettime(ac) := tt
    END;

on_alarm(ac) =
    PRE ac : setalarmclocks
    THEN
      alarmstate2(ac) := ringing
    END;

off_alarm(ac) =
    PRE ac : ringingalarmclocks
    THEN
      alarmstate2(ac) := quiet
    END;
```

```
  cancel(ac) =
      PRE ac : setalarmclocks
      THEN
        alarmstate2(ac) := alarmoff ||
        alarmsettime := {ac} <<| alarmsettime
      END

END
```

One drawback is that more complex proof obligations arise in the second form of representation (involving inverse images of functions, and functional overriding, rather than simple theorems regarding sets). In addition it is not possible to simultaneously update the state function of two different instances of a type:

$$
\begin{array}{l}
\textbf{invalid(ii)} \;=\\
\quad \texttt{PRE}\ \textbf{ii} \in \textbf{doors2} \wedge \textbf{doors2} \neq \{\}\\
\quad \texttt{THEN}\\
\quad\quad \texttt{ANY}\ \textbf{dd}\\
\quad\quad \texttt{WHERE}\\
\quad\quad\quad \textbf{dd} \in \textbf{doors2} \wedge \textbf{dd} \neq \textbf{ii}\\
\quad\quad \texttt{THEN}\\
\quad\quad\quad \textbf{doorstate(dd)} := \textbf{dooropen} \;||\\
\quad\quad\quad \textbf{doorstate(ii)} := \textbf{doorclosed}\\
\quad\quad \texttt{END}\\
\quad \texttt{END}
\end{array}
$$

This is an invalid operation because the same variable **doorstate** appears more than once on the left hand side of the expanded substitution

$$
\begin{array}{rl}
\textbf{doorstate}, \textbf{doorstate} := & \textbf{doorstate} \oplus \{ \textbf{dd} \mapsto \textbf{dooropen} \},\\
& \textbf{doorstate} \oplus \{ \textbf{ii} \mapsto \textbf{doorclosed} \}
\end{array}
$$

7.5 Ship Load Planning

This case study involves the analysis in OMT of a system which supports the creation of loading sequences for bulk carriers to ensure that safe hull stress limits are not exceeded during loading and unloading operations. It is based on some aspects of an application within the Afrodite ESPRIT project, which is using VDM^{++} as a specification language to enhance the dependability of load planners. The B AMN language was applied to the problem by the authors as a test of the integrated approach to formal and structured techniques. It was found that the formal specification process assisted in clarifying and simplifying the original data model, in addition to providing more precise and detailed semantic information. The development was pursued to the point where realistic animations of the entire system could be given, but implementation in code was not undertaken.

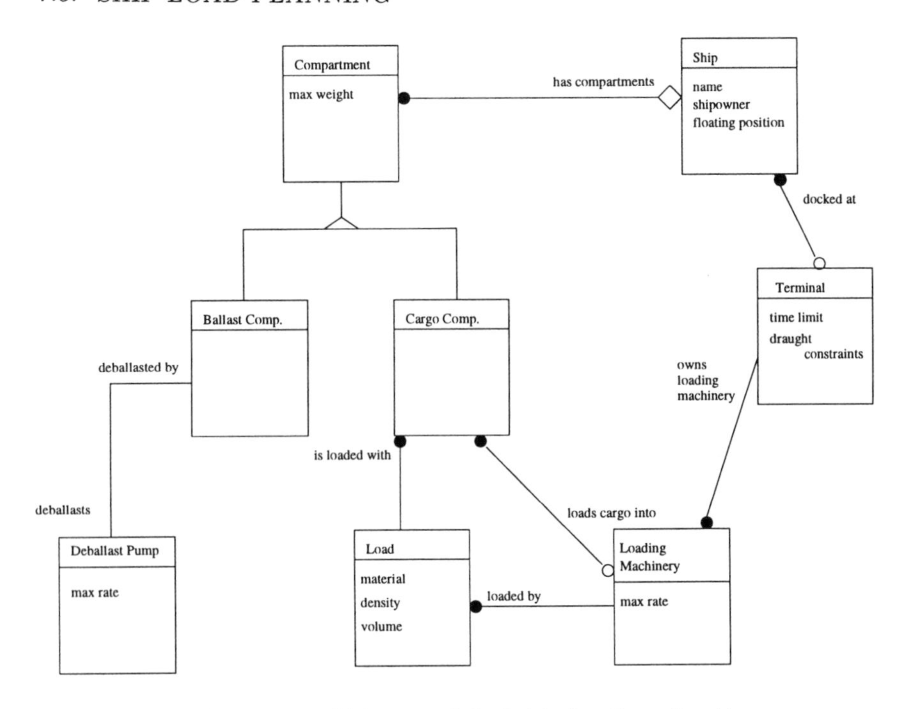

Figure 7.4: Object model of ship-loading situation

An extract from the object model of the domain of the system is given in Figure 7.4.

From this object model we can begin to outline the data components of the modules of the specification, with each entity type being represented by a machine. The subtypes Cargo Compartment and Ballast Compartment are included in the representation of Compartment. Although it is possible for a compartment to act, at different times in its history, as either a ballast or cargo compartment, at any particular moment in time it is considered to be only acting in one role in this model.

The situation is highly dynamic (ships arrive at ports, are aligned with terminals, and compartments are loaded or deballasted, ballast pumps are switched on or off, and so forth). Thus the OMT dynamic models of the entities involved are critical for meaningful formalisation.

The dynamic model of Deballast Pump is shown in Figure 7.5.

Transitions of the dynamic model will correspond quite closely to operations of the formalised entity. Indeed, attributes of the transition events will become, in most cases, parameters of the corresponding operations (these operations will also have an additional parameter which denotes the object upon which the transition is acting).

As a result of analysis of these two models, we have the following initial specification of the Deballast Pump entity:

```
MACHINE
```

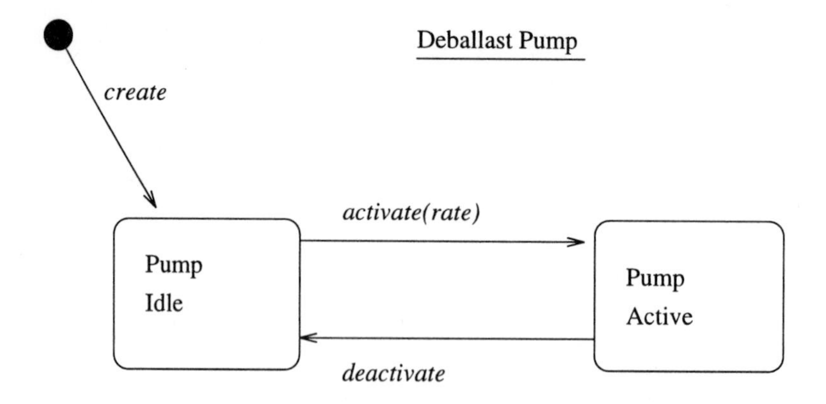

Figure 7.5: Dynamic model of Deballast Pump

```
  Deballast_pump(DEBALLASTPUMPS)
SEES
  Bool_TYPE
SETS
  DBP_STATE = {pump_idle, pump_active}
VARIABLES
  deballast_pumps, max_db_rate, dbp_state
DEFINITIONS
  DBRATE == 0..1000
INVARIANT
   deballast_pumps <: DEBALLASTPUMPS  &
   max_db_rate : deballast_pumps --> DBRATE &
   dbp_state : deballast_pumps --> DBP_STATE
INITIALISATION
  deballast_pumps, max_db_rate, dbp_state := {}, {}, {}
OPERATIONS

  ok, dbp <-- create_deballast_pump(mr) =
      PRE mr : DBRATE
      THEN
        IF deballast_pumps /= DEBALLASTPUMPS
        THEN
          ANY oo
          WHERE oo : DEBALLASTPUMPS - deballast_pumps
          THEN
            deballast_pumps := deballast_pumps \/ {oo} ||
            dbp_state(oo) := pump_idle ||
            dbp := oo ||
            max_db_rate(oo) := mr ||
            ok := TRUE
          END
        ELSE
          ok := FALSE
```

```
        END
      END;
```

Activate a specified pump with a given pumping rate, which must be less than the maximum rate allowed for the pump:

```
  activate_pump(dbp,rate) =
      PRE dbp : deballast_pumps &
          rate : DBRATE &
          rate <= max_db_rate(dbp) &
          dbp_state(dbp) = pump_idle
      THEN
        dbp_state(dbp) := pump_active
      END;

  deactivate_pump(dbp) =
      PRE dbp : deballast_pumps &
          dbp_state(dbp) = pump_active
      THEN
        dbp_state(dbp) := pump_idle
      END

END
```

The **create_deballast_pump** operation is, to an extent, an artifact of the formal specification, rather than the domain model. It serves to introduce a new ballast pump representation into the database which is managing information about the ship loading situation. It is particularly important in supporting the animation of the formal specification, and hence in validating it against the user requirements.

The dynamic model for loading machines is very similar to that for ballast pumps, and the abstract specification of loading machines is analogous to that for ballast pumps:

```
MACHINE
  LoadMachinery(LMACHINERY)
SETS
  LMSTATE = {idle, operating}
VARIABLES
  load_machinery,
  max_load_rate,
  load_mach_state
DEFINITIONS
  LRATE == 0..1000
INVARIANT
  load_machinery <: LMACHINERY &
  max_load_rate : load_machinery --> LRATE &
  load_mach_state : load_machinery --> LMSTATE
INITIALISATION
  load_machinery,
  max_load_rate, load_mach_state  := {}, {}, {}
```

```
OPERATIONS

  lm <-- create_load_mach(mlr) =
          PRE mlr : LRATE
          THEN
            IF load_machinery /= LMACHINERY
            THEN
              ANY oo
              WHERE oo : LMACHINERY - load_machinery
              THEN
                lm := oo ||
                load_machinery := load_machinery \/ {oo} ||
                max_load_rate(oo) := mlr ||
                load_mach_state(oo) := idle
              END
            END
          END;

  activate_load_mach(ll,lrate) =
          PRE lrate : NAT & ll : load_machinery &
              lrate <= max_load_rate(ll) &
              load_mach_state(ll) = idle
          THEN
            load_mach_state(ll) := operating
          END;

  deactivate_load_mach(ll) =
          PRE ll : load_machinery &
              load_mach_state(ll) = operating
          THEN
            load_mach_state(ll) := idle
          END

END
```

At the next level in the system, we have a model of compartments. At present we will only model compartments which remain permanently in the role of a ballast or cargo compartment, and which do not change their role over time. Thus the dynamic model of compartments can be split into two models, for ballast and cargo compartments separately. Part of the model for ballast compartments is shown in Figure 7.6.

In this figure the notation **transition/operation** denotes a state transition **transition** of ballast compartments which involves a transition **operation** on the ballast pump associated with the ballast compartment. Only half of the circular life-cycle of a ballast pump is shown here (the other half would define one or two transitions to specify the ballasting of a compartment). The dynamic model for cargo compartments is similar, with an additional transition **assign_loading_mach** which is activated only in the **Unloaded** state, and which is a self-transition on this state. The transitions **load_compartment**

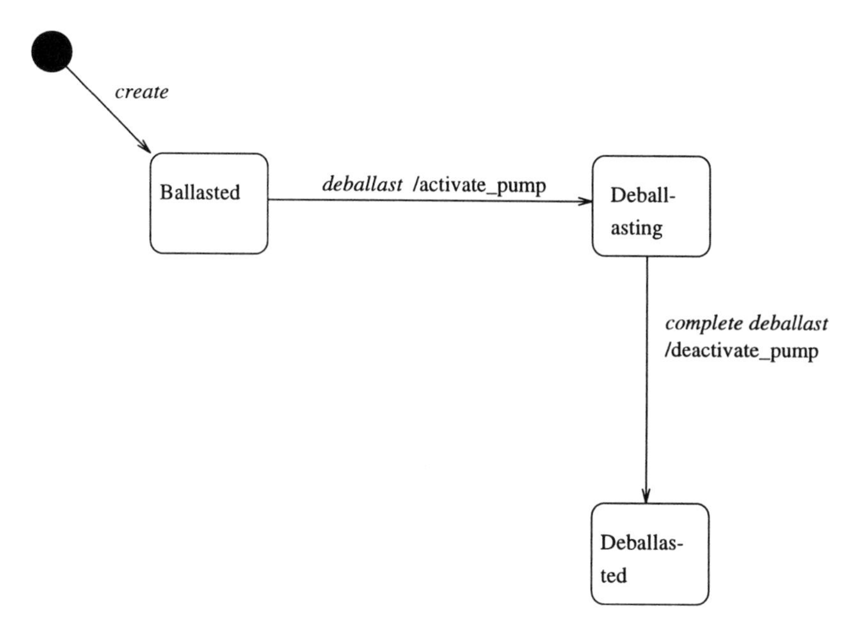

Figure 7.6: Dynamic model of Ballast Compartment

and **complete_loadc** act in a similar way to **deballast** and **complete_deballast**, and invoke operations **activate_load_mach** and **deactivate_load_mach** respectively on the load machinery associated with the compartment.

Thus the Compartment entity is formalised by the following machine:

```
MACHINE
  Compartment(COMPARTMENT,DEBALLASTPUMP,LMACHINERY)
SEES
  Load, Bool_TYPE
INCLUDES
  Deballast_pump(DEBALLASTPUMP),
  LoadMachinery(LMACHINERY)
PROMOTES
  create_deballast_pump,
  create_load_mach
VARIABLES
  cargo_compartments,
  ballast_compartments,

  unloaded_compartments, loading_compartments,
  loaded_compartments,

  ballasted_compartments, deballasting_compartments,
  deballasted_compartments,
```

```
  compartments, is_loaded_with,
  c_ballast_pump,
  c_max_weight, loading_mach
DEFINITIONS
  DBRATE == 0..1000;
  LRATE == 0..1000;

  DURATION == 0..5000
INVARIANT
  compartments <: COMPARTMENT &
  cargo_compartments <: compartments &
  ballast_compartments <: compartments &

  cargo_compartments /\ ballast_compartments = {}  &

  unloaded_compartments <: cargo_compartments &
  loading_compartments <: cargo_compartments &
  loaded_compartments <: cargo_compartments &

  unloaded_compartments /\ loading_compartments = {} &
  loading_compartments /\ loaded_compartments = {} &
  unloaded_compartments /\ loaded_compartments = {} &

  ballasted_compartments <: ballast_compartments &
  deballasting_compartments <: ballast_compartments &
  deballasted_compartments <: ballast_compartments &

  ballasted_compartments /\ deballasting_compartments = {} &
  deballasted_compartments /\ deballasting_compartments = {} &
  ballasted_compartments /\ deballasted_compartments = {} &

  is_loaded_with : cargo_compartments --> LOAD &

  c_ballast_pump : ballast_compartments --> DEBALLASTPUMP &
  c_max_weight : compartments --> 0..1000000 &
  loading_mach : cargo_compartments +-> LMACHINERY

INITIALISATION

  compartments, cargo_compartments,
  ballast_compartments, is_loaded_with,
  ballasted_compartments, deballasted_compartments,
  deballasting_compartments,
  unloaded_compartments, loading_compartments,
  loaded_compartments, c_ballast_pump,
```

```
  c_ballast_pump,
  c_max_weight, loading_mach  :=  {}, {}, {}, {},
       {}, {}, {}, {},
       {}, {}, {}, {}, {}

OPERATIONS

  ok, cc <-- create_cargo_compartment(mw) =
      PRE
          mw : 0..1000000
      THEN
        IF COMPARTMENT /= compartments
        THEN
          ANY oo, ld
          WHERE oo : COMPARTMENT - compartments & ld: LOAD
          THEN
            compartments := compartments \/ {oo} ||
            cargo_compartments := cargo_compartments \/ {oo} ||
            cc := oo ||
            is_loaded_with(oo) := ld ||
            unloaded_compartments :=
                     unloaded_compartments \/ {oo} ||
            c_max_weight(oo) := mw ||
            ok := TRUE
          END
        ELSE
          ok := FALSE
        END
      END;

  set_load(cc,ld) =
      PRE cc : unloaded_compartments &
          ld : loads
      THEN
        unloaded_compartments := unloaded_compartments - {cc} ||
        loaded_compartments := loaded_compartments \/ {cc} ||
        is_loaded_with(cc) := ld
      END;

  ok, cc <-- create_ballast_compartment(mw,dbp) =
      PRE
          mw : 0..1000000 & dbp: deballast_pumps
      THEN
        IF COMPARTMENT /= compartments
        THEN
          ANY oo
```

```
        WHERE oo : COMPARTMENT - compartments
        THEN
          compartments := compartments \/ {oo} ||
          ballast_compartments := ballast_compartments \/ {oo} ||
          cc := oo ||
          c_ballast_pump(oo) := dbp ||
          ballasted_compartments :=
                   ballasted_compartments \/ {oo} ||
          c_max_weight(oo) := mw ||
          ok := TRUE
        END
      ELSE
        ok := FALSE
      END
    END;

add_ballast_pump(cc,dbp) =
    PRE dbp : deballast_pumps &
        cc : ballast_compartments
    THEN
      c_ballast_pump(cc) := dbp
    END;

assign_loading_mach(cc,mach) =
   PRE cc : cargo_compartments &
       cc : unloaded_compartments &
       mach : load_machinery - ran(loading_mach)
   THEN
     loading_mach(cc) := mach
   END;

  /* Initiate the deballasting of a compartment:  */

deballast(cc,rate) =
    PRE cc : ballast_compartments &
        cc : ballasted_compartments &
        rate : DBRATE &
        rate <= max_db_rate(c_ballast_pump(cc))
    THEN
      activate_pump(c_ballast_pump(cc),rate) ||
      deballasting_compartments :=
               deballasting_compartments \/ {cc} ||
      ballasted_compartments :=
               ballasted_compartments - {cc}
    END;
```

```
  load_compartment(cc,lrate) =
      PRE cc : cargo_compartments &
          cc : unloaded_compartments  &
          cc : dom(loading_mach) &
          lrate : LRATE &
          lrate <= max_load_rate(loading_mach(cc))
      THEN
        loading_compartments := loading_compartments \/ {cc} ||
        unloaded_compartments := unloaded_compartments - {cc} ||
        activate_load_mach(loading_mach(cc),lrate)
      END;

   /* Complete the deballasting of a compartment:  */

  complete_deballast(cc) =
      PRE cc : ballast_compartments &
          cc : deballasting_compartments &
          c_ballast_pump(cc) : deballast_pumps &
          dbp_state(c_ballast_pump(cc)) = pump_active
      THEN
        deballasted_compartments :=
                    deballasted_compartments \/ {cc} ||
        deballasting_compartments :=
                    deballasting_compartments - {cc} ||
        deactivate_pump(c_ballast_pump(cc))
      END;

  complete_loadc(cc) =
      PRE cc : cargo_compartments &
          cc : loading_compartments
      THEN
        loaded_compartments := loaded_compartments \/ {cc} ||
        loading_compartments := loading_compartments - {cc} ||
        deactivate_load_mach(loading_mach(cc))
      END

END
```

Finally, we can analyse the possible transitions and states which a ship may undergo during the loading process. An initial model is shown in Figure 7.7. In this diagram there are transitions, for example, **start_loading**, which have conditions attached to them (**Exists Compartments in unloaded and ballasted**). These transitions are only enabled in the case that the transition condition is true. For example, the loading operation can only start (in our simplified model) if there is at least one compartment which is ready to be loaded, and a (different) compartment which is ready to be deballasted. The most significant state here is the **Loading and Deballasting** state. In this state one ship compartment is being loaded whilst another is being deballasted. We can perform five transitions in this state. The first two, **deballast_new_compartment** and **load_new_compartment**, do not change the state, but involve respectively:

- the selection of a new ballasted compartment, and the initiation of the deballasting of this compartment, together with the completion of the deballasting of the compartment which was being deballasted prior to the transition being triggered;
- the selection of a new unloaded compartment, and the initiation of the loading of this compartment, together with the completion of the loading of the compartment which was being loaded prior to the transition being triggered.

A ship can leave this state in one of three ways:

- *safety_violation*: a safety violation occurs, which is modelled by an unspecified boolean function **unsafe_load_condition** (defined in **Ship_Sensors**) returning a true value (the calculation of this function from sensor reports and loading information is one of the central functions of the monitoring component of the load planner);
- *constraint_violation*: a timing constraint or other (non-safety related) failure in the loading operation has occurred, and the loading is deemed to have been unsuccessful;
- *complete_load*: there is no further compartment to be loaded or deballasted, and there has been no safety or constraint violation, so that the loading is deemed to have completed successfully.

In either of the last two cases the ship is considered fit to leave port. The transitions are considered to have priority in the order given above, ie, the **safety_violation** transition is always taken if it is enabled by its condition.

The corresponding formal specification is then given by:

```
MACHINE
  Ship(SHIP,shipsize,LOAD,COMPARTMENT,DEBALLASTPUMP,LMACHINERY)
CONSTRAINTS
  shipsize >= 1 & shipsize <= 11  /* the number of compartments */
SEES
  Bool_TYPE, Ship_Sensors
EXTENDS
  Load(LOAD,LMACHINERY),
  Compartment(COMPARTMENT,DEBALLASTPUMP,LMACHINERY)
SETS
  SHIP_STATE = {at_sea, docking, docked,
                loading_and_deballasting,
                incompletely_loaded, unsafely_loaded,
```

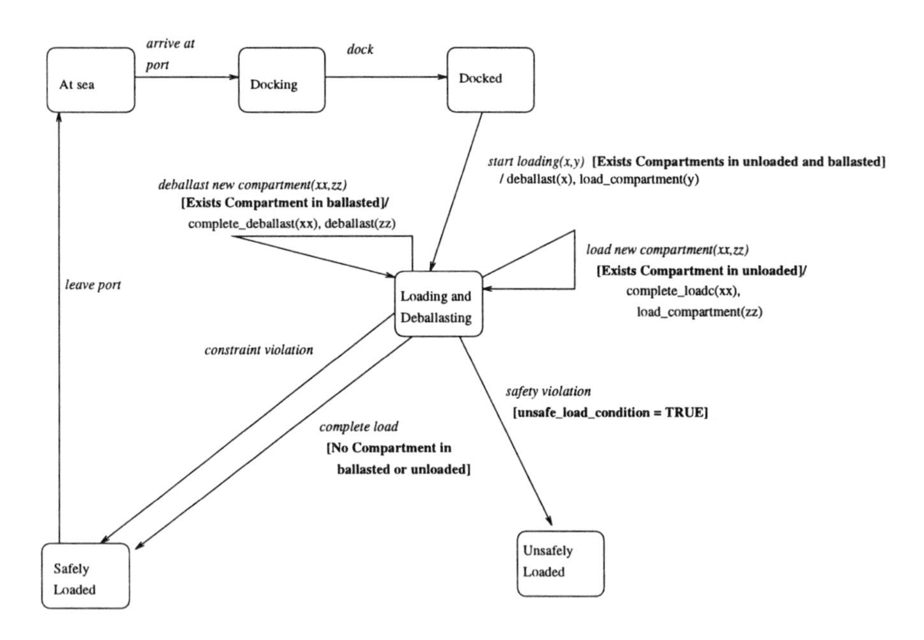

Figure 7.7: Dynamic model of ship

```
                  safely_loaded}
CONSTANTS
  safe_loading_sequence
PROPERTIES
  safe_loading_sequence : SHIP*COMPARTMENT*COMPARTMENT*LRATE*DBRATE -->
                             BOOL
```

This function provides a check that a proposed loading and deballasting operation is safe. The details of this test are deferred at present.

```
VARIABLES
  ships,
  s_compartments,
  ship_state, floating_position
DEFINITIONS
  LRATE == 0..1000;
  DBRATE == 0..1000;
  DRAUGHT == 400..4000
INVARIANT
  ships <: SHIP &
  s_compartments : ships --> (1..shipsize +-> COMPARTMENT) &
  ship_state : ships --> SHIP_STATE &
  floating_position : ships --> DRAUGHT
INITIALISATION
  ships,
  s_compartments,
  ship_state, floating_position  :=  {}, {}, {}, {}
```

```
OPERATIONS

  ok, ss <-- create_ship(fp) =
      PRE fp : DRAUGHT
      THEN
        IF ships /= SHIP
        THEN
          ANY oo
          WHERE  oo : SHIP - ships
          THEN
            ships := ships \/ {oo} ||
            ss := oo ||
            s_compartments(oo) := {} ||
            floating_position(oo) := fp ||
            ship_state(oo) := at_sea ||
            ok := TRUE
          END
        ELSE
          ok := FALSE
        END
      END;

  add_compartment(ss,numb,cc) =
      PRE ss : ships &
          numb : 1..shipsize &
          numb /: dom(s_compartments(ss)) &
          cc : compartments - ran(s_compartments(ss))
      THEN
        s_compartments(ss)(numb) := cc
      END;

  arrive_at_port(ss) =
    PRE ss : ships & ship_state(ss) = at_sea
    THEN
      ship_state(ss) := docking
    END;

  dock_safely(ss) =
    PRE ss : ships &
        ship_state(ss) = docking
    THEN
      ship_state(ss) := docked
    END;
```

For simplicity we assume that only one compartment is being loaded at any point in a loading operation, and that only one is being deballasted.

```
  ok <-- start_loading(ss) =
    PRE ss : ships &
        ship_state(ss) = docked
    THEN
```

```
    ship_state(ss) := loading_and_deballasting ||
    ANY xx, yy, lrate, dbrate
    WHERE
      xx : ran(s_compartments(ss)) &
      yy : ran(s_compartments(ss)) &
      xx /= yy &
      xx : ballast_compartments &
      yy : cargo_compartments &
      xx : ballasted_compartments &
      yy : unloaded_compartments &
      yy : dom(loading_mach) &
      lrate : LRATE & dbrate : DBRATE &
      dbrate <= max_db_rate(c_ballast_pump(xx)) &
      lrate <= max_load_rate(loading_mach(yy))
    THEN
      deballast(xx,dbrate) ||
      load_compartment(yy,lrate) ||
      ok := safe_loading_sequence(ss,xx,yy,lrate,dbrate)
    END
  END;

ssafe <-- monitor_ship_condition(ss) =
  PRE ss : ships
  THEN
    ssafe := safe_load_condition(ss)
  END;

load_new_compartment1(ss,xx,yy) =
  PRE ss : ships &
      ship_state(ss) = loading_and_deballasting &
      xx : ran(s_compartments(ss)) & yy : ran(s_compartments(ss)) &
      xx /= yy &
      xx : ballast_compartments &
      yy : cargo_compartments &
      xx : deballasting_compartments &
      yy : loading_compartments
  THEN
    complete_loadc(yy)
  END;

new <-- load_new_compartment2(ss,xx) =
  PRE ss : ships &
      ship_state(ss) = loading_and_deballasting &
      xx : ran(s_compartments(ss)) &
      xx : deballasting_compartments
  THEN
    ANY zz, lrate
    WHERE zz : ran(s_compartments(ss)) &
          zz /= xx &
          zz : cargo_compartments &
```

```
        zz :  unloaded_compartments &
        lrate : LRATE &
        zz : dom(loading_mach) &
        lrate <= max_load_rate(loading_mach(zz))
    THEN
      load_compartment(zz,lrate) ||
      new := zz
    END
  END;

deballast_new_compartment1(ss,xx,yy) =
  PRE ss : ships &
      ship_state(ss) = loading_and_deballasting &
      xx : ran(s_compartments(ss)) &
      yy : ran(s_compartments(ss)) &
      xx /= yy &
      xx : ballast_compartments &
      yy : cargo_compartments &
      xx : deballasting_compartments &
      yy : loading_compartments
  THEN
    complete_deballast(xx)
  END;

new  <--  deballast_new_compartment2(ss,yy) =
  PRE ss : ships &
      ship_state(ss) = loading_and_deballasting &
      yy : ran(s_compartments(ss)) &
      yy : cargo_compartments &
      yy : loading_compartments
  THEN
    ANY zz, dbrate
    WHERE zz : ran(s_compartments(ss)) &
          zz : ballast_compartments &
          zz /= yy &
          zz : ballasted_compartments &
          dbrate : DBRATE &
          dbrate <= max_db_rate(c_ballast_pump(zz))
    THEN
      deballast(zz,dbrate) ||
      new := zz
    END
  END;

complete_loading(ss) =
  PRE ss : ships &
      ship_state(ss) = loading_and_deballasting  &
      !cc.(cc : ran(s_compartments(ss)) =>
                cc /: unloaded_compartments &
                cc /: ballasted_compartments)
```

```
    THEN
      ANY xx, yy
      WHERE
        xx : ran(s_compartments(ss)) &
        yy : ran(s_compartments(ss)) &
        xx : cargo_compartments &
        xx : loading_compartments &
        yy : ballast_compartments &
        yy : deballasting_compartments
      THEN
        complete_deballast(yy) ||
        complete_loadc(xx)
      END ||
      ship_state(ss) := safely_loaded
    END;

  constraint_violation(ss,xx,yy) =
    PRE ss : ships &
        ship_state(ss) = loading_and_deballasting &
        xx : ran(s_compartments(ss))
        & yy : ran(s_compartments(ss)) &
        xx /= yy &
        xx : ballast_compartments &
        yy : cargo_compartments &
        xx : deballasting_compartments &
        yy : loading_compartments
    THEN
      complete_loadc(yy) ||
      complete_deballast(xx) ||
      ship_state(ss) := incompletely_loaded
    END;

   safety_violation(ss) =
     PRE ss : ships &
         ship_state(ss) = loading_and_deballasting
     THEN
       IF (safe_load_condition(ss) = FALSE)
       THEN
         ship_state(ss) := unsafely_loaded
       END
     END;

   leave_port(ss) =
      PRE ss : ships &
          ship_state(ss) : {incompletely_loaded, safely_loaded}
      THEN
        ship_state(ss) := at_sea
      END

END
```

For completeness we also model the interaction between a ship and the terminals at which it may be loaded. We provide an outline formal model of a Terminal in the following machine:

```
MACHINE
  Terminal(TERMINAL,LMACHINERY)
VARIABLES
  terminals,
  owns_loading_machinery,
  time_limit,
  draught_constraint
DEFINITIONS
  DURATION == 0..5000;
  DRAUGHT == 400..4000
INVARIANT
  terminals <: TERMINAL &
  owns_loading_machinery : terminals --> FIN(LMACHINERY) &
  time_limit : terminals --> DURATION &
  draught_constraint : terminals --> DRAUGHT

INITIALISATION

  terminals, owns_loading_machinery,
  time_limit, draught_constraint  :=  {}, {}, {}, {}

OPERATIONS

  tt <-- create_terminal(tlim,dcons) =
      PRE tlim : DURATION &
          dcons : DRAUGHT
      THEN
        IF terminals /= TERMINAL
        THEN
          ANY oo
          WHERE  oo : TERMINAL - terminals
          THEN
            tt := oo ||
            terminals := terminals \/ {oo} ||
            owns_loading_machinery(oo) := {} ||
            time_limit(oo) := tlim ||
            draught_constraint(oo)  :=  dcons
          END
        END
      END;

  add_loading_mach(term,lmach) =
      PRE term : terminals & lmach : LMACHINERY &
          lmach /: owns_loading_machinery(term)
      THEN
        owns_loading_machinery(term) :=
```

```
            owns_loading_machinery(term) \/ {lmach}
      END

END
```

Finally, we can define an overall specification of the loading system, suitable for animation:

```
MACHINE
  LoadingSystem(SHIP,TERMINAL,LMACHINERY,
                COMPARTMENT,DEBALLASTPUMP,LOAD)
EXTENDS
  Ship(SHIP,LOAD,COMPARTMENT,
       DEBALLASTPUMP,LMACHINERY),
  Terminal(TERMINAL,LMACHINERY)
SEES
  Bool_TYPE
VARIABLES
  docked_at
INVARIANT
  docked_at : ships +-> terminals  &

  !ss.(ss : ships  &
       ship_state(ss) = loading_and_deballasting  =>
                        #tt.(tt : terminals & docked_at(ss) = tt))

INITIALISATION
  docked_at := {}
OPERATIONS

  bb <-- allowed_to_dock(ss,tt) =
    PRE ss : ships & tt : terminals
    THEN
      IF floating_position(ss) <= draught_constraint(tt)
      THEN
        bb := TRUE
      ELSE
        bb := FALSE
      END
    END;

 new_dock_safely(ss,tt) =
    PRE ss : ships & tt : terminals &
        ship_state(ss) = docking &
        floating_position(ss) <= draught_constraint(tt)
    THEN
      docked_at(ss) := tt ||
      dock_safely(ss)
    END;

 new_leave_port(ss,tt) =
```

```
    PRE ss : ships & tt : terminals &
        docked_at(ss) = tt &
        ship_state(ss) : {safely_loaded, incompletely_loaded}
    THEN
      docked_at := docked_at - {ss |-> tt} ||
      leave_port(ss)
    END

END
```

The overall structure of the system is shown in Figure 7.8.

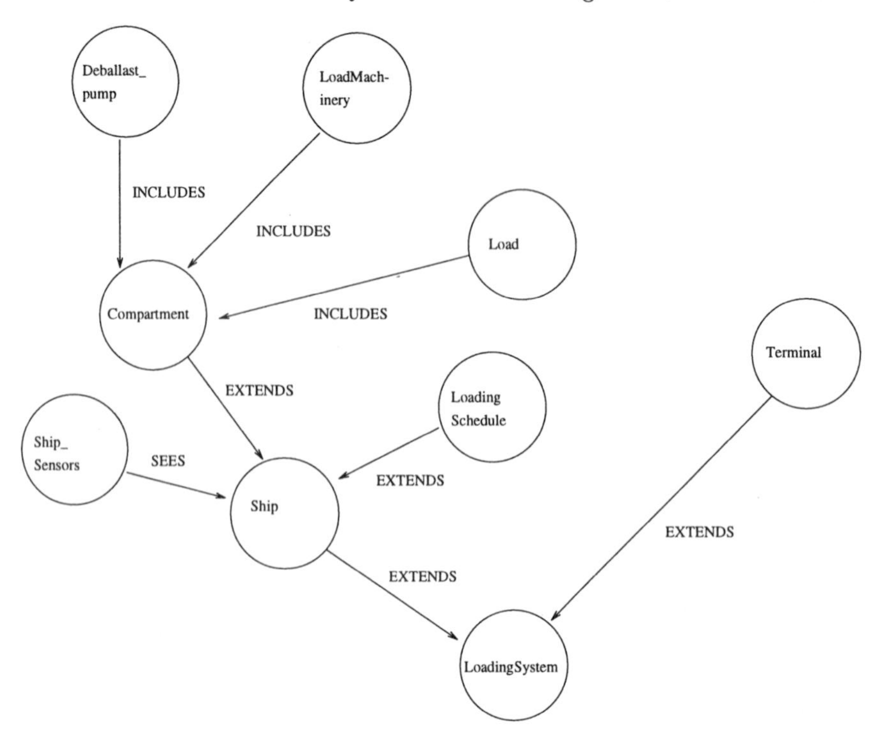

Figure 7.8: Ship loading specification architecture

Note that there are correspondences between the form of structuring mechanism which is used in the specification architecture, and the nature of the relationship between components of the system being specified. For example, there is a `SEES` access to **Ship_Sensors** by **Ship**, which restricts the **Ship** component to read-only access to the state of **Ship_Sensors**. In the corresponding real-world situation ship sensors act as an input to the ship-loading system and their results are not controlled by this system. In contrast, an `INCLUDES` access mechanism allows the including component to control the included component via the operations of the latter (and indeed, this relationship is then exclusive to the controlling component). Thus, `SEES` can be used to represent a monitoring relationship and `INCLUDES` or `EXTENDS` a control relationship.

7.5.1 Results

An initial specification of 890 lines of B AMN was written from the data and dynamic models of the system and used to animate the system. The specification was proved to be internally correct.

Some treatment of safety aspects of the system based on its formal specification can already be performed. For example, we could place an invariant in the **LoadingSystem** machine which required that every ship which is recorded as being in the **loading_and_deballasting** state is actually docked at a terminal:

$$\forall\ ss.(ss \in \mathbf{ships} \wedge \mathbf{ship\_state}(ss) = \mathbf{loading\_and\_deballasting} \Rightarrow \exists\ \mathbf{tt}.(\mathbf{tt} \in \mathbf{terminals} \wedge \mathbf{docked\_at}(ss) = \mathbf{tt}))$$

Proof of this invariant would assure us that the formal specification was a realistic model of the situation, and that any software which was formally developed from the specification would not allow such anomolous situations to occur.

In general, a load monitoring system can lead to a hazard if an unsafe ship condition exists which is not signalled as such by the system. The inverse problem, that a false signal of an unsafe condition is given, is of less concern from a safety standpoint, however it is certainly undesirable from a functional standpoint.

There can be various reasons why a failure to signal an unsafe ship condition can occur:

1. numeric overflows in the software;
2. numeric inaccuracies in the software;
3. memory overflows (exhausting space);
4. memory errors (attempting to access a non-existent address);
5. hardware errors;
6. errors in formalising rules and algorithms which are used to determine if a hazardous ship condition exists;
7. errors in these rules and algorithms themselves;
8. errors in sensors which supply information to the software;
9. human errors in data entry;
10. human errors in system set up.

Formal treatment of fault types 1, 2, 3 and 4 can be undertaken for B AMN developments, as described in [Haughton and Lano, 1994]. Fault type 6 can be minimised by use of animation and formal proof of validation properties of the system (such as the integrity constraint given above). The present specification is however too incomplete and insufficiently detailed for such analysis to be performed.

The structuring of the specification can provide some information on the propagation of failures through the system. For example, since **Compartment** INCLUDES **Deballast_pump** and **LoadMachinery**, and only promotes the creation operations from these machines, we know that control over the operation of existing deballast pumps and loading machinery can only be exercised in conjunction with operations on compartments. Thus, failures in the control of these components can be traced initially to failures in the **Compartment** module (assuming that the structure of the final implementation mirrors the specification structure). In general, an erroneous state in a machine **M** can be propagated directly to any machine which accesses it by means of a SEES, USES, INCLUDES or EXTENDS clause. More specifically, a numeric overflow failure in an operation **op** will be due either to a numeric overflow failure within its own code, in its input parameters, or in the results of operations that it invokes. Thus the failure can be decomposed into failures of operations which **op** depends upon.

7.6 Formalisation of Dynamic Models: A Lift System

This system is a simple control mechanism for an (unspecified) set of lifts and floors. The entities of the system are **Lift**, **Door**, **Button**. We also envisage the possibility that **Floor** may be regarded as an entity type, rather than a primitive unstructured type (since it may possess an attribute giving the floor number in addition to any special characteristic of the floor). A lift may have a *current floor* (the one it is on, if it is not moving or not in a decomissioned state), and it may have a *destination floor* (the floor it is moving to or has been requested to move to). These are *not* total attributes, since they only make sense for certain lift *states*. Floors and lift buttons are in 1-1 correspondence, as are lifts and doors. However, we would usually navigate from a lift to its associated door and from a button to its associated floor, since lifts and buttons will be the initiators or sources of actions which involve these associated objects.

The static data model of the system is therefore as shown in Figure 7.9, using OMT notation.

The notation expresses that for every instance of the **Lift** entity type, there is one or zero instances of the **Floor** entity type associated via the relation **current_floor**, and for every instance of the **Floor** entity type, there are many (zero or more) instances of the **Lift** entity type associated via the relation **current_floor**.

The dynamic model notation of OMT [Rumbaugh *et al.*, 1991] uses a version of Harel statecharts [Harel, 1987] to express the dynamics of a system. A simple example, including the use of synchronisation between subsystems using communication from one dynamic model to another, is given in Figures 7.10

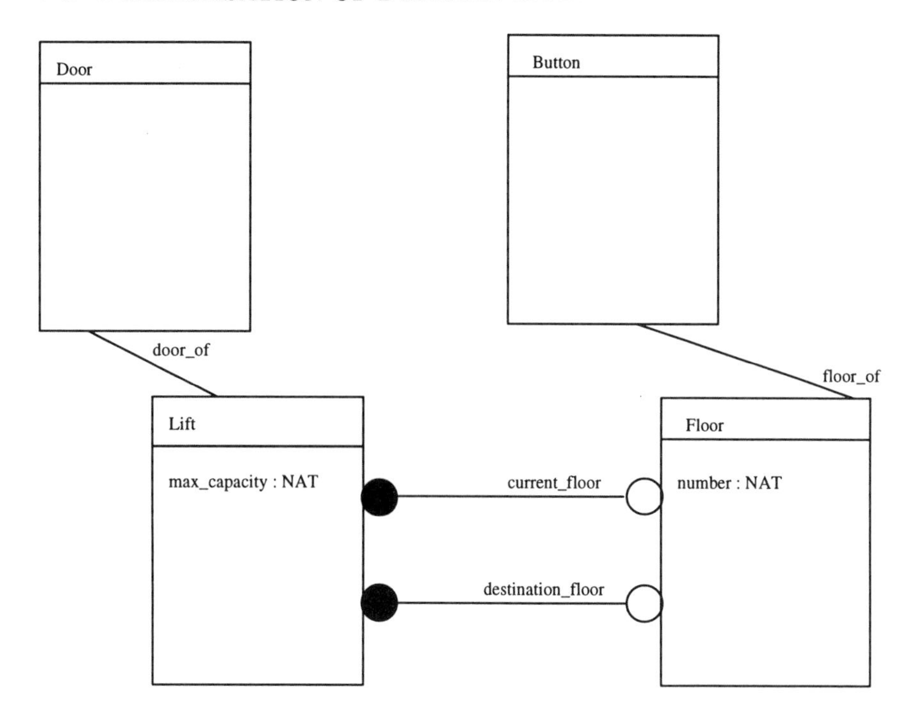

Figure 7.9: Static data model of lift system

and 7.11 to specify the dynamic behaviour of lifts and buttons.

Although lifts will usually be physically associated with a particular floor, we do not want to consider that each floor comprises a different state for a lift – since such a set of states would then be different in each particular lift configuration. Instead, we choose to consider only three essentially distinct states.

The dynamic model for doors is shown in Figure 7.11.

Note that the preconditions for transitions between states may be quite complex, and can refer to the set of existing instances of the class concerned (and to those sets of instances in a given state).

As an example the precondition

$$\mathbf{destination\_floor}^{-1}(\!|\ \{\mathbf{ff}\}\ |\!) = \varnothing$$

of the **request_lift(ff)** event expresses the condition

$$\neg\ \exists \mathbf{ll}.(\mathbf{ll} \in \mathbf{moving\_lifts} \wedge \mathbf{destination\_floor}(\mathbf{ll}) = \mathbf{ff})$$

That is, no other lift is currently heading for the floor **ff** whose button issued the request.

Each machine representing a dynamically modifiable set of instances of an entity type will possess both a set representing the set of all *possible* instances of the object class type, and a variable which is the subset of this set representing

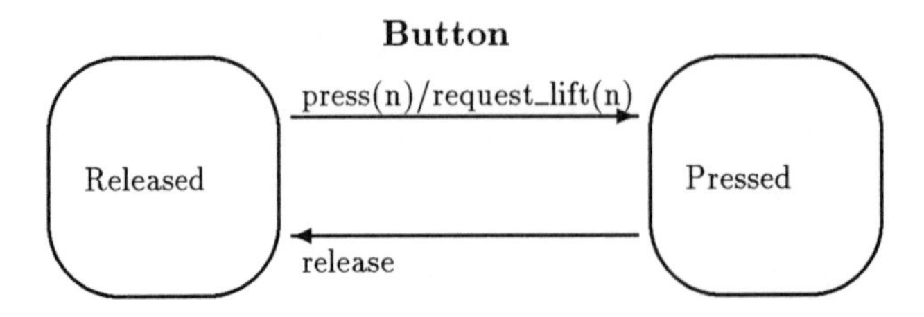

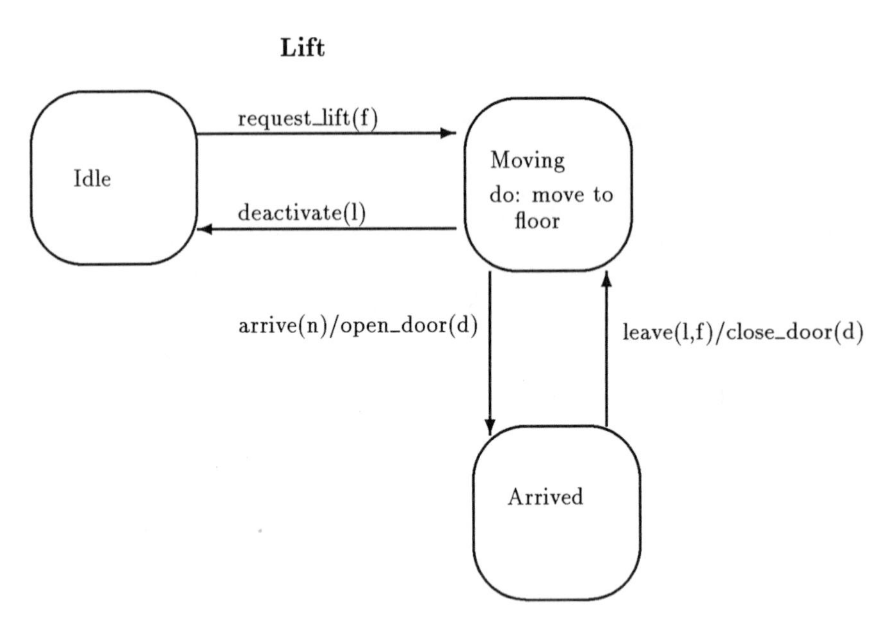

Figure 7.10: Dynamic models for Lift and Button

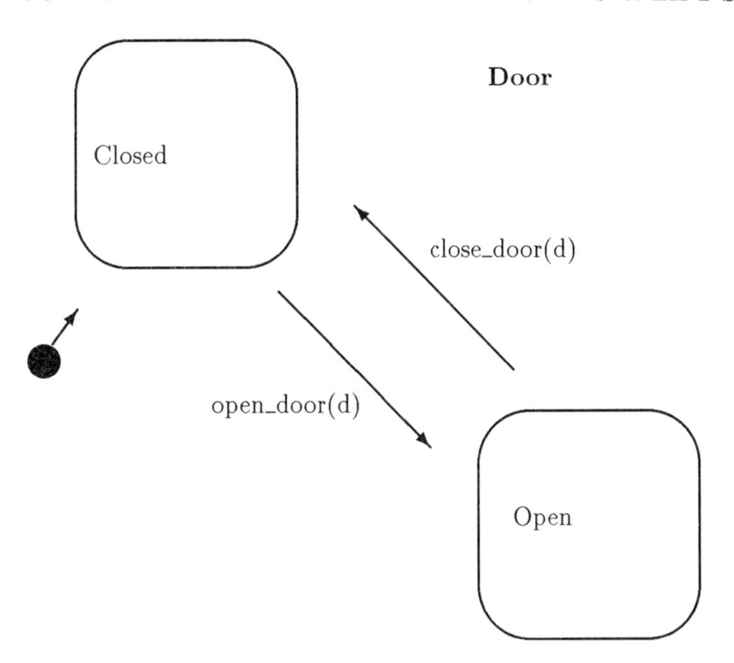

Figure 7.11: Dynamic model for Door

the existing instances of the type. The set may be defined internally to the machine in the SETS clause, in order to be instantiated in an implementation of the machine. Alternatively, in the case that the machine will be included in other machines, with the set of possible instances being shared between several machines, or instantiated to a specific value, it must be defined as a parameter of the machine. In this case we wish to make the lift system entirely configurable by an external user, and so we choose to make the sets of possible instances of a type suitable parameters of the machines representing the types.

7.6.1 From Analysis to Specification

Floors

The static data model leads to the definition of four outline machines. The simplest is the machine representing floors (we omit the number representing the number of the floor):

```
MACHINE
  Floor(FLOOR)
CONSTANTS
  ground_floor
PROPERTIES
  ground_floor : FLOOR
```

```
END
```

This machine encapsulates the set of possible floors, and a special floor (which is not used in the present specification, but is provided with a view to reuse of the system in more sophisticated specifications, such as a system where the 'default' destination floor is the ground floor). Because this is a static entity type (new members are not dynamically added or deleted) we only have a given set **FLOOR** and not a set of existing entities **floors** $\subseteq$ **FLOOR**.

There are no proof obligations for internal consistency of this machine.

Doors

The machine representing doors encapsulates three sets:

- **doors** – representing the instances of the door entity;
- **open_doors** – representing the instances of doors in the Open state;
- **closed_doors** – representing the instances of doors in the Closed state.

There are no other variables of the machine since doors have no actual attributes.

We must impose the condition that the sets of open and closed doors are disjoint and make up the whole of the set of doors – this is the same condition that is normally expected for disjoint subtypes of an entity type.

The outline specification is therefore:

```
MACHINE
  Door(DOOR)
VARIABLES
  doors, open_doors, closed_doors
INVARIANT
  doors <: DOOR &
  open_doors <: doors &
  closed_doors <: doors &

  open_doors \/ closed_doors = doors &
  open_doors /\ closed_doors = {}
INITIALISATION
  doors, open_doors, closed_doors := {}, {}, {}
OPERATIONS

  dd <-- create_door =
      PRE
        doors /= DOOR
      THEN
        ANY oo
        WHERE
```

```
            oo : DOOR - doors
        THEN
            dd := oo ||
            doors := doors \/ {oo} ||
            closed_doors := closed_doors \/ {oo}
        END
      END;

  add_door(dd) =
      PRE dd : DOOR - doors
      THEN
        doors := doors \/ {dd} ||
        closed_doors := closed_doors \/ {dd}
      END;

  open_door(dd) =
       PRE
          dd : doors
       THEN
          open_doors := open_doors \/ {dd} ||
          closed_doors := closed_doors - {dd}
       END;

  close_door(dd) =
       PRE
          dd : doors
       THEN
          closed_doors := closed_doors \/ {dd} ||
          open_doors := open_doors - {dd}
       END

END
```

The initial state of a door is **Closed**, which is expressed in the **create_door** operation.

Buttons

The treatment of this is similar to the Door entity. It has the inclusion relations:

```
MACHINE
   Button(BUTTON)
SEES Floor
  ....
END
```

Lifts

Similarly, for the Lift entity type:

```
MACHINE
  Lift(LIFT)
SEES Floor
USES Door
VARIABLES
     lifts, moving_lifts, arrived_lifts, idle_lifts,
     current_floor, destination_floor, door_of
INVARIANT
     lifts <: LIFT &
     arrived_lifts <: lifts &
     moving_lifts <: lifts &
     idle_lifts <: lifts &

     arrived_lifts /\ moving_lifts = {} &
     idle_lifts /\ arrived_lifts = {} &
     idle_lifts /\ moving_lifts = {} &

     arrived_lifts \/ idle_lifts \/
             moving_lifts = lifts &

     current_floor : arrived_lifts --> FLOOR &
     destination_floor : moving_lifts --> FLOOR &
     door_of : lifts >-> doors

INITIALISATION
     lifts, arrived_lifts,
     idle_lifts, moving_lifts, current_floor,
     destination_floor, door_of   :=
                              {}, {}, {}, {}, {}, {}, {}

OPERATIONS
     add_lift(ll,dd) =
         PRE ll : LIFT &
             ll /: lifts &
             dd : doors &
             dd /: ran(door_of)
         THEN
           idle_lifts := idle_lifts \/ {ll} ||
           lifts := lifts \/ {ll} ||
           door_of(ll) := dd
         END;

     request_lift(ff) =
```

```
    PRE ff : FLOOR &
        (idle_lifts /= {}  or
         destination_floor~[{ff}] /= {})

         /*" #(kk).(kk:moving_lifts |
                destination_floor(kk) = ff) ) "*/

    THEN
      IF destination_floor~[{ff}] /= {}
      THEN
        skip   /* there is already a lift
                  serving that floor */
      ELSE
        ANY ll
        WHERE ll : idle_lifts
        THEN
          idle_lifts := idle_lifts - {ll} ||
          moving_lifts := moving_lifts \/ {ll} ||
          destination_floor(ll) := ff
        END
      END
    END;

arrive(ll,ff) =
   PRE ff : FLOOR &
       ll : moving_lifts &
       destination_floor(ll) = ff
   THEN
     moving_lifts := moving_lifts - {ll} ||
     destination_floor :=
        destination_floor - {ll |-> ff} ||
     arrived_lifts := arrived_lifts \/ {ll} ||
     current_floor(ll) := ff
   END;

leave(ll,ff) =
   PRE ll : arrived_lifts &
       ff : FLOOR &
       ff /= current_floor(ll)
   THEN
     arrived_lifts := arrived_lifts - {ll} ||
     moving_lifts := moving_lifts \/ {ll} ||
     current_floor := {ll} <<| current_floor  ||
     destination_floor(ll) := ff
   END;
```

```
    deactivate(ll) =
       PRE ll : moving_lifts
       THEN
         moving_lifts := moving_lifts - {ll} ||
         destination_floor := {ll} <<| destination_floor ||
         idle_lifts := idle_lifts \/ {ll}
       END
END
```

The attribute

door_of : **lifts** $\rightarrowtail$ **doors**

is recorded as a total injection from lifts to doors. That is, different lifts must have different doors, but it is possible for doors to exist without an associated lift, contrary to the static data model. The problem with directly expressing the original one-to-one correspondence is that pairs of lifts and doors would need to be created together, together with a record of their association via **door_of**. Thus the sets of entity instances for these two types would need to be defined as variables in the same machine (since at the abstract machine level, we are not allowed to use sequential composition to compose operations from INCLUDED machines).

This is a very simple model, in which a lift can only arrive at a floor for which it is destined. Moreover, it must be given a definite new destination when it leaves a floor. We can enhance this model to include an attribute **actual_floor**. Alternatively, we can include an operation **move_to_floor** which represents the action (or activity) of state **Moving**:

```
    move_to_floor(ll,ff) =
        PRE
          ll : moving_lifts &
          ff = destination_floor(ll)
        THEN
          skip
        END;
```

Lift System

Synchronisation between subsystem components is performed by multiple generalised substitutions of actions of individual components in a machine which includes each component. We could therefore define:

```
MACHINE
  LiftSystem(BUTTON,LIFT,DOOR,FLOOR)
INCLUDES
  Button(BUTTON),
  Floor(FLOOR),
  Lift(LIFT),
```

```
  Door(DOOR)
PROMOTES
  release, create_door
OPERATIONS

  add_lift_and_door(ll,dd) =
        PRE ll : LIFT &
            dd :  doors &
            ll /: lifts &
            dd /: ran(door_of)
        THEN
          close_door(dd) ||
          add_lift(ll,dd)
        END;

  add_button_and_floor(nn,ff) =
        PRE
          nn : BUTTON &
          nn /: buttons &
          ff : FLOOR &
          ff /: ran(floor_of)
        THEN
          add_button(nn,ff)
        END;

  press_and_request_lift(bb) =
        PRE
          bb : buttons  /* and there is a lift to request */
          &
          (idle_lifts /= {}  or
              destination_floor~[{floor_of(bb)}] /= {})
        THEN
          press(bb) ||
          request_lift(floor_of(bb))
        END;

  arrive_and_open_door(ll,ff) =
        PRE
          ll : moving_lifts &
          ff : FLOOR &
          destination_floor(ll) = ff
        THEN
          arrive(ll,ff) ||
          open_door(door_of(ll))
        END ;
```

```
leave_and_close_door(ll,ff) =
     PRE
       ll : arrived_lifts &
       ff : FLOOR &
       ff /= current_floor(ll)
     THEN
       leave(ll,ff) ||
       close_door(door_of(ll))
     END;

deactivate_and_close_door(ll) =
     PRE ll : moving_lifts
     THEN
       close_door(door_of(ll)) ||
       deactivate(ll)
     END


END
```

In this composed system, each operation which combines operations from included machines must reproduce each precondition of every operation of which it is composed.

7.6.2 Applications of Formalisation

The formalisation of the semantics of statecharts which has been performed by the above translation can assist in establishing safety properties of the systems specified by these charts.

For example, if we were required to show that a lift can never be moving with its doors open, then we could formalise this assertion as the additional invariant clause of **LiftSystem**:

$$\forall \mathbf{ll}.(\mathbf{ll} \in \mathbf{moving\_lifts} \;\Rightarrow\; \mathbf{door\_of}(\mathbf{ll}) \in \mathbf{closed\_doors})$$

An attempt to establish this will identify that this can be proved, provided that

$$\forall \mathbf{ll}.(\mathbf{ll} \in \mathbf{idle\_lifts} \;\Rightarrow\; \mathbf{door\_of}(\mathbf{ll}) \in \mathbf{closed\_doors})$$

In refinements of the specification, these properties would also be required to hold in the refined state, so that the implemented code would satisfy this safety property.

Conclusions

This chapter has described specific techniques to assist in the correct formalisation of user and business requirements upon a system, based on systematic translation processes from structured, diagrammatic models of a system to formal models in the B AMN specification language. This approach also has benefits in improving the structure of the formal specifications, in comparison to an approach in which formalisation is performed without prior analysis. Maintainability and reusability in particular should be enhanced, since there are stronger links between the domain terminology and concepts and the specification elements. The increased traceability of requirements through the formal specification and development also improves the assessability of systems.

Chapter 8

Introducing Object-Oriented Concepts into a Net-Based Hierarchical Software Development Process

8.1 Introduction

The software development life cycle commonly consists of three phases: requirement analysis, system design, and system implementation. Object-Oriented (OO) development, versus functional structure approaches based on top-down functional decomposition, has features of both top-down analysis and design, and bottom-up design and implementation [Yang, 1994]. In the phases of top-down analysis and design, solutions to integrate and match the OO technology with the functional decomposition techniques, will allow and encourage the reuse of the existing knowledge and products (e.g. CASE tools). This paper presents an integrated model of the Object-Oriented (OO) paradigm and Channel/Agency (CA) Nets [Reisig, 1987]. The main purpose of this net-based model is to facilitate requirement analysis and to sketch the design by introducing OO concepts from the beginning, i.e. at the top level of a hierarchical and formal method based software development life cycle, called the PROOFS life cycle [van der Aalst *et al.*, 1993]. Such a model permits a continuous and systematic transition from informal descriptions of an OO system to formal specifications through a specific semi-formal way .

In order to encourage and promote the use of formal techniques in the development of Heterogeneous Distributed Applications (HDA), the ESPRIT II PROOFS project aimed at the development of a tool set supporting a development method suitable for more and more complex HDA. By emphasising the integrated use of existing well known formalisms and tools, the so-called PROOFS method [van der Aalst *et al.*, 1993], based on the S-CORT method [Hildebrand and Treves, 1990], consists of a set of formalisms and a relevant methodology, including Channel/Agency (CA) nets [Reisig, 1987], Petri Nets (PNs), High-level Petri Nets like Coloured Petri Nets (CPNs) [Jensen, 1987] and net-based functional language ExSpect [van Hee *et al.*, 1990], and data model NIAM [Verheijen and van Backkum, 1982]. On the other hand, it is observed that the object-orientation technique is widely used, often informally, in the domain of HDA due to its feasible mechanism to capture a model of the real world. One of the topics thus proposed in PROOFS has been to introduce the OO paradigm into the framework of the PROOFS method. This work had the meaning both in bringing advanced OO techniques into traditional development methods and in introducing various powerful formal methods into the OO development.

Object-orientation, as a newly born technique in the 1980's, has widely emerged in almost every domain of software engineering, ranging from requirement analysis (OOA [Coad and Yourdon, 1990]), system design (OOD [Booch, 1991, Meyer, 1988, Ward, 1989]), to programming language (OOP: Smalltalk, C++, SPOKE [Alcatel ISR., 1995], etc.). This advanced technique has also attracted much attention, over the last several years, from the specification field. In this field, interests have focused on introducing OO concepts into existing well-known models such as Petri nets [Battiston and Cindio, 1993, Bastide *et al.*, 1993, Lakos and Keen, 1991], VDM [Durr and van Katwijk, 1992], Z [Carrington *et al.*, 1989] and SDL [Moller Pedersen *et al.*, 1987], etc. Our study in this paper is much inspired by these efforts, and is streamlined by the challenge launched in PROOFS.

The integration of the OO paradigm into PROOFS is twofold: with the adopted models and with the software development process (SDP). These two aspects of the integration could be independent when we think that each model is autonomous and can be used in different ways with regard to the SDP, but they are closely related when we discuss them in the defined PROOFS development life-cycle where multi-models are used in a hierarchical transformation way due to various systematic links [van der Aalst *et al.*, 1993]. This paper focuses on an OO integration model during the top-level phases of the development process and its link with lower-level models.

The Channel/Agency nets (CA nets) have been generally proposed in [Reisig, 1987], and they have been formally defined in [Treves, 1992, Agoulmine, 1994]. Briefly, a CA net is a semi-formal net-like model for preliminary specifications in which a system is considered as a directed graph of active and passive components, named *Agencies* and *Channels* respectively. They are proposed to be used for the initial and semi-formal sketches and specification at the requirement analysis, then for the preliminary architecture design phases of the development life cycle.

However, even though CA nets are supported by part of Petri net theory, and have a similar graphical representation as Petri nets, they are not fully formal, because CA nets do not have any marking (the quantity and the type of information is not specified) and the links between channels and agencies are only abstract relations. Consequently, the interpretation of a CA net has a semantic ambiguity. Feasible solutions to clarify these semantics are around defining different classes of CA nets to make them specific and adapted to different application domains. As one of the solutions, integrating object-oriented concepts into the CA net will highly reduce the ambiguity degree when some specific and relevant semantic and syntactic rules are introduced into CA nets to fit with the OO applications development [Yang, 1994].

In this paper, we propose an object oriented CA net model, called O-CA, in order to achieve a suitable CA net model adaptable to the description of object-oriented systems. We define an O-CA net to be a CA net which obeys a set of specific rules and has a clear semantic interpretation with OO concepts and constraints. The way to construct O-CA nets is to map OO concepts into CA net components and techniques (e.g. refinement, splitting, abstraction, embedding, etc.) by giving reasonable specific semantics to CA nets. A proper step by step method which depicts the use of this model is proposed in the context of PROOFS life cycle. Furthermore, a short study on how to transform O-CA nets into PNs and CPNs is also presented as guideline of linking this high level specification model to the lower level specification models during the hierarchical development process. Finally an example is given for illustrating how to construct an O-CA net model for a distributed Document Conferencing Application.

8.2 Channel/Agency Nets

8.2.1 General Presentation

The Channel/Agency (CA) net model is a semi-formal and graphical language for the requirement capture and system specification. It is based on a restricted number of basic concepts that allow an easy system description.

Graphically, a CA net is made of squares (□), circles (○), arrows (←, →) and relevant labels:

- An *agency* (a square) represents an **active component** which produces or consumes, transports or changes information or items.
- A *channel* (a circle) represents a **passive component** which stores information or items between the agencies.

- An ***arrow*** represents a **relation** between two components, which indicates a potential information flow or information access.
- A ***label*** (string) identifies its attached component.
- An agency can only be connected to channels and visa verse, i.e. any component is not allowed to be connected to any other one of the same kind.

Due to the fact that components of a CA net are able to be refined, split, embedded and sectioned [Reisig, 1987], specification using CA nets is modular, top-down and hierarchical, so that CA nets are suitable for an use in a top-down analysis and design like those proposed in [Reisig, 1987] and the PROOFS method [van der Aalst *et al.*, 1993].

8.2.2 Techniques for the Use of CA Nets

Techniques for the use of CA nets in software engineering have been proposed in [Reisig, 1987]. Among them, we are mainly interested by the techniques of abstraction and refinement and the techniques of embedding and sectioning. The techniques of refinement have been classified in [Treves, 1992].

Techniques of abstraction and refinement

A ***refinement*** consists in replacing a single element of a net by a sub-net of an arbitrary size. The reverse process is the ***abstraction*** of a net, where a sub-net is substituted by a single component.

Relevant notions, called *port* and *boarder*, have been introduced in order to achieve these processes to be coherent.

We call ***input port*** an input linked component to component and ***output port*** an output linked component to considered component. An input port may also be an output port. We call ***boarder*** the set of components of a sub-net, associated to a refinement of the considered component, connected to the input and output ports of the considered component.

The refinement process relative to a component can be divided into several phases:

- Identification of the sub-components which will substitute the component to be refined.
- linking of these sub-components according to the detailed specification.
- linking of the sub-CA net through its boarder with its input and output ports.

Fig. 8.2 illustrates a refinement of an agency given in Fig. 8.1.

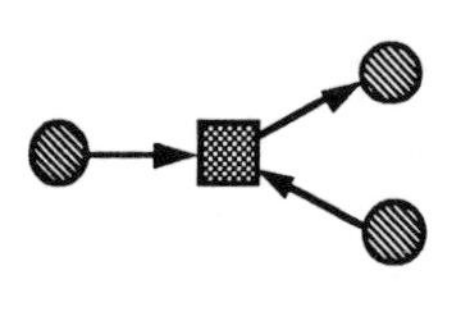

Figure 8.1: A simple CA

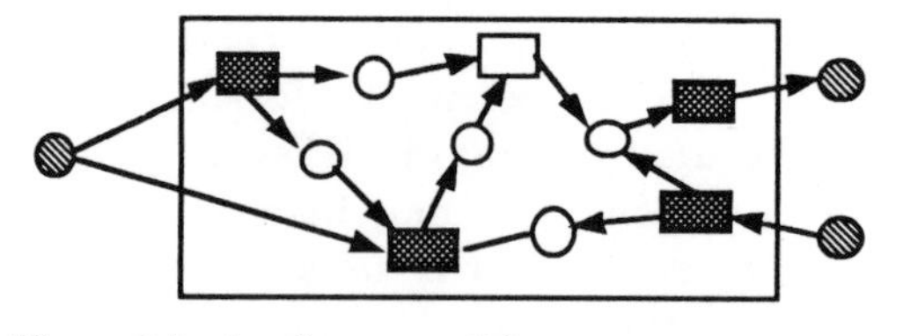

Figure 8.2: A refinement of the where each ■ is a boarder, while each ● is a port

A refinement is called a ***splitting*** when the sub-net is reduced to its boarder. A splitting is devoted especially for the expression of the distribution of the information within a channel, e.g. the definition of the different types of messages to be considered.

Techniques of embedding and sectioning

An ***embedding*** means to introduce additional components to a given net. The reverse of embedding is called ***sectioning***.

Embedding permits the reuse of already given components, while refinement allows the detailing of information within a given component.

Considering the graphical representation, an embedding M of a net N is shown by just adding the newly introduced elements of M to N. There is no convention on how to distinguish the previous elements of N from the newly introduced ones.

8.3 General Object Concepts and Characteristics

We will pay attention to the following principal object concepts and characteristics. The relevant terms are based on [Booch, 1991].

Object

An object may generally be described as an entity that:

- has a ***state*** which denotes the value the object carries.
- is characterised by the ***operations*** (***methods***) it provides and that it requires (or uses) from other objects;
- is an ***instance*** of a given ***class***;
- is identified by a name (or identity);
- has a restricted ***visibility*** to other objects;
- may be viewed either by its specification or by its implementation.

State and attributes

The state of an object is related to as a set of valued attributes of the object. The state may change over time.

Operations and the "use" relation among objects

The operations provided or required by an object determine the object's behaviours which change the state of the object.

An object may provide operations which can be invoked (used) by other objects and may also use operations of other objects. The invocation contains a number of operation arguments. The effect of an operation is that the object exhibits a defined part of its behaviour, changes to a new value and reports results containing a number of arguments.

Objects interact by invoking operations, and we call such an interaction a ***"use" relation***. An object is said to "use" another object if the former requires one or more of the operations provided by the latter.

Interfaces

There are two kinds of interfaces according to the input or output functions. The ***provided interface*** defines the operations provided and the associated arguments and types, whereas the ***required interface*** lists the operations and types required from other objects.

Compositions and the "include" or "be-a-part-of" relation between objects

An object can be internally structured into a combination of some other objects called ***component objects***. The operations of an object can be based on operations of the component objects.

An object A is said to "include" another object A1, or A1 is said to "be-a part-of" A, when A is a combination of a set of component objects, and A1 is one of the component ones.

Class

A ***class*** regroups objects that have the same static and dynamic descriptions. This concept defines what attributes the objects have and what the operations of the objects are, which types of arguments they require, and what are the restrictions on operation invocations. All static knowledge about an object is specified in its class. A class is a definition and a template to enable the creation or instantiation of new objects.

An instance is an object created from a class. The class describes behaviour and information structure of the instance, while current state of the instance is defined by the operation performed on the instance.

Inheritance and the "be-a" relation between classes

Inheritance is a relationship between classes, wherein one subclass shares the structure or behaviour defined in one (single inheritance) or more (multiple inheritance) other super-classes. Inheritance defines a "be-a" hierarchy between classes.

The effect of applying this relation is that the operations of the super-class (including the operation invocation messages, their response messages and behaviour of these operations in the particular super-class) are combined with these operations additionally defined in the subclass. In consequence, the operations of the set of super-class are implicitly available at the interface of the subclass, although additional operations can be provided in the subclass.

Encapsulation and visibility

Encapsulation is synonymous to information hiding, i.e., it hides the internal structure of an object as well as the implementation of its operations.

The ***visibility*** of an object depends on one's view. As every object has an hidden part and a visible part, an object may be viewed through two different ways: from the outside and from the inside. Whereas the outside view of an object serves to capture the abstract behaviour of the object, the inside view indicates how that behaviour is implemented.

Abstraction focuses upon the outside view of an objects and encapsulation prevents users from seeing its inside view [Booch, 1991].

8.4 The O-CA Net Model

A very important problem encountered in the PROOFS project is how to enable the OO users and developers to enter into the original non OO life-cycle, reversibly to make the PROOFS method reusable for OO application development. Comparing with other models (Petri Nets, Coloured Petri Nets, etc.), CA nets have more natural features of integrating OO concepts, due to its high abstract and general specification power. An object-oriented CA net model [Yang, 1994] can provide means of high level description to the OO application, and its transformation into Petri Nets or other high-level Petri Nets (CPNs [Jensen, 1987], ExSpect [van Hee *et al.*, 1990], etc..).

The O-CA net model is dedicated as an applicable one for the high-level specification of systems architecture and the description of relationships between OO components (e.g. objects, classes, interfaces and operations). It permits a smooth "link" from the informal description of an OO system to the formal specification via a specific semi-formal way where the main concepts of the OO paradigm are introduced at the stages of requirement analysis and architecture design.

In the following, the description of OO concepts with CA nets is presented. The approach considered for this purpose is the integration of the OO paradigm into the CA nets model. Specific rules, which extend the rules for the use of CA nets during the design process, applicable to this new model are then given.

8.4.1 Integration Methods

Generally speaking, the task of integrating a model A to a model B can be seen as an attempt to maintain as much as possible B's philosophy, and introduce a constrained set of the new A's syntactic or semantic concepts upon B.

We define an O-CA net to be a CA net which obeys a set of specific rules and has a clear semantic interpretation with OO concepts and constraints. The way to construct a O-CA net is to map the OO concepts to a CA net components and techniques (e.g. refinement, splitting, abstraction, embedding, etc.) in the sense that will provide the required semantics to the CA net. To achieve such a model which supports the object-oriented modelling, based on the major OO concepts and characteristics described in section 8.3, we give an adaptation between the OO paradigm and CA nets, according to the modelling requirement. As CA nets are syntactically limited to specify only process aspects of a system versus its data aspects, the state and attributes of an object are not considered for the moment in O-CA nets.

Before achieving this, let us first keep in mind that an agency is active, and can be interpreted as an event, an entity, a process, a function, an operation, a service, or a dependency, etc., and that a channel is passive, and can be interpreted as a condition , a link, an information, or a statement, etc.. Based on these elementary principles, an agency will be an object, an interface, or an operation, whereas a channel will be an interaction relation between two objects, or an input/output message.

8.4.2 Description of Object Concepts in O-CA Nets

Object in O-CA nets

(a) Object

An object is expressed by an agency. Such an agency may be refined by introducing its contents (interfaces, operations, component objects, and relevant links between them).

(b) Operations

As an active component, an operation belonging to an object is expressed by an agency which always appears once the object has been refined.

(c) Interfaces

To ensure an object internal information hiding (encapsulation), its interfaces, represented by agencies, are introduced to be the only visible "paths" to or from any external objects. As operations are referenced by the interfaces of the object, operations and the interfaces are always linked together by channels.

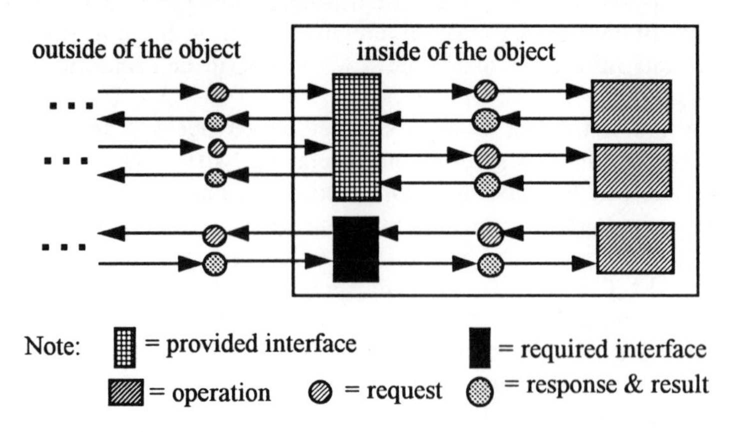

Figure 8.3: Interfaces, operations and links between them

As shown in Fig. 8.3, an object, its operations and interfaces are represented in an O-CA net as follows: a provided interface (resp. required interface) is represented by an agency which is internally linked to the invoked (resp. invoking) operation agency and which is externally linked to other objects via the request and response & result channels.

Relations among objects in O-CA nets

(a) "use" relation

The "use" relation between two objects is expressed by a channel. Furthermore, via a channel between two object agencies, the interaction of the two objects can be described as in Fig. 8.4. The plain arrows denote the invocation whereas the dotted ones represent the results report of the "use" relationship. Note that often, in short, we only use undirected arcs to describe some examples with which we only want to show the existence of a use relation between two objects rather than the use direction, as those presented later in Fig. 8.7, Fig. 8.8, etc.

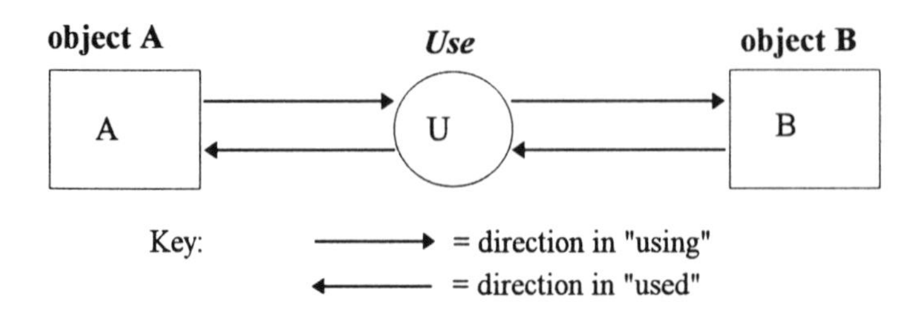

Figure 8.4: Objects and the "use" relation between them

Each interaction between objects is carried out by means of invoking operations. This concept can be presented by splitting the "use" channel into several sub-channels, each corresponding to a special operation invocation (see Fig. 8.5).

An operation invocation includes, commonly, a request message from the using object to the used one, and also a message containing the response and the result of the invocation from the used object to the using one. As indicated in Fig. 8.6, for instance, the operation invocation U1 in Fig. 8.5 can be split by introducing its request and the response & result.

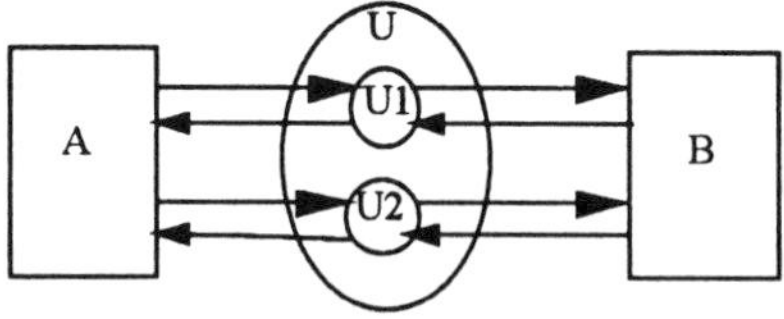

Figure 8.5: Splitting the use relation U: operation invocations U1, U2

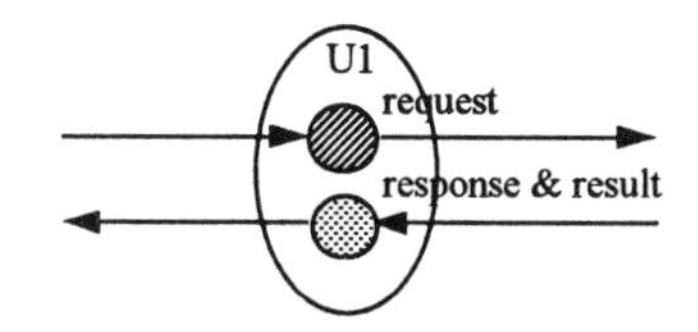

Figure 8.6: Splitting an operation invocation: operation request and response & result

(b) "include" or "be-a-part-of" relation (composition)

Besides its operations and interfaces, an object may include also several component objects. To describe the "include" or "be-a-part-of" relation, we again use the refinement techniques: an object is refined by introducing its component objects and at the same time, as proposed above, its interfaces and operations. This kind of "zoom-in", and "zoom-out" pictures expresses the "include", and reversibly the "be-a-part-of" relation.

Let us consider a simple example: an object A includes three component object A1, A2, and A3, as well as two operations and one interface; between A1 and A2, A2 and A3, there exists a "use" relation; between operations, component objects and interface, there are some other internal relations. Fig. 8.7 gives the O-CA net related to a refinement of the object A.

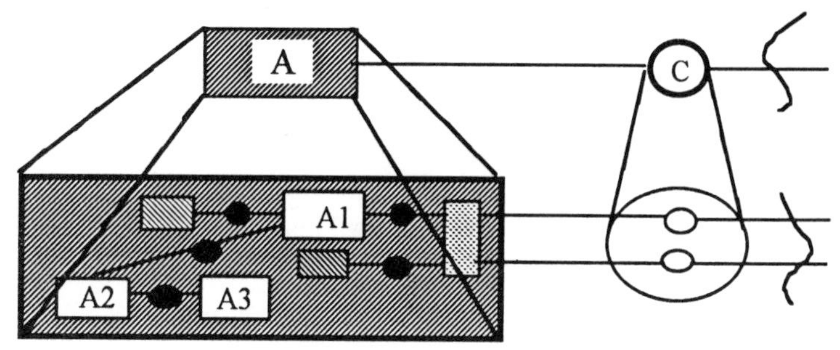

Figure 8.7: A refinement of object A and a relevant splitting of channel

Class and the relations between classes in O-CA nets

(a) Class

As a class of objects represents an abstraction of a set of objects which share a common structure and a common behaviour, classes and relations between different classes can be represented by a similar CA net as that of its instances, i.e. objects. For example, a class is represented by an agency, and the use relation between two classes is represented by a channel, etc.. In addition, the components of a class are all considered as generic types or classes, e.g. the interface type, operation type, component class etc.. Naturally, an interface type, an operation type, or a component class is represented by an agency respectively.

(b) "Be-a" relation (inheritance)

The "be-a" relation enables the inheritance of behaviour specified for the super-classes into the subclass, and it may be extended consistently in the subclass.

Based on the fact that one subclass agency 's refinement consists of all the components of its super-class and the other extra components which make it specific, the principle of "embedding and sectioning" [Reisig, 1987] can be used to describe this relation. Embedding permits introduction of additional components (including agencies, channels and relevant arcs) to a given net. After a super-class 's CA net (including its interfaces, operations, relations with other classes) has been completed, its subclass's CA net can be easily obtained by adding the extended operations and relevant interfaces according to the embedding principle. Fig. 8.8 gives an example of the proposed expression of inheritance in O-CA net: subclass S inherits its super-class P's properties (interface, operations, links) and with its own additional properties (interface, operations and links, see those **bold** squares, cycles and arcs).

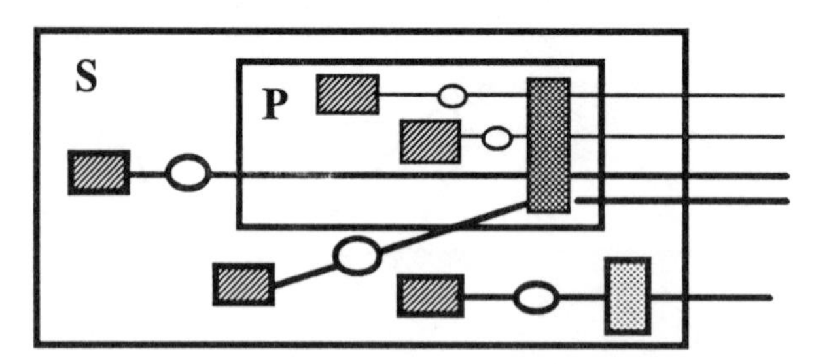

Figure 8.8: An example of an inheritance of the class P

(c) Instantiation of class

An object is an instance of its class. Instantiation of a class is to instantiate the class's static structure as well as class's behaviour.

Based on an O-CA net of a class A and the use relation between A and another class B, the instantiations of the class A can be explained by splitting. The way of splitting the class A depends on the way how the created instances of the class A use the operations of the class B: we can imagine that B's behaviour appears differently with respect to each individual instance of A.

We can suppose that each instance has an unique identifier which correspond to an index of the class, for example object A(1), object A(2), etc. There are two kinds of instantiations with respect to the object index: the index independent instantiation and the index dependent instantiation. In the first case, all instances of the class A uses all operations defined in the used class B. In the second case, all instances do not use necessarily all operations defined in B. In Fig. 8.9 and Fig. 8.10, we illustrate the two kinds of instantiations by two different O-CA nets respectively.

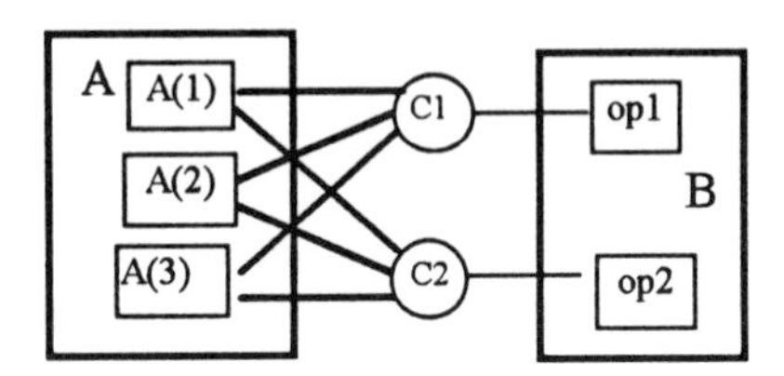

Figure 8.9: Index independent of A with respect to B

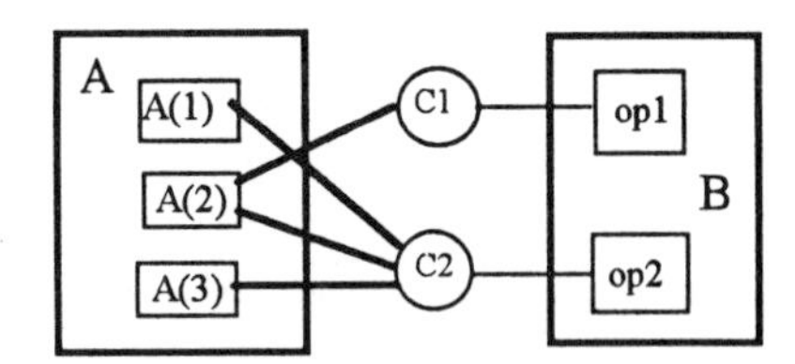

Figure 8.10: Index dependent of A with respect to B

We can interpreter the two different nets as follows: In the first case (Fig. 8.4-7), each instance, or object, is a copy of its class in the level of structure and behaviour and it shares the use relation C between the class A and the class B, i.e. all A(1), A(2) and A(3) can use all operations in B. In the second case (Fig. 8.4-8), each object is not only a identified or indexed copy (A(1), A(2) or A(3)) of the class with the same structure and behaviour as the class, it has also an identified or indexed use relation to the used class B with respect to the operations it uses in B. For example, A(1) uses only op2 in B, A(3) uses only op1, but A(2) can use both op1 and op2.

In fact, a class to be instantiated is just like a template which may be parameterised by objects [Booch, 1991]. Instantiation by parameterizing a class concerns therefore the data aspects (e.g. the index of an object) of the class. As we has mentioned that the O-CA nets have syntactical limit for modelling data aspect, to enforce O-CA nets by formal CA nets (CA+ nets [Agoulmine, 1994]) is highly necessary. We will discuss this point in a future study.

Object visibility and encapsulation in O-CA nets

We can abstract the main ideas of all this section by assessing the way and the power of O-CA nets for modelling object visibility and encapsulation.

Visibility of objects is shown both inside objects and outside objects. In O-CA nets, an object is expressed by an agency which can be refined by introducing its internal components such as its component objects (agencies) , its operations and its operation interfaces (agencies), and the links between the components (channels). This point shows an inside visibility of objects. Fig. 8.4-9 illustrates this idea in its right picture.

On the other hand, outside visibility of an object can be shown in different abstraction levels along the refinement process of the O-CA net. For instance, from the abstract "use" relation between two objects at the top level of the O-CA hierarchical net model to the precise operation invocation relation at the end of the refinement of the model, one can gradually see first the using and the used object, then the interfaces of both objects, and then, through the interface, the concerned operations which provide services or request services. Fig. 8.11, in its left picture, shows a moment when one can see the interface which lists the available operations and receives requests and sends responses.

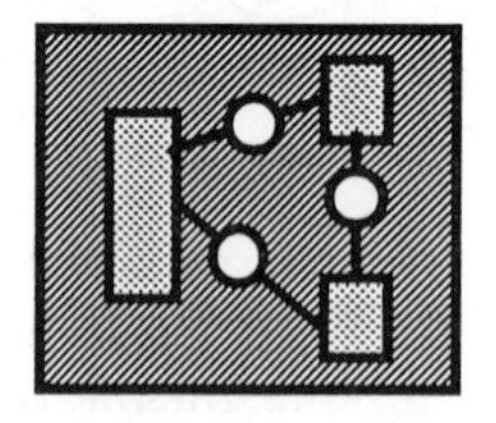

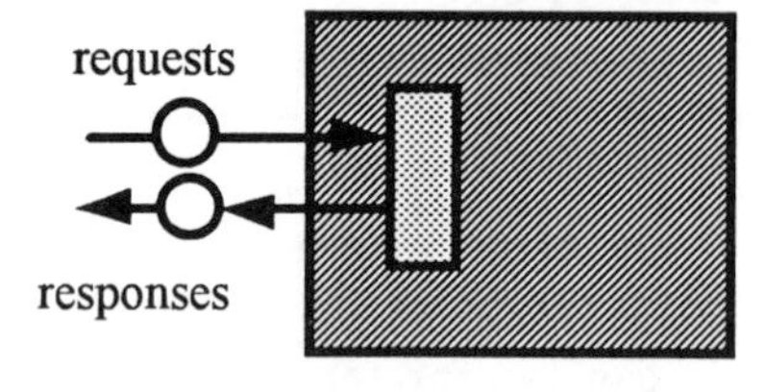

An inside view of the object

An outside view of the object

Figure 8.11: Inside view and outside view of an object

CA nets are by no means intended to describe detailed object design and implementation information, which are rather the subject of Petri Nets (PNs). However the way towards PNs from CA nets is very direct, when one "opens" the encapsulation and enriches the O-CA model by introducing more detailed specification of the OO system, especially its dynamic behaviours (for being derived into PNs), and data aspects (for being derived into Coloured Petri Nets (CPNS)). So it is clear that CA nets are faced both by end-users or designers in the phases of requirement analysis and sketch design, whereas PNs-like models are faced by designers in the phases of detailed system design. An important fact is that the latter can be smoothly derived from the former [Agoulmine, 1994].

8.4.3 Structure Rules and Transformation Rules of O-CA Nets

Structure rules define the CA components and their construction, and transformation rules include refinement, splitting, abstraction, and embedding rules as described in section 8.2.2., each of which defines a transformation operation of the CA net. These rules are rather syntactic than semantic.

Special syntactic and semantic constraints may make O-CA nets be specialised CA nets, where special rules are necessary both at syntactic level and semantic level.

Before defining new rules, we first recall some defined rules and definitions for general CA nets [Reisig, 1987, Agoulmine, 1994]:

Definition 0.1

A **minimal** refinement (or splitting) of a component X of a net N is a refinement (or a splitting) such that number of links of the boarder of the refinement of X with the ports of X is preserved with respect the links of X with its ports.

Definition 0.2

A maximal refinement (or splitting) of a component X of a net N is a refinement (or splitting) such that each sub-component X(i) of the boarder of the refinement of X is linked to all the ports of X with respect the direction of the links of X with its ports.

Rule 0.1

Two components of the same type cannot be linked. An **O-CA net** respects additionally the following rules and constraints.

Structure rules

Rule 1.1

At the **top level** (eg. the first step) of a specification, classes and "use" relations between them are only identified. Classes are represented by agencies and the "use" relation by a channel connected to these agencies, as given in Fig. 8.4.

As a consequence of various refinements and splitting, we have:

Rule 1.2

Any channel is connected at least with two agencies at **any level** of the specification.

Agency refinement rules

Rule 2.1

Two kinds of agencies are allowed. One is named non-terminal agency which can be refined, while another named terminal agency which cannot be refined again, but may be split.

Rule 2.2

Agencies representing classes or sub-classes are non-terminal. Agencies representing operations and interfaces are terminal.

Rule 2.3

The agencies at the **top level** are non-terminal ones.

Rule 2.4

At any level of the specification, except the top-level one, all kind of agencies may appear.

Rule 2.5

A specification is **terminated** when all the agencies are terminal.

Channel and agency splitting rules

Rule 3.1

An **index independent instantiation** is described with a maximal splitting, while an **index dependent instantiation** cannot be described with a maximal splitting.

Rule 3.2

No refinement rule is applicable for a channel, as a channel is only considered as a use relation or a communication message.

Net abstraction rules

Rule 4.1

A **net abstraction** is restricted to the reverse operation of a refinement or a splitting previously done.

Net sectioning rule

Rule 5.1

The net sectioning rule is not used in our purpose.

8.5 O-CA Nets and PROOFS Method

The PROOFS method [van der Aalst *et al.*, 1993], based on S-CORT [Hildebrand and Treves, 1990] and supported by multi-formalisms including CA nets, High-level Petri Nets like Coloured Petri nets (CPN) [Jensen, 1987] and ExSpect [van Hee *et al.*, 1990], NIAM data model and supporting tools, is justified as a development method covering every phase of the software life cycle of the development of large and heterogeneous distributed systems, from requirement analysis to maintenance. The method consists of the following five stages in which Petri nets are used in an omnipresent manner:

- P0: Preliminary study of the communication system (requirement analysis).

- P1: Definition of the communication system architecture.
- P2: Specification and check of communications between the communication system.
- P3: Definition of the active components of the communication system.
- P4: Implementation, integration and operation of the communication system.

Following on from the P0 stage, an approximate solution in terms of CA nets is proposed in order to set down the orientations of the system and define what can be carried out and what could be reused in an existing situation (protocols, services, etc.). During P1, CA nets are systematically used to identify the elements forming the system to be designed and describe the architecture of the system in a modular way, while NIAM is used for data conceptual modelling. By P2, the functional, logical, physical and quantitative aspects of the systems are described in detail by means of a partially formal modular method, based on CA net derived PNs, CPNs or ExSpect nets. On the basis of the P2 specification of each protocol, P3 focus on the modelling of the system elements by using the net-based functional language ExSpect. During the final stage P4, code generation and system maintenance based on the formal models produced during previous stages are carried out.

8.5.1 Use of O-CA Nets

The use of O-CA nets focus on the "top-level object relations", "object logical architecture" and "object communications" of an OO system, which correspond to some tasks of P0, P1 and P2 in the PROOFS life cycle, i.e. the high level and general description with CA nets before the detailed system description with high-level PNs. The following method of using O-CA nets is corresponding to these parts of descriptions.

Step 1 Identification of the "top" classes of objects and the use relations among them with agencies and channels: a class of objects is a non-terminal agency, whereas a use relation is a channel, each channel with an input directed arc from the using class and an output directed arc to used class.

Step 2.a Refinement of a non-terminal class agency by presenting each of its component classes as a non-terminal agency, each of its operations as a terminal agency and both provided interfaces and required interfaces as split-able terminal agencies. Linking the operations with the relevant interfaces. Between two non-terminal sub-agencies, if there is a use relation, an internal channel and relevant directed arcs should be added.

Step 2.b Embedding a class agency for the inheritance of a sub-class.

Step 3 Splitting each use channel by introducing each operation invocation as a sub-channel and linking them to the relevant interface agency within the two relevant agencies.

Step 4 Instantiation of classes by splitting.

Step 5 Verifying the completeness of the obtained O-CA net and correcting it by reviewing the design scenario. And finally reducing the size of the net by abstracting some unnecessary components [Berthelot and Roucairol, 1980]. If the net is non connected, there should be logical error. Often we will find the problem due to some non splitable channels which have been incorrectly split.

Step 2.a, step 2.b and step 3 should be iteratively used until there is no non-terminal agencies. With each step, suitable net selections should be adopted. Step 5 guarantees the logical completeness of the finally obtained O-CA net.

After each step, i.e. each transformation of the O-CA net, the proper labels attached to the derived new components should be presented.

Comparing with the PROOFS method, first with step 1, corresponding to the task P0, we can present the top-level object classes and the use relations among them, i.e.. a rough OO general communication structure of the OO system. Then the actions from step 1 to step 4 produce an object logical architecture specification.

We have seen, O-CA nets are used to describe both objects and classes. In the case of classes, every component in an O-CA net represents classes or types (object classes, interface types, operation types and message types), and the Net embedding rule (rule 5.1) can be used in the step 2.b a sub-class by embedding the O-CA specification of its inherited super-class or super-classes (multiple inheritance).

8.5.2 Link from O-CA Nets to PNs and CPNs

Being semi-formal, a CA net-like model is considered a bridge between informal models, e.g. natural languages, and formal models especially like Petri nets-like model, e.g. PNs, CPNs.

The PROOFS method, based on the CA nets derived from P0 to P2, is claimed then to produce systematically more detailed models, e.g. the PNs, or CPNs (or ExSpect) during the P2, and P3 stages. Here, as a basis for our further study, we shortly present an intended method to link O-CA nets to PNs and CPNs.

The principle of the link includes the following points:

- for ordinary PNs, i.e. Place/Transition nets:
  (1) A given O-CA net gives a basic net structure of its derived PN, a more detailed net structure can always be achieved by continuously refining the O-CA net components until the components are terminal.

(2) A derived PN can then be obtained by introducing the dynamic aspect specifications into the based net structure, practically, i.e. by allocating tokens into the net structure. The dynamic aspects deal with communication, concurrence and synchronisation features.

- for CPNs

(3) Besides above points, the data aspects concerning colours (types, values) of tokens and assessments or functions of the transitions should be defined.

As we know, a so-called ***net structure*** is in fact a directed graph with two kinds of nodes, places and transitions, interconnected by arcs (only nodes with different types can directly connected by arcs). In general, a CA net is a static Petri net structure in which active components, i.e. agencies, correspond to transitions in Petri nets, passive components, i.e. channels, correspond to places in Petri nets, and arcs are also arcs in Petri nets. A CA net may be a Petri net minus tokens. It is worth indicating that the real connection from CA nets to PNs/CPNs are carried out after all possible refinements, i.e. (1), are done.

A terminal agency representing an operation in the O-CA net can be then interpreted by a special sub-PN or a functional language like ExSpect by further steps according to the top-down specification method.

In [Agoulmine, 1994], two extended CA net models, named CA+ and CA++, are proposed to facilitate the transformation of CA nets to PNs and furthermore CPNs. The idea is to consider CA nets as a basic model of the system design and to enrich it first with a language expressing the dynamics of the system, secondly with Abstract Data Types. The derived model then can be directly translated firstly the ordinary PNs and then the CPNs.

As O-CA nets are regrouped as a sub-class of generic CA nets, the CA+, and CA++ nets are also suitable to O-CA nets to achieve points (2) and (3).

8.6 Example

In this section we give an example of using the step by step method proposed in the section 8.5.1 to build an O-CA net of a small part of a Document Conferencing Application (DCA).

In a distributed DCA environment, there are one editor-in-chief and several participants working to produce a document. We call the object EDesk a logical representation of the person editor-in-chief, and an object PDesk as a logical representation of a participant, and the object Document as a document processor. Naturally, the EDesk and PDesk *use* the Document to produce a document together. Fig. 8.12 shows the step 1 of the specification: the top-level object classes.

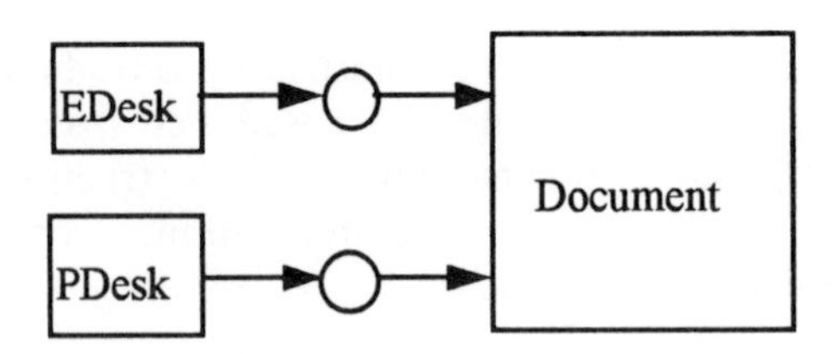

Figure 8.12: The top-level objects in a part of the DCA

We suppose that a Document *includes* a Header, several Chapters and a Bibliography, and a *provided interface* linked to all of these *component objects*. A PDesk has two *operations*: read and write1, and two relevant *required interfaces*. The operation read means to read one of any parts of the document (the Header, every Chapter and the Bibliography). The operation write1 means to write the Bibliography and the Chapter which one PDesk is responsible for (for example: PDesk(i) can only write1(i) the Chapter(i) and the Bibliography).

Each part of the Document has two *provided operations* get, and put which can be invoked (*used*) by PDesk and EDesk via the *provided interfaces* linked by the channels G (read->get), P1 (write1->put1) and P2 (write2->put2).

A partial O-CA net describing the refinement of the class PDesk, the refinement of the class Document and the operation invocation from the PDesk to the Document is shown in Fig. 8.13 as follows: a PDesk can read all parts of the Document, but can only write one Chapter and the Bibliography. The read operation must invoke the get operation of the used component object to show the concerned part of the document and, in parallel, the write1 operation has to invoke the put operation of the used component object to input new text into the relevant Chapter or the Bibliography of the Document. Fig. 8.13 is not immediately obtained from Fig. 8.12, in fact several steps of refinement and splitting are used: step 2.a (PDesk, Document) and step 3 (links between PDesk and Document).

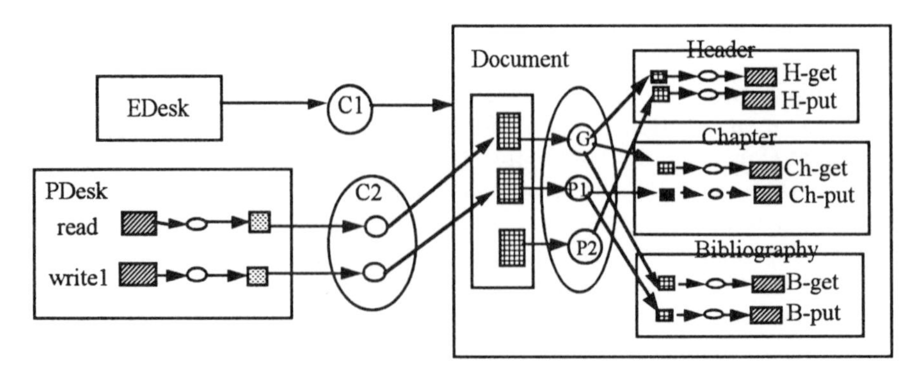

Figure 8.13: A partial O-CA net concerning the refinements of PDesk and Document

Moreover, EDesk is a sub-class of PDesk, i.e. the EDesk *is-a* PDesk, but acts a more important role than a PDesk: it can write (or consolidate) all parts of the Document rather than only some parts. So we can obtain the partial O-CA net concerning EDesk, with the step2.b, by first inheriting the PDesk refinement then adding the additional write2 operation which permit to write the reset parts that are not allowed by write1. In consequence, with the step 3, the relevant links (channels and arcs) are also added into the newly O-CA net. Fig. 8.14 depicts a globally refined O-CA net of the system after the two steps.

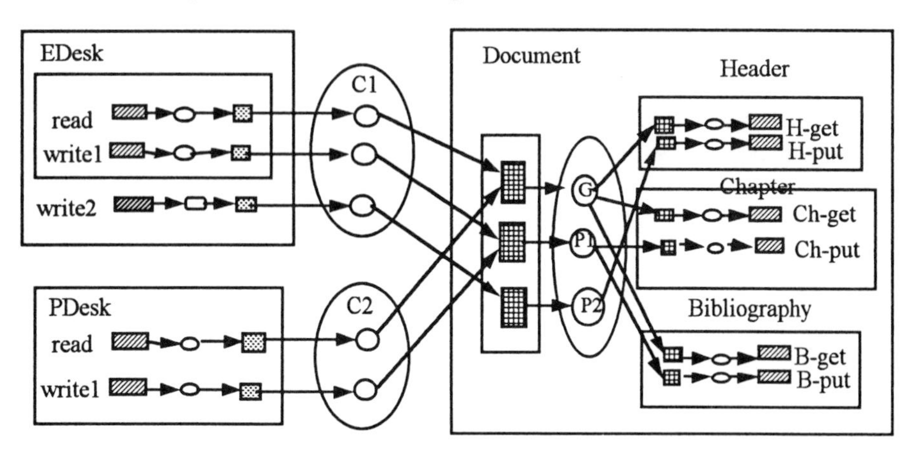

Figure 8.14: An entire O-CA net concerning the class EDesk, PDesk and Document

To illustrate instantiations of classes, we suppose that two objects PDesk(i) and PDesk(j) *are instanciated* from the class PDesk, and the objects Chapter(i) and Chapter(j) *are instanciated* from the class Chapter. Other objects like EDesk, Header and Bibliography, which are unique in the system, *are instanciated* from their classes EDesk, Header and Bibliography respectively. According to the rule 3.2, with step 4, the O-CA net of all the instanciated objects can be obtained as shown in the Fig. 8.15. We remind that each PDesk(i) is allowed to write only Chapter(i) and the Bibliography.

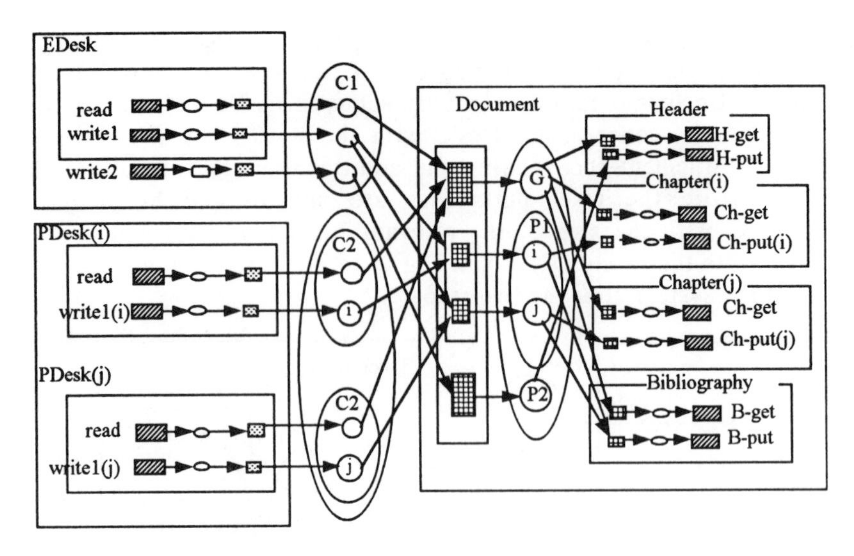

Figure 8.15: Instantiations of classes

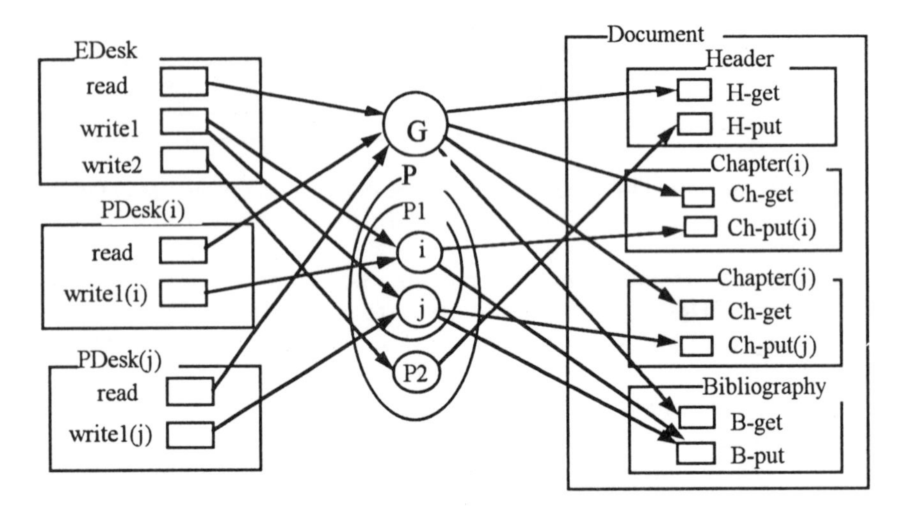

Figure 8.16: The simplified O-CA net

In the final step, we reduce the size of the net as Fig. 8.16 shows, in which all the *interface* agencies are reduced from a view of the net simplification. The improved O-CA net in Fig. 8.16 can be seen as a static Petri net structure, and can be further enriched and transformed into PN and CPN by using the method proposed in 8.5.2.

8.7 Conclusion

Making a model, especially a formal model, from scratch is one of the most difficult task of software development. In this paper, we have shown the feasibility of introducing OO concepts to facilitate this very important task in a Petri nets based development process.

An Object Oriented Channel/Agency net model, named O-CA nets, has been proposed. An O-CA net, as a specific sub-class of CA net, is a CA net containing semantic interpretations that corresponds to OO concepts. In addition, O-CA nets provide the relevant rules defining the structure and the transformation of the model associated to the OO concepts.

This model includes basic OO concepts, i.e. the concepts of objects, object classes, and the three important relations among objects or classes (use, is-a-part-of, and is-a). This model can be used during the system requirement analysis and logical preliminary design phases in a Petri nets based life cycle like PROOFS to describe the objects (resp. classes) and the "use" relationships among them, the hierarchical composition structure of a compound object, the invocations from any "using" objects to the operations offered by the relevant "used" objects, the communication messages of invocations, the instantiations of classes, and the inheritances of classes.

Finally a short study on how to transform O-CA nets into PNs and CPNs is presented. The idea is to enrich O-CA nets by introducing the dynamic behaviours, then the data aspects of the OO system respectively. It is also claimed by the proposed method that operations of objects, as terminal components in the O-CA net model can be substituted by an sub-PN or interpreted by a function specification like ExSpect in the next stage of detailed system design in the development process. This study will be a hint or guideline for our further studies on how to use CA+ nets and CA++ nets to enrich O-CA nets for the purpose of system data modelling and dynamic modelling.

Acknowledgements

The authors gratefully acknowldge some financial support for this research from the CCE under contract EP 5342 ESPRIT II PROOFS.

Part IV

Formal Foundations of Object Technology

Chapter 9

Design Structures for Object-Based Systems

9.1 Introduction

Temporal logic has been used with great success since [Pnueli, 1977] as a formalism for supporting the specification of the behaviour of concurrent, reactive systems. However, it is not enough to have a convenient formalism in which to specify and verify the properties of systems. In order to make the specification effort manageable, we need ways of modularising it. That is, we need formal building blocks out of which complex systems may be built, formal mechanisms for interconnecting previously specified components, as well as disciplines for decomposing and organising specifications in terms of such building blocks and interconnection mechanisms. For instance, when specifying a producer-consumer-buffer system, we should be able to start from separate specifications of each of these components, possibly "reusing" existing specifications, and interconnect them in such a way that the whole system behaves as intended.

Likewise, when developing components that are intended to be used as modules in different environments (i.e. that can be "reused"), we should not have to worry about the specific systems with which they will be required to interact. (That is, we should not and cannot depend on "engineering omniscience".) For instance, when specifying a buffer, we should not have to define in advance the clients (producers and consumers) of the services of the buffer. Instead, in order to use a previously specified component as a "server" in the context of a system, we should be able to define the *interfaces* via which each "client" has access to the services being provided by the "server".

Therefore, we should recognise that the modularisation of system development requires *design structures* that have not been accounted for in the traditional logical view of specification as applied to concurrent and reactive systems. On the one hand, connections between temporal logic and process languages do exist (e.g. [Barringer *et al.*, 1985] for CSP [Hoare, 1985]), and notions of module have been proposed for concurrent programs [Lamport, 1983], but these languages are much closer to programming than to specification. They direct us to very specific ways of process combination (explicit control flow), failing to support looser forms of interconnection such as client-server relationships which are crucial for achieving higher levels of interoperability, maintainability and reusability in systems design. More recent proposals like Lamport's Temporal Logic of Actions [Lamport, 1994] have incorporated logical mechanisms for controlling change within systems, but they fail to provide higher levels mechanisms for the modularisation of the specifications themselves, namely because they rely on a "shared variable" paradigm and on a global name space. In this sense, our aim is to bring into the specification of reactive systems modularisation principles based on formal design structures that are more akin to, say, object-oriented development [Meyer, 1988]. On the other hand, attempts at marrying object orientation and reactive systems development already exist [e.g. Järvinen *et al.*, 1990], but without the blessings of an underlying formal logic as a semantic domain.

The work that we shall report in this paper provides such a foundation. It integrates the flexibility of object-oriented design techniques (in a broad sense), the formal semantics that is associated with temporal logic, and algebraic techniques for modularising specification akin to those which have been proposed for Abstract Data Types (ADTs) [e.g., Ehrig and Mahr, 1985, Goguen, 1986, Sannella and Tarlecki, 1988]. The key idea, which we have put forward in [Fiadeiro and Maibaum, 1992], is to adopt theories instead of formulae as building blocks for component specification, using interpretations between theories (theory morphisms) as a means for specifying interconnections between components in the spirit of the categorial approach to systems theory proposed in [Goguen and Ginali, 1978].

However, although theories and interpretations between theories have proved to provide an adequate semantic domain for discussing system specification and design, actual specification/design modules are better accounted for in terms of particular *(standard) structures of theories*, as already identified for ADT Specification [Ehrig and Mahr, 1990]. Our purpose in this paper is to extend the framework put forward in [Fiadeiro and Maibaum, 1992], which we shall briefly recall in Sections 2 and 3, with design structures for modularisation. A design structure consists of a kernel theory that specifies the behaviour of a component, a collection of theories specifying the interfaces via which the component is connected to the other components of the system, and another theory (the body) which connects the interfaces with the kernel. We shall analyse different types of interfaces, corresponding to different kinds of component interconnection. In Section 4 we discuss action calling (as proposed in [Sernadas *et al.*, 1990]), a non-symmetric synchronisation mechanism in which the occurrence of an action implies the occurrence of another one. In Section 5 we discuss object calling, a generalisation of the previous mechanism in which an action synchronises with a transaction of another object.

9.2 Temporal Specification of Objects

In order to make the paper self-contained, we shall begin by summarising the temporal logic of objects which we have defined in [Fiadeiro and Maibaum, 1992]. As a temporal logic, this formalism is quite standard. In fact, the ability to support the modularisation techniques that we shall be discussing is, up to some extent, independent of the specific temporal logic that we use.

An object as a component of a reactive system is a dynamic entity which possesses a state that can be changed only through a circumscribed set of actions (– this being the normal informal definition of *encapsulation*). Besides changing the state of the component, these actions also provide the means for the component to interact with other components. Hence, the language of an object must provide symbols that account for observations of the state of the object (which we call attribute symbols), and symbols that account for the actions that the object may perform. Such collections of attribute and action symbols are grouped into what we call the *signature* of the object. Besides these two categories of symbols, we must also account for the data types that are to be used as a universe for the definition of the domain and codomain of attributes and of action parameters:

Definition: An *object signature* is a triple (Σ,A,Γ) where

- S=(S,W) is a data signature in the sense of [Ehrig and Mahr, 1985], i.e. a set S (of sort symbols) and a $S^*\times S$-indexed family W (of function symbols).
- A is a $S^*\times S$-indexed family (of attribute symbols).
- Γ is an S^*-indexed family (of action symbols).

The families Ω, A, and Γ are assumed to be disjoint and finite (i.e., there are only a finite number of function, attribute and action symbols).

A semantic interpretation structure for an object signature corresponds to a Kripke structure for discrete, linear temporal logic, i.e. it will provide a mapping that gives the values taken by the attributes at each instant, and a mapping that gives the actions that will occur at each instant. It also includes a Σ-algebra that interprets the data types:

Definition: A *θ–interpretation structure* for a signature θ=(Σ,A,Γ) is a triple $(\mathcal{U},\mathcal{A},\mathcal{G})$ where:

- $\mathcal{U}$ is a Σ-algebra;
- $\mathcal{A}$ maps $f \in A_{<s_1,\ldots,s_n>,s}$ to $\mathcal{A}(f): s_{1\mathcal{U}} \times \ldots \times s_{n\mathcal{U}} \times \mathbb{N}_0 \rightarrow s_{\mathcal{U}}$;
- $\mathcal{G}$ maps $g \in \Gamma_{<s_1,\ldots,s_n>}$ to $\mathcal{G}(g): s_{1\mathcal{U}} \times \ldots \times s_{n\mathcal{U}} \rightarrow \wp(\mathbb{N}_0)$.

Notice the state dependence of $\mathcal{A}$ and $\mathcal{G}$. Encapsulation, which we henceforth refer to as *locality*, leads to a restricted class of interpretation structures which we shall take as models for object signatures:

Definition: Given an object signature θ=(Σ,A,Γ) and a θ-interpretation structure $(\mathcal{U},\mathcal{A},\mathcal{G})$, let

$$\mathcal{E}=\{i \in \mathbb{N}_0 \mid i \in \mathcal{G}(g)(a_1,\ldots,a_n)$$

for some

$$g \in \Gamma_{<s_1,\ldots,s_n>}, (a_1,\ldots,a_n) \in s_{1\mathcal{U}} \times \ldots \times s_{n\mathcal{U}}\}.$$

That is, $\mathcal{E}$ consists of those instants during which an action of the object occurs; these instants denote state transitions in which the object will be engaged. These state transitions are said to be *witnessed* by θ. The θ-interpretation structure $(\mathcal{U},\mathcal{A},\mathcal{G})$ is said to be a *locus* iff, for every instant $i \notin \mathcal{E}$ and every $f \in A$,

$$\mathcal{A}(f)(i) = \mathcal{A}(f)(i+1).$$

In other words, attributes remain unchanged during non witnessed transitions.

Attribute symbols generate terms whose denotation varies in time (non-rigid designators) and action symbols generate predicates of a linear and discrete temporal language [Goldblatt, 1987] which we shall use to describe objects. An object description is just a theory presentation in this logic:

Definition: An *object description* is a pair (θ,Φ) where θ is an object signature and Φ is a (finite) set of θ-formulae (the axioms of the description). A model of a description is a θ-locus that makes all the formulae of Φ valid.

Rather than give an exhaustive definition of this language, which can be found in [Fiadeiro and Maibaum, 1992], we shall introduce it via an example. For instance, the behaviour of a producer (as in a producer–consumer–buffer system) can be described as follows:

PRODUCER

data sorts: ITEM, BOOL
data functions: boolean operations

attribute symbols:	action symbols:
current: ITEM	produce(ITEM)
waiting: BOOL	store(ITEM)

axioms:

p0) usual axioms for the booleans
p1) produce(i) $\to$ **X**current = i
p2) produce(i) $\to$ **X**waiting = true
p3) store(i) $\to$ **X**waiting = false
p4) **BEG** $\to$ waiting = false
p5) produce(i) $\to$ waiting = false
p6) store(i) $\to$ current = i
p7) store(i) $\to$ waiting = true
p8) waiting=true $\to$ **F**store(current)

Intuitively, *current* denotes the last item produced by the producer, and *waiting* indicates whether the producer is waiting to store it. These are attributes (non-rigid designators) as they will likely be the subject of change effected by actions.

The sort symbol ITEM stands for the items that are produced. Although we shall not explicitly address genericity and parameterisation, this sort can be thought of as being "generic", and the signature to be "parameterised" by this sort symbol.

The chosen action symbols account for the production and storage of items. Notice that there is no reference to the buffer in which the produced items will be stored. We shall see later (section 3) how we can establish a relationship between a producer and a buffer by requiring them to synchronise on the store actions.

The symbol **X** is the temporal operator "next", or "tomorrow" (recalling that we are using linear discrete time, with an initial instant). Hence, at an instant *w*, **X**current denotes the value taken by *current* at time w+1. (As in [Manna and Pnueli, 1991], we are allowing the next operator to be applied both to formulae and terms.) On the other hand, an action predicate is satisfied at every instant where the action occurs. Hence, p1 asserts that producing item i sets *current* to i. Likewise, p2 asserts that a produce action sets *waiting* to *true*. These axioms are assumed to hold at every moment during the life of the object (and for every assignment of values to the variables).

Axiom p4 is an initialisation condition: **BEG** is satisfied only at the first instant ot time, so p4 asserts that the initial value of *waiting* is *false*.

These axioms account for the effects of the actions on the attributes. With respect to safety and liveness requirements, p5 asserts that a produce action can only occur when *waiting* is *false*, and p8 requires that, when *waiting* is *true*, a store of the current item will eventually occur (**F** is the temporal operator "eventually").

Summarising, typical formulae that are used for describing objects are of the form:

1. action$\wedge$condition $\rightarrow$ **X**condition

Such axioms specify the effects of the actions on the attributes.

2. action $\rightarrow$ condition

Such axioms specify restrictions on the occurrence of the actions and are the basis of safety properties for the object.

3. condition $\rightarrow$ **F**action

Such axioms specify requirements on the occurrence of the actions and are the basis of the liveness properties for the object. All conditions are first-order formulae in the language of the attributes and data types.

There is a further property of every object description. This is the property that reflects the locality condition: for every signature $\theta = (\Sigma, A, \Gamma)$,

$$\Big(\bigvee_{g \in \Gamma} (\exists x_g) g(x_g)\Big) \vee \Big(\bigwedge_{a \in A} (\forall x_a)(\mathbf{X}a(x_a) = a(x_a))\Big)$$

is a logical axiom for any specification (θ, Φ). (A similar localisation mechanism has been explored in Lamport's Temporal Logic of Actions [Lamport, 1994].) Indeed, it states that, at every instant, either one of the action predicates is satisfied (meaning that one of the actions of the object occurs), or the attributes will remain invariant. For instance, the locality condition for the producer is

$$(\exists i{:}ITEM)produce(i) \vee (\exists i{:}ITEM)store(i) \vee$$
$$(\mathbf{X}current{=}current \wedge \mathbf{X}waiting{=}waiting)$$

Such a requirement imposes a discipline on the way in which we are allowed to identify system components and, thus, its formalisation is an important part of any framework designed to support modularity in specification. The formalisation of the object-oriented notion of locality within the temporal logic approach to specification has been discussed in some detail in [Fiadeiro and Maibaum, 1992]. Other forms of localisation have also been considered in [Ryan *et al.*, 1991].

We end this section with the description of consumers and buffers. When we abstract from the consumer activity the behaviour that is of interest in relation to the producer-consumer-buffer system, we find out that it is, in a sense, the "mirror image" of the production activity:

CONSUMER

data sorts: ITEM, BOOL
data functions: boolean operations

attribute symbols:	action symbols:
current: ITEM	consume(ITEM)
waiting: BOOL	extract(ITEM)

axioms:

c0) usual axioms on the booleans
c1) extract(i) $\rightarrow$ **X**current = i
c2) extract(i) $\rightarrow$ **X**waiting = true
c3) consume(i) $\rightarrow$ **X**waiting = false
c4) **BEG** $\rightarrow$ waiting = false
c5) extract(i) $\rightarrow$ waiting = false
c6) consume(i) $\rightarrow$ current = i
c7) consume(i) $\rightarrow$ waiting = true
c8) waiting=true $\rightarrow$ **F**consume(current)

We should stress that the use of the same symbols in both signatures (PRODUCER and CONSUMER) does not indicate (or hint at) any relationship between the described objects. The use (meaning) of symbols is local to the signature. As we shall explain in section 3, any relationship must be explicitly stated via morphisms.

It is easy to see that we obtain the overall sequential behaviour that could be expected: a consumer extracts an item and must wait until that item is consumed before extracting another item.

With respect to the buffer, we could specify its behaviour as follows:

BUFFER

data sorts: ITEM, list(ITEM)
data functions: operations on lists

attribute symbols:	action symbols:
queue: list(ITEM)	put(ITEM)
	get(ITEM)

axioms:

b0) usual axioms specifying lists
b1) put(i) $\rightarrow$ **X**queue=queue•i
b2) get(i) $\rightarrow$ $(\exists q)$(queue=(i•q) $\wedge$ **X**queue=q)

That is, the buffer stores items on a list and releases them in a FIFO discipline. Notice that such a description of a queue differs considerably from the temporal specification given in [Manna and Pnueli, 1991]. The difference is that we use here abstract data types for modelling the attribute structure of objects. Naturally, it is possible that the abstract data types that are used are not part of the intended implementation environment. This just means that the object description has to be refined accordingly, e.g. as discussed in [Fiadeiro and Maibaum, 1994].

Finally, we should stress that it is more the organisation of specifications into theories than the actual logic that is used in the specifications that we want to stress in the paper. Lamport has shown (e.g. [Lamport, 1994]) that a too liberal use of standard temporal logic (the one we have used above) can lead to specifications that are not implementable. We feel, however, that this is a problem to be solved by carefully defining a specification language and method and giving them a temporal semantics that can be proved to be implementable (once the target implementation language and paradigm are defined – see [Fiadeiro and Maibaum, 1995] for further discussion on this topic). We do not wish to put forward any form of temporal logic as a specification language but, rather, to show that there are formal mechanisms available for supporting the modularisation constructs that are typical of object-based methods.

9.3 Interconnecting Objects

The cornerstone of the modular approach to reactive system specification that we have been developing is the ability to start from separate descriptions of how components behave and interconnect these descriptions in order to define how components interact within a system (and form new, larger grain, components).

From a formal point of view, and following the direction initiated by J. Goguen [e.g. Goguen and Ginali, 1978], Category Theory provides a neat formalisation of component interconnection as a means of defining complex systems: "given a category of widgets, the operation of putting a system of widgets together to form a super-widget corresponds to taking a (co)limit of the diagram of widgets that shows how to interconnect them". Particularising to our case, having developed a collection of theories that act as descriptions of objects, we can describe a complex system by assembling as nodes of a diagram as many instances (components) of these theories (component types) as required, and by using morphisms between them, in order to establish the required interconnections.

As structure-preserving mappings, morphisms establish the relationship that must exist between two object descriptions so that one of them may be considered as a component of the other. Formally, this corresponds to the notion of interpretation between theories as defined in [Fiadeiro and Maibaum, 1992]:

Definition: Given two object signatures $\theta_1=(\Sigma_1,A_1,\Gamma_1)$ and $\theta_2=(\Sigma_2,A_2,\Gamma_2)$, a morphism σ from θ_1 to θ_2 consists of

- a morphism of algebraic signatures σ_υ: $\Sigma_1 \rightarrow \Sigma_2$;
- for each $f{:}s_1,\ldots,s_n \rightarrow s$ in A_1 an attribute symbol $\sigma_\alpha(f){:}\sigma_\upsilon(s_1),\ldots,\sigma_\upsilon(s_n) \rightarrow \sigma_\upsilon(s)$ in A_2;
- for each g: $s_1,\ldots,s_n$ in Γ_1 an action symbol $\sigma_\gamma(g)$: $\sigma_\upsilon(s_1),\ldots,\sigma_\upsilon(s_n)$ in Γ_2.

A mapping (translation) between the formulae of each signature can be associated with a signature morphism, providing for each formula f of θ_1 its translation $\sigma(f)$ as a formula of θ_2. Based on these notions of signature morphism and formula translation, we define a morphism between two descriptions (θ_1,Φ_1) and (θ_2,Φ_2) as follows:

Definition: Given theory presentations (θ_1,Φ_1) and (θ_2,Φ_2), a signature morphism σ: $\theta_1 \rightarrow \theta_2$ is a morphism between the theory presentations iff

- $\Phi_2 \vdash_{\theta_2} \sigma(p)$ for every $p \in \Phi_1$
- $\Phi_2 \vdash_{\theta_2} \sigma((\bigvee_{g\in\Gamma_1} (\exists x_g)g(x_g)) \vee (\bigwedge_{a\in A_1} (\forall x_a)(\mathbf{X}a(x_a) = a(x_a))))$

Notice that, besides requiring that every axiom of the source theory be translated to a theorem of the target specification, it is necessary to require that the locality axiom of the source signature is also a theorem of the target theory. The reason is that, for the morphism to capture the notion of the domain specification being a sub-component, the encapsulation of the sub-component must be preserved.

Theorem: Theory presentations and their morphisms constitute a finitely co-complete category.

The sub-object through which two objects are made to interact may be as simple as an action. Indeed, it is easy to isolate an action as an "atomic" object, corresponding to a channel as traditional in parallel program design:

ITEM-CHANNEL

data sorts: BOOL, ITEM
data functions: boolean operations

attribute symbols:	action symbols:	axioms:
Ø	a(ITEM)	Ø

Such a channel can be used to connect two objects. These connections are established through morphisms. The morphisms identify two actions, one belonging to each object, specifying a synchronisation point in the lives of the two objects. For instance, the interaction between a producer and a buffer may be established by requiring them to synchronise at each pair (store(i),put(i)). With this in mind, we have to define the morphisms that connect **ITEM-CHANNEL** to the descriptions PRODUCER and BUFFER:

store: ITEM-CHANNEL → PRODUCER
put: ITEM-CHANNEL → BUFFER

<u>data signature</u>: inclusion	<u>data signature</u>: inclusion
<u>actions</u>: a $\mapsto$ store	<u>actions</u>: a $\mapsto$ put

The fact that the objects thus interconnected interact by sharing the identified actions (i.e., communicate along the specified channel) results from the fact that the pushout of the two morphisms identifies the two action symbols. Hence, in the resulting system, there is only one action corresponding to the communication between the two objects. See [Fiadeiro and Maibaum, 1992] for further details.

A channel of the same type may be used to connect a consumer and a buffer, although through different morphisms:

extract: ITEM-CHANNEL → CONSUMER
get: ITEM-CHANNEL → BUFFER

<u>data signature</u>: inclusion	<u>data signature</u>: inclusion
<u>actions</u>: a $\mapsto$ extract	<u>actions</u>: a $\mapsto$ get

A system consisting of a producer, a consumer and a buffer interacting in this way may be specified through a diagram (in the sense of category theory) where components are nodes and the arrows (morphisms) indicate how the components are interconnected:

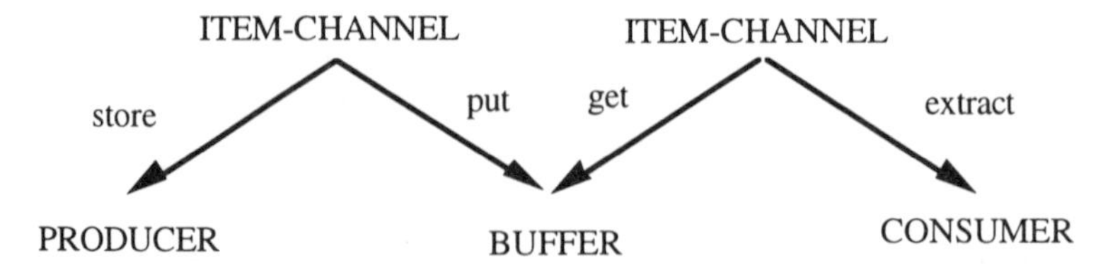

That is, the diagram itself provides a specification of the complex system, its logical configuration as a structure of interconnected components. In fact, because the category of temporal theories defined above is well behaved (is finitely co-complete), such a diagram itself denotes a theory which is the minimal one that contains the specifications of the components and respects the specified interaction, in the sense that only one action is obtained for each synchronisation pair identified by the morphisms. This theory can be seen as the decription of the global system as a module in the sense that it can be used itself as a component of a more complex system. See [Fiadeiro and Maibaum, 1992] for a detailed account of the calculation of colimits in this category.

9.4 Interfacing: Action Calling

The nature of the interconnection that is established, as in the example above, is quite simple: by relating actions in two different objects we are saying that the two are to be identified when considered as actions of the global system. That is, the two objects are required to synchronise in order to perform it.

What would then happen if we were required to connect two producers to the same buffer? Clearly, we cannot say that the put action of the buffer is to be shared with the store action of the second producer: as sharing means synchronisation, by transitivity we would be requiring the two producers to synchronise in order to store an item, which is not at all what we intend. What we want is to say that both producers are "clients" of the buffer, but use it independently.

An analogy with electronic components is perhaps useful. If we regard a computer as a server and we want to connect another terminal to the computer, we have to assign it to a free port. We can regard the action symbols of an object description as ports, and the interconnecting objects (such as channels above) as cables, the morphisms specifying the ports to which the cables are to be connected. (The parameters of the action symbol provide for the data that is to be exchanged during synchronisation.) These interconnections via ports are such that every other object connected to a given port synchronises in the communication. Hence, if we want two objects to communicate independently with a third one, each via the same operation, we must make sure that they are assigned different ports, albeit of the same "type".

This discussion seems to suggest that theory presentations per se are not the right design structures that we wish to have as specification modules. Rather, via object descriptions such as that of the buffer, we would like to specify particular forms of behaviour that we would wish to encapsulate, any "client" of the components thus specified needing to be connected via adequate dedicated interfaces. How can we then define such interfaces?

First of all, it seems clear that interaction should not be achieved by synchronising with the primitive actions of the objects (such as the buffer) but by "calling" them [Sernadas *et al.*, 1990]. That is, instead of saying that store(i) and put(i) should always synchronise, which can be seen to correspond to the requirement store(i) $\leftrightarrow$ put(i),

we would like to state that store calls the "method" put of the buffer. This situation corresponds instead to the requirement

store(i) $\rightarrow$ put(i),

which breaks the symmetry of the former synchronisation constraint. Indeed, this formula merely requires store actions to "subsume" put actions. In this way, if we want to have two producers P_1 and P_2 interacting separately with the buffer, we can specify

P_1.store(i) $\rightarrow$ put(i)
P_2.store(i) $\rightarrow$ put(i)

where by P_n.store we are renaming the corresponding actions of the producers.

In summary, any object should provide its services via actions that "call" its internal actions (which we shall call *methods*). As a design structure corresponding to a buffer we should thus have (1) the description BUFFER that specifies the behaviour of the buffer, (2) interface actions through which its methods may be called, and (3) axioms that state how the interface actions call the methods.

These considerations lead us to the following definition:

Definition: A *design structure* consists of

- an object description KER (its kernel);
- an extension BDY (for body) of KER such that $A_{BDY}=A_{KER}$
- a collection of interfaces, each of which consists of a channel description, i.e. a theory presentation ITF whose signature contains only one action symbol, and a morphism ι_{ITF}: ITF $\rightarrow$ BDY connecting the channel to the body presentation.

Let us discuss in more detail each of these points. On the one hand, the theory presentation BDY is an extension of KER that contains the definitions of the various interface actions. The requirement that $A_{BDY}=A_{KER}$ means that we are not allowed to extend the state of the kernel. The need for new attributes (e.g. for enabling or disabling interface actions) should lead to an extension of the kernel by aggregating new objects. Indeed, the kernel of a design structure may be a complex object reflecting the fact that the services that are provided are manyfold. We shall comment further below on the nature of typical axioms with which BDY extends KER.

On the other hand, because extensions are morphisms, the extension must be locality preserving, as seen in Section 3. That is, the locality formula of the kernel

$$\Big(\bigvee_{g\in\Gamma_{kernel}}(\exists x_g)g(x_g)\Big) \vee \Big(\bigwedge_{a\in A_{kernel}}(\forall x_a)(\mathbf{X}a(x_a) = a(x_a))\Big)$$

must hold for the body (notice that the translation involved in the extension is an identity). This requires that the new actions that we are introducing (the interface actions) subsume some of the kernel actions (unless they leave the attributes unchanged, which in general is not what is intended). Hence, there is a necessary binding between the actions that we declare in the interfaces and the actions that we provide in the kernel. Such bindings are established in the body.

Each such design structure can be depicted in terms of a diagram:

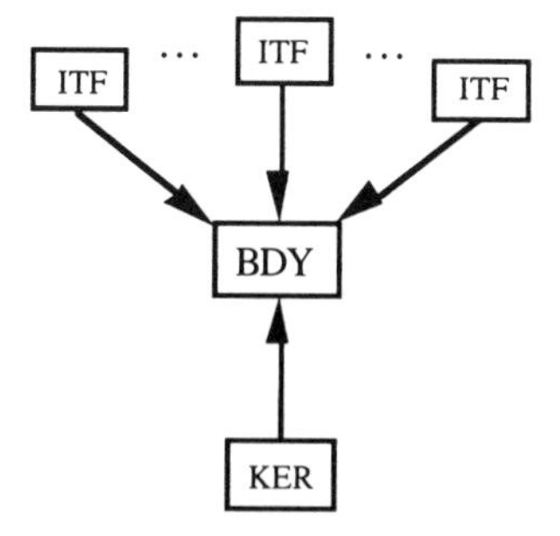

An example has already been discussed: if we want to build a system that consists of two producers and a consumer interacting via a buffer, we must provide the buffer with separate interfaces through which we may connect its "clients". The relevant design structure would then consist of

- KER: the theory presentation BUFFER
- ITF: three interfaces
  P_1: {P_1.store(ITEM)}
  P_2: {P_2.store(ITEM)}
  C: {C.extract(ITEM)}
- BDY: the theory presentation that extends KER with
  - the new action symbols P_1.store, P_2.store and C.extract
  - the axioms
    P_1.store(i) $\rightarrow$ put(i)
    P_2.store(i) $\rightarrow$ put(i)

$$C.extract(i) \rightarrow get(i)$$

The body of a design structure may contain many types of information. For instance, closure conditions are usually associated with the design structure stating that only the specified interface actions may call the kernel's methods. Mutual exclusion between the interface actions may also be required. Indeed, in the case of the buffer, we could have the following axioms in the body of the design structure:

$$\neg(P_1.store(i) \wedge P_2.store(i))$$

stating that access to the buffer via the two store interfaces is mutually exclusive, and

$$get(i) \rightarrow C.extract(i)$$
$$put(i) \rightarrow P_1.store(i) \vee P_2.store(i)$$

stating that the put and get methods are only called via the given interface actions. These axioms are essential when we want to make the server private to the specified clients, in the sense that any extension of the design structure in order to accomodate more clients is forced to be done through existing interfaces. (If this is not what is required, the closure condition would have to be modified in order to preserve logical consistency.)

Let us now discuss how to connect a client to the server. Clearly, all we have to do is link the client to one of the interfaces of the server through a morphism. That is, all we have to do is identify (via a morphism) the action symbol in the client that corresponds to that of the interface (i.e. to connect the ports)and make sure that the axioms of the interface are theorems of the client (i.e. that the conditions for the interface to be connected are met). In the case of two producers and one consumer connected to the same buffer we would have:

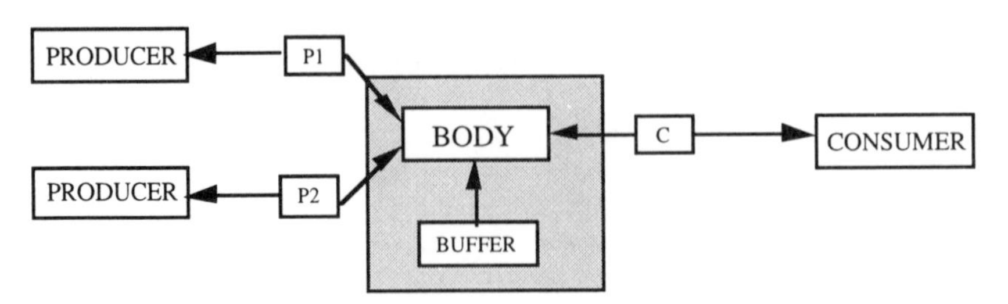

Naturally, the client itself may be the kernel of another design structure (as each description can be seen as a trivial design structure per se, the morphisms involved being the identity), leading to the following general pattern of interconnections between design structures through which we specify systems:

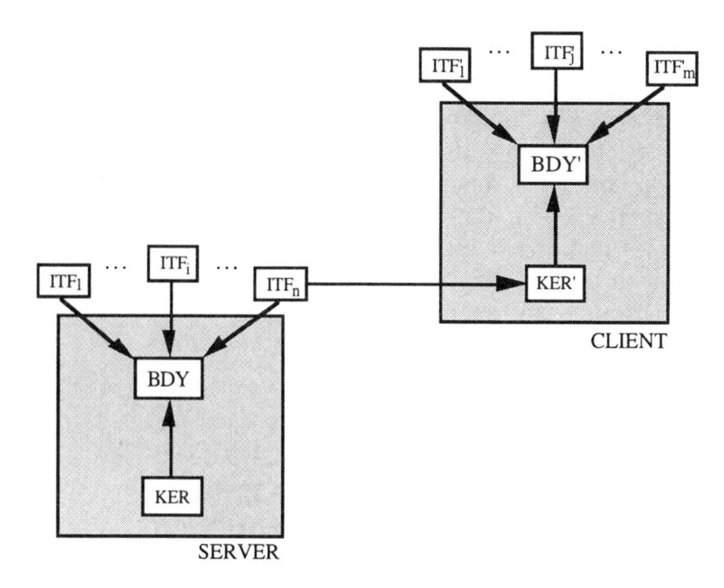

Because these are diagrams in the category of object descriptions (see Section 3), the resulting system can again be described by the colimit of its configuration diagram.

In fact, colimits can be used for defining operations between design structures. For instance, connections such as the one above allow us to aggregate the two design structures to form a new one:

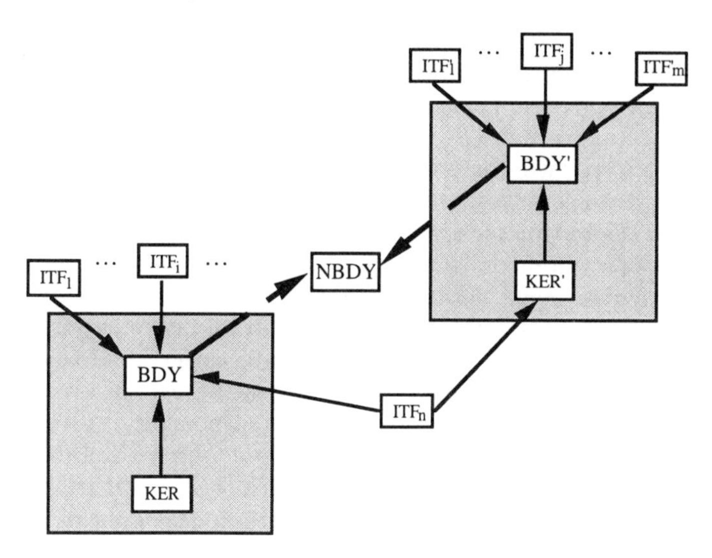

The body of the new design structure is obtained from the pushout of the two morphisms that connect the two bodies BDY and BDY' via the interface ITF_n (where the second morphism is the composition of the ones through KER'). The kernel of the new design structure is given by the sum of the given ones. Its interfaces are the ones which "remain", the connecting morphisms being obtained by composition.

Design structures can also be built through inheritance. When we inherit the kernel of a design structure into another kernel (which establishes a morphism between them), the interfaces of the former kernel may also be used as interfaces for the inheriting object as a kernel itself (of another design structure). The body of the new design structure is obtained through the pushout of the inheritance morphism and the body extension of the old design structure:

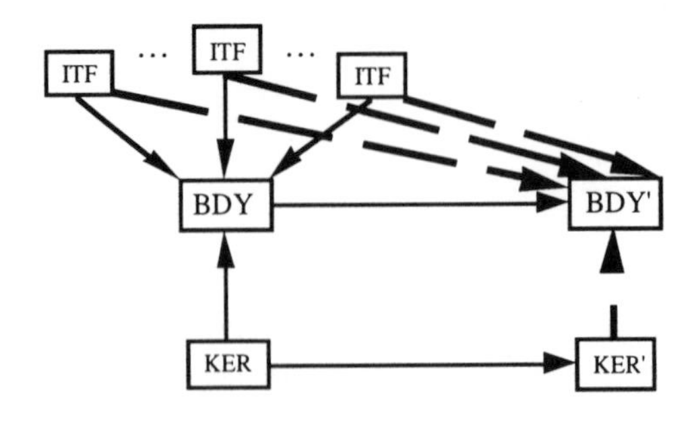

9.5 Abstraction: Object Calling

In the previous sections, we have illustrated how components may be separately described and later on interconnected in order to describe more complex systems. The possibility to establish interfaces via which different objects interact with a same object not by action sharing but by action calling allows for more flexible configurations, namely for supporting client-server architectures.

However, even this looser form of interaction can be too restrictive. The proposed formal model of component description and interconnection relies on a synchronous, multiprocessor architecture in which, at each execution step, several actions may be performed concurrently. In order to support stepwise development, this architecture can be as abstract as required in the sense that the granularity of the actions is not fixed by the intended implementation environment. Naturally, all objects at a given layer of abstraction have the same action granularity: the granularity that is given by the (abstract) processors. It is this uniform granularity that justifies the use of the temporal operator **X** (next). Hence, the model proposed in the previous sections does not support, directly, interaction between objects which have different action granularities.

For instance, in the producer-consumer-buffer example that we have been discussing, the description we gave of a producer may be seen as a "view" which abstracts from a more complex activity of production the actions that are relevant for the system that we are interested in putting together (transmission of the produced items to consumers via a buffer). In order to get the whole picture, we might use the description FACTORY of an object responsible for production. There is, however, no reason for assuming that the action of producing an item is atomic in FACTORY. It may well correspond to some complex activity, in which case it is certainly better described as a sub-activity of FACTORY, i.e. via a sub-object. In this section, we generalise the design structures introduced in Section 4 in order to allow for actions to call sub-objects of other objects and not just (atomic) actions.

The idea, which we have put forward in [Fiadeiro and Maibaum, 1994] for supporting stepwise refinement of object descriptions, is to let the interfaces of the design structures to be more complex theories (denoting transactions) than just channel descriptions (i.e. single actions) and extend the interconnection mechanisms to allow for actions to be mapped to such transactional interfaces.

Such interfaces have to be of a special kind so that we can abstract from their models occurrences of (higher level) actions. The idea is to identify in each interface points ("concrete" actions) that mark the beginning and the end of the occurrence of the action being abstracted from that interface. (Such start and end points have been used in the literature, namely [Aceto and Hennessy, 1989] in the context of Process Algebra.) Hence, a design structure for abstraction is such that:

- Each interface ITF is a theory which includes two distinguished actions beg_{ITF} and end_{ITF}, and an attribute in_{ITF}:BOOL, as well as the following theorems

  $BEG \rightarrow \neg in_{ITF}$

  $beg_{ITF} \rightarrow \neg in_{ITF} \wedge (\mathbf{X} in_{ITF} \vee end_{ITF})$

  $end_{ITF} \rightarrow (in_{ITF} \vee beg_{ITF}) \wedge \mathbf{X} \neg in_{ITF}$

  $beg_{ITF} \vee end_{ITF} \vee (in_{ITF} \leftrightarrow \mathbf{X} in_{ITF})$

  $in_{ITF} \rightarrow \mathbf{F} end_{ITF}$

  These theorems capture the fact that the attribute in_{ITF} is true only between occurrences of beg_{ITF} and end_{ITF}, and is local to these two actions. Furthermore, once the transaction has started, its end is required.

- The body BDY extends the kernel KER with the interface actions and with the attributes in_{ITF} for each interface I_{ITF}.

Notice that interaction by action calling is a particular case of this construction. Indeed, it is possible to have an interface I_a to consist of just one action a: we just have to pick beg_a=end_a=a. In this case, in_a will always be false.

Although we do have a generalisation of the notion of interface discussed in section 4, the morphisms which we used to interconnect interfaces and objects are not generalised that easily. Indeed, the structure of the category itself fixes a granularity for the actions, preventing the more general interconnection between actions (propositions) and transactional interfaces (theories). That is, such interconnections take us out of the category.

However, there is a way of solving this problem. Given a design structure with transactional interfaces, we can abstract from it an object description (theory) with actions that correspond to the interfaces and, hence, of the right granularity.

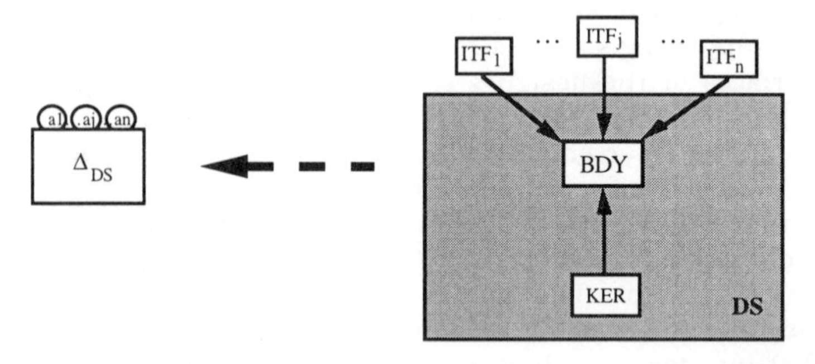

Definition: Let DS=(KER,BDY,$(ITF_a)_{a\in IDX}$) be a design structure with transactional interfaces. Let KER=(Σ,ATT_K,ACT_K). We define the signature θ_{DS} of the abstracted object as follows: θ_{DS}=(Σ,ATT_K,IDX) where IDX is indexing the interfaces.

Let us now define the possible behaviours of the abstracted object. Given an interpretation structure ($\mathcal{U}$,$\mathcal{A}_B$,$\mathcal{G}_B$) for the body BDY, we can abstract a set of interpretation structures for θ_{DS}. Each abstraction consists of:

1) a function $\rho: \omega \rightarrow 2^{\omega}$ (from abstract to concrete "time") such that for every $i\in\omega$,
   a) $\rho(i)$ is an interval [beg(i),end(i)]
   b) beg(i)<beg(i+1) and end(i)<end(i+1)
   c) for every beg(i+1)$\leq$m<end(i) and attribute $f\in ATT_K$,
   $\mathcal{A}_B(f)(m)=\mathcal{A}_B(f)(m+1)$.

That is to say, we identify for each abstract step the interval of concrete steps that realise it. For flexibility and generality, we allow for such intervals to overlap but require that attributes remain invariant during intersections.

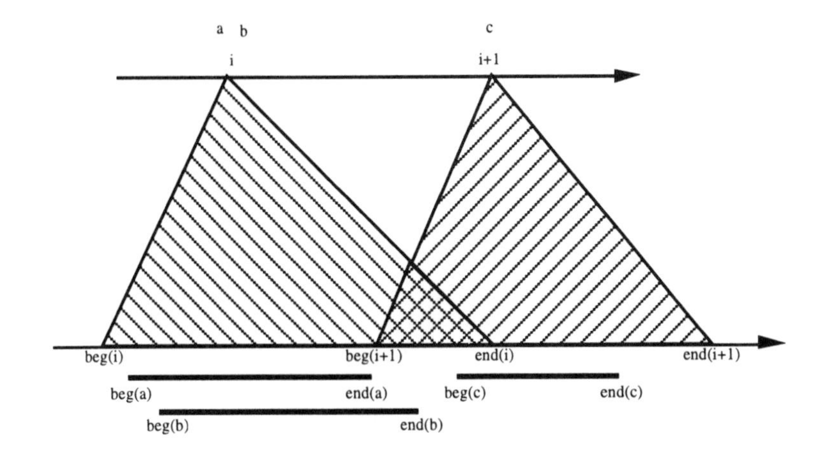

2) a function α: I $\rightarrow 2^{\omega}$ (from abstract actions to abstract "time") such that, for every a$\in$I and i$\in\alpha$(a), there exist exactly one j$\in\rho$(i) and one k$\in\rho$(i), j$\leq$k, such that
   a) j$\in \mathcal{G}_B$(beg$_a$)
   b) k$\in \mathcal{G}_B$(end$_a$)
   c) $\mathcal{A}_B$(f)(m)=$\mathcal{A}_B$(f)(m+1) for every attribute f$\in$ATT$_K$ and beg(i)$\leq$m<j (i.e. no action may start once the values of the attributes have changed)
   d) $\mathcal{A}_B$(f)(m)=$\mathcal{A}_B$(f)(m+1) for every attribute f$\in$ATT$_K$ and k<m$\leq$end(i) (i.e. after the end of an action the values of the attributes cannot change until the end of the interval).

That is to say, the occurrence of an action at (abstract) time i corresponds to the execution of the interface for that action once and only once during the interval ρ(i).

3) if j$\in \mathcal{G}_B$(c) for some action c of an interface I$_a$ then there is i$\in\omega$ such that j$\in\rho$(i) and i$\in\alpha$(a). That is to say, actions of the interfaces occur only as part of executions of the abstract actions.

Definition: The θ_{DS}-interpretation structure $(\mathcal{U}, \mathcal{A}_{(\rho,\alpha)}, \mathcal{G}_{(\rho,\alpha)})$ abstracted through (ρ,α) is defined as follows:

- $\mathcal{S}_{(\rho,\alpha)}(f)(i)=\mathcal{S}_B(f)(beg(\rho(i)))$
- $\mathcal{A}_{(\rho,\alpha)}=\alpha$

That is to say, attributes take at abstract time i the value that they take at the beginning of the interval $\rho(i)$, and the abstracted actions occur in the intervals during which the corresponding interfaces execute.

Proposition: each interpretation structure abstracted in this way is a locus for θ_{DS}, i.e. it satisfies the locality condition for θ_{DS}.

Having defined the signature and the models of the abstracted object description, we can now discuss the properties (theorems) which characterise the behaviour of the abstracted object. What we obtain through this process of abstraction is not a description in the strict sense, i.e. it is not a finite axiomatisation. (This is similar to what happens with a derivation in ADT-specification.) But, using the axiomatisation of the body, we are able to derive the properties of the abstracted object, which is enough, for instance, to prove that it provides an interpretation of another object. This derivation can be done via the following inference rules, where Δ_{DS} stands for the abstracted description:

- change: $$\frac{BDY \vdash (beg_a \wedge \varphi \rightarrow \mathbf{F}(end_a \wedge \mathbf{X}\psi))}{\Delta_{DS} \vdash (a \wedge \varphi \rightarrow \mathbf{X}\psi)}$$

- safety: $$\frac{BDY \vdash (beg_a \rightarrow \varphi)}{\Delta_{DS} \vdash (a \rightarrow \varphi)}$$

- liveness: $$\frac{BDY \vdash (\varphi \rightarrow \mathbf{F}beg_a)}{\Delta_{DS} \vdash (\varphi \rightarrow \mathbf{F}a)}$$

Each of these rules corresponds to one of the classes of formulae that are used to describe the behaviour of objects as discussed in Section 2. In particular, φ and ψ denote first-order formulae in the language of the attributes and data types. The first rule states that the effects of the occurrence of an abstracted action are recovered at the end of the corresponding transaction. Notice the role of conditions 2c and 2d in establishing the correctness of this rule: we must make sure that the values of the attributes at the start of the interval are still the same at the start of the action, and that their value at the end of the action will be the same at the end of the interval. The second and third rules transfer the safety and liveness requirements on the abstracted actions to the beginning of the corresponding transaction (which, recall, has a termination liveness requirement).

It is now possible to synchronise the abstracted object with the other objects in the system using the categorial techniques that we described in section 3.

9.6 Concluding Remarks

In this paper, we have proposed a formal notion of design structure with which we want to support modularity in the temporal logic approach to reactive system specification and thus achieve higher levels of interoperability, maintainability and reusability in system design. Basically, a design structure comprises a theory presentation that acts as a description of the behaviour of a component, and a collection of theory presentations that correspond to the interfaces via which other components may interact with it. A third theory presentation (the body) specifies the way the interfaces are connected to the kernel and synchronised with each other. We showed in particular how clientship could be formalised as a primitive for building the specification of complex systems out of the specifications of its components.

Relationships between components, namely interconnections between servers and their clients, were modelled via theory morphisms (interpretations between theories), thus following Goguen's work on General Systems Theory [Goguen and Ginali, 1978]. Furthermore, because the category of theories and interpretations between theories is finitely cocomplete [Fiadeiro and Maibaum, 1992], a complex system can be described by the theory which results from the colimit of the diagram which depicts its (logical) configuration as a structure of simpler, interacting components.

As defined in the paper, interconnection is established in a synchronous discipline by having actions of different objects synchronised. In this setting, we showed how different levels of interaction can be modelled between components. We analysed simple interconnections corresponding to action sharing, interconnection via action interfaces, which corresponds to action calling as in [Sernadas *et al.*, 1990], and interconnection via object interfaces, which corresponds to object calling. In this latter case, we relate two different levels of granularity, abstracting actions from transactions. These more flexible forms of object interconnection are essential for supporting object-oriented disciplines of system development, namely reusability. Indeed, if ever a market of software components is to be established, interfacing mechanisms will be necessary in order to interconnect already existing components to a given system, namely in order to bring the components to the same level of action granularity.

Acknowledgements

This work was partially supported by the Esprit BRAs 6071 IS-CORE (Information Systems: Correctness and Reusability), 8035 MODELAGE (A Common Formal Model of Cooperating Intelligent Agents), the HCM Scientific Network CHRX-CT92-0054 MEDICIS (Methodology for the Development of Computer System Specifications), and the Eureka project ESF (DTI UK). Discussions with Félix Costa, Stephen Goldsack, Peter Hartel, Stuart Kent and Pedro Inácio are also gratefully acknowledged.

Chapter 10

Interconnection of Object Specifications

We present a very simple account of interconnections of systems of distributed, concurrent, interacting objects. We give an abstract definition of object class specifications, and show how these may be composed into larger systems in a way that captures complex objects and parallel composition with synchronisation. The distributed autonomy of objects is one of the key concepts in object-orientation, and we hope that this simple treatment of parallel composition will clarify the most important aspects of the hierarchical structure of distributed systems of objects.

Of particular interest is our use of commutativity properties to give a *truly concurrent* model of parallelism in distributed systems, in contrast to *interleaving* models of parallelism. If a and b are atomic events that occur independently in distributed components of a system, then in an interleaving model of the parallel composition of a and b, either a happens before b, or b before a. In our approach, we simply deny the relevance of any temporal ordering to the execution of independent events a and b by asserting that 'a *then* b' is the same as 'b *then* a'. In other words, independence of events is expressed by commutativity properties.

The use of commutativity properties to capture true concurrency dates back at least as far as the work of Mazurkiewicz (for example [Mazurkiewicz, 1984]). Monteiro and Pereira [Monteiro and Pereira, 1986] explicitly use limits of monoids to obtain commutativity properties for a sheaf theoretic model of concurrency. Our approach to interconnections of objects is similar in many respects to Monteiro and Pereira's sheaf theoretic approach, and also relies on work by Goguen on categorical systems theory (see [Goguen, 1970, Goguen, 1972b, Goguen, 1975b]; applications of this work to object-orientation can be found in Goguen [Goguen, 1992], Ehrich *et al.* [Ehrich *et al.*, 1991], Wolfram and Goguen [Wolfram and Goguen, 1992], and Cîrstea [Cîrstea, 1995]).

We consider objects to have a local, hidden state, aspects of which may be observed by means of *attributes*, and which may be updated by *methods*. Thus, an object class is specified by declaring a set of attributes, a set of methods, and stating how the methods affect the values of the attributes. This is the basis of the hidden sorted algebra approach to object specification [Goguen, 1991b, Goguen and Diaconescu, 1994, Goguen and Malcolm, 1994, Malcolm and Goguen, 1994]. Hidden sorted algebra is a variety of many sorted algebraic specification which uses hidden sorts to model the hidden local states of objects, and visible sorts to model the data values that objects manipulate.

Goguen and Diaconescu [Goguen and Diaconescu, 1994] have proposed a construction called the 'independent sum' to capture the parallel connection with synchronisation of systems of interacting objects. A goal of the present chapter is to investigate in a simple and abstract way some of the ideas underlying the independent sum construction. Let us illustrate this construction by means of an example (for technical details, see Goguen and Diaconescu, *loc. cit.*).

The following specification defines a class of cells which store natural numbers; the specification imports a module called `DATA`, which we assume defines the data types used in the object class specification, in this case at least the sort `Nat` of natural numbers.

```
obj X is pr DATA .
  hsort  h .
  op  init : -> h .
  op  getx : h -> Nat .
  op  putx : h Nat -> h .
  var H : h .    var N : Nat .
  eq  getx(init)  =  0 .
  eq  getx(putx(H,N))  =  N .
endo
```

The local state of these cells is represented by the hidden sort `h`, and each cell has an assignment method `putx` to update its state and an attribute `getx` which returns the value currently held in the cell. The constant `init` of sort `h` represents the initialisation, or creation, of the cell.

Goguen and Diaconescu consider such a specification as describing the behaviour of a cell object, and introduce a construction called 'independent sum' for composing systems of interacting objects out of such specifications. For example, suppose the above specification defines the behaviour of a cell `X`, and suppose the cell `Y` is defined by a similar specification, with all `x`'s replaced by `y`'s. Then the behaviour of the system consisting of `X` and `Y` operating in parallel is defined by the following specification:

```
obj X||Y is pr DATA .
  hsort  h .
  op  init : -> h .
  op  getx : h -> Nat .
  op  putx : h Nat -> h .
```

```
    op  gety : h -> Nat .
    op  puty : h Nat -> h .
    var H : h .   vars M N : Nat .
    eq  getx(init)  =  0 .
    eq  getx(putx(H,N))  =  N .
    eq  gety(init)  =  0 .
    eq  gety(puty(H,N))  =  N .
    eq  getx(puty(H,N))  =  getx(H) .
    eq  gety(putx(H,N))  =  gety(H) .
    eq  putx(puty(H,M),N)  =  puty(putx(H,N),M) .
endo
```

Note that while `h` represents the state of `X` in the first specification, in the second specification `h` represents the product of the states of `X` and `Y`, i.e., it represents the state of the composite system. The first four equations in the specification `X||Y` are taken directly from `X` and from `Y`; the final three equations state that `X` and `Y` operate independently: assignment to `X` doesn't affect the value of `Y`'s attribute, and vice-versa. The final equation is particularly relevant to subsequent sections: the independence of `X` and `Y` is captured by stating that assignment to `X` *commutes* with assignment to `Y`.

The independent sum $X\|Y$ represents the parallel connection of X and Y, synchronised on `init` (i.e., $X\|Y$ has just one initialisation, which synchronously initialises X and Y). The process can be repeated to give a hierarchical construction of large systems. For example, if we have a third cell, Z, then we could form $Y\|Z$, which we could then conjoin with $X\|Y$ to give $(X\|Y)\|_Y(Y\|Z)$. The subscripted Y indicates that the common part Y is to be shared, that is, the independent sum will not make duplicate versions of the object Y. This last composite object might be thought of as the pushout of the inclusions of Y into $X\|Y$ and $Y\|Z$; in fact, it is not a pushout, though Goguen and Diaconescu give a universal characterisation for the independent sum construction.

In the following sections, we investigate other universal ways of capturing parallel connection with synchronisation. Section 10.1 gives some preliminary technical background; Section 10.2 discusses a basic approach to interconnections of simple classes of processes; Sections 10.3 and 10.4 discuss interconnections of object and systems of objects, respectively. A key idea in what follows is the use of commutativity properties to capture true concurrency, similar to the commutativity axioms in the independent sum.

Acknowledgements

The research reported in this chapter has been supported in part by the CEC under ESPRIT-2 BRA Working Groups 6071, IS-CORE (Information Systems COrrectness and REusability) and 6112, COMPASS (COMPrehensive Algebraic Approach to System Specification and development), and a contract under the management of the Information Technology Promotion Agency (IPA), Japan, as part of the Industrial Science and Technology Frontier Program 'New

Models for Software Architectures', sponsored by NEDO (New Energy and Industrial Technology Development Organization).

10.1 Some Preliminaries

In order to make our account reasonably self-contained, we begin with brief summaries of some technical notions used in the sequel. The main concepts we require are: categories, functors, limits, monoids, right actions of monoids, and sheaves. These are introduced in the following subsections; readers already familiar with the concepts can skip to Section 10.2.

10.1.1 Categories

Category theory allows an attractively abstract treatment of many constructions in mathematics and computer science. We give only a few basic definitions here; good introductions can be found in [Barr and Wells, 1990, Goguen, 1991a, Lane, 1971, Pierce, 1991].

A **category** C consists of the following:

- a class $|\mathsf{C}|$ of **objects**;
- a class $\|\mathsf{C}\|$ of **morphisms** (sometimes called 'arrows');
- two maps $\partial_0, \partial_1 : \|\mathsf{C}\| \to |\mathsf{C}|$, which give, respectively, the **source** and **target** of a morphism;
- for each $c \in |\mathsf{C}|$, a distinguished morphism 1_c called the **identity**, with $\partial_0(1_c) = c = \partial_1(1_c)$;
- a partial operation $\_;\_ : \|\mathsf{C}\| \times \|\mathsf{C}\| \to \|\mathsf{C}\|$, called **composition**, which is defined on (f, g) when $\partial_1(f) = \partial_0(g)$, in which case $\partial_0(f\,;g) = \partial_0(f)$ and $\partial_1(f\,;g) = \partial_1(g)$.

Moreover, it is required that the following axioms be satisfied:

- Identities are neutral elements of composition, in that for each $c \in |\mathsf{C}|$ and $f, g \in \|\mathsf{C}\|$,

  $$1_c\,;f = f \quad \text{and} \quad g\,;1_c = g$$

  whenever these compositions are defined;
- composition is associative, in that for all $f, g, h \in \|\mathsf{C}\|$,

  $$f\,;(g\,;h) \;=\; (f\,;g)\,;h$$

  whenever both sides are defined.

For $f \in \|\mathsf{C}\|$ we write $f : a \to b$ iff $\partial_0(f) = a$ and $\partial_1(f) = b$. The same information can be presented pictorially:

$$a \xrightarrow{f} b$$

and if $g : b \to c$, then the composite $f\,;g : a \to c$ can be depicted:

$$a \xrightarrow{f} b \xrightarrow{g} c$$

Moreover, if $h : c \to d$, then the equation $f\,;(g\,;h) = (f\,;g)\,;h$ can be depicted thus:

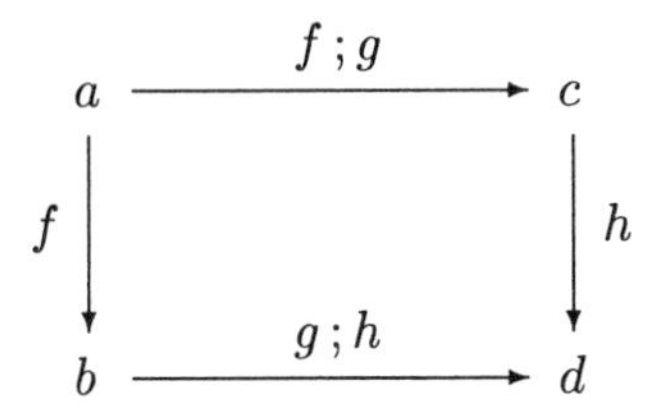

Such a picture is referred to as a **commuting diagram**: this means that any two paths in the diagram that start and end at the same points are equal.

The 'objects' of a category are different from the 'objects' of object-orientation; if there is any risk of confusing the two, we will refer to the former as 'category objects', and the latter simply as 'objects'.

An important example of a category is Sets, whose objects are sets, and whose morphisms are functions (or, more precisely, triples (A, f, B) where A and B are sets and f is a map from A to B; then $\partial_0(A, f, B) = A$ and $\partial_1(A, f, B) = B$). Composition of morphisms is just composition of functions. (We write composition in 'diagrammatic order', i.e., in the same direction as the arrows. The more usual notation for $f\,;g$ is $g \circ f$.)

Another example is the category that arises from a partial order. If $(C, \leq)$ is a partial order, then we can construct a category, which we also call $(C, \leq)$, whose objects are the elements of C, and whose morphisms are pairs of elements (a, b) such that $a \leq b$. In particular, there is a morphism $(a, b) : a \to b$ iff $a \leq b$. If $a \leq b$ and $b \leq c$, then the composite $(a, b)\,;(b, c)$ is $(a, c) : a \to c$.

Given any category C, its **opposite** category C^{op} is obtained by turning around the arrows of C. Specifically, the objects of C^{op} are the objects of C, and the morphisms are those of C, but $f : a \to b$ in C^{op} iff $f : b \to a$ in C. For example, if $(C, \leq)$ is the category arising from a partial order, then $(C, \leq)^{\mathrm{op}}$ is the category arising from the partial order $(C, \geq)$, where $c \geq c'$ iff $c \leq c'$.

An example of a very artificial category is the category whose objects are 0, 1, and 2, with identity morphisms for each of these objects, and with a morphism $l : 1 \to 0$ and another $r : 2 \to 0$. This category can be pictured as a diagram:

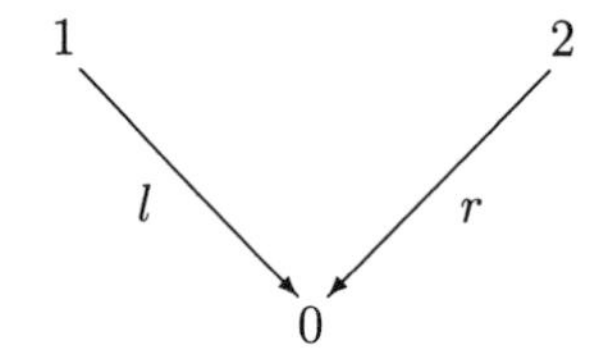

The identity morphisms have been omitted from this diagram. In following sections we discuss the relationship between this category and synchronisation: for future reference, let us call this category Sync. We might think of Sync as denoting a system consisting of two objects, 1 and 2, which are synchronised on a common subobject, 0. In fact, this rather strange example of a category can be thought of as a pictorial representation of such a system, and an instance of such a system would be obtained by instantiating the category objects 0, 1 and 2 with objects, and instantiating the arrows l and r with morphisms between those objects which express the relevant subobject relationship. Such instantiations may be achieved by *functors*.

10.1.2 Functors and Natural Transformations

Given two categories C and D, a **functor from C to D** is a pair $F = (|F|, \|F\|)$, where $|F|$ maps objects of C to objects of D, i.e., $|F| : |\mathsf{C}| \to |\mathsf{D}|$, and $\|F\|$ maps morphisms of C to morphisms of D in such a way that:

- for all $f \in \|\mathsf{C}\|$, if $f : a \to b$ then $\|F\|(f) : |F|(a) \to |F|(b)$;
- $\|F\|(1_a) = 1_{|F|(a)}$; and
- if $f : a \to b$ and $g : b \to c$ are morphisms of C, then $\|F\|(f\,;g) = \|F\|(f)\,;\|F\|(g)$.

We write $F : \mathsf{C} \to \mathsf{D}$ to denote that F is a functor from C to D.

From now on, if F is a functor we will write F for both $|F|$ and $\|F\|$, and Fx instead of $F(x)$, for x an object or an arrow, so that, for example, the main features of functors can be summarised concisely as follows.

- for all $f \in \|\mathsf{C}\|$, if $f : a \to b$ then $Ff : Fa \to Fb$;
- $F\,1_a = 1_{Fa}$; and
- if $f : a \to b$ and $g : b \to c$ are morphisms of C, then $F(f\,;g) = Ff\,;Fg$.

An example of a functor is the *identity* functor, $I : \mathsf{Sets} \to \mathsf{Sets}$, defined by $Ia = a$ for each set a, and $If = f$ for each function f. Another example is $\mathit{List} : \mathsf{Sets} \to \mathsf{Sets}$, defined by $\mathit{List}\,a = a^*$, the set of lists over the set a, and $\mathit{List}\,f = f^*$, the function that applies $f : a \to b$ componentwise to a list in a^*, giving a list in b^* as result.

Given two functors $F, G : \mathsf{C} \to \mathsf{D}$, a **natural transformation** η from F to G is a family $\eta_a : F(a) \to G(a)$ of morphisms of D indexed by objects $a \in |\mathsf{C}|$, such that for all morphisms $f : a \to b$ of C, the following diagram commutes.

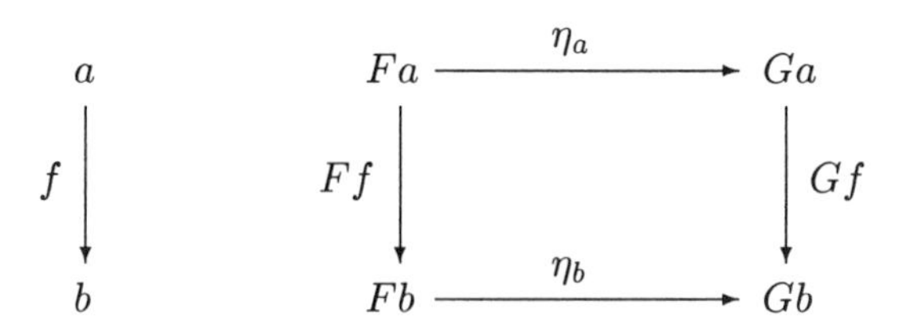

That is, $Ff\,;\eta_b = \eta_a\,;Gf$.

For example, the family of functions η_a that take $x \in a$ to the singleton list $\langle x \rangle \in a^*$ is a natural transformation from I to *List*.

10.1.3 Limits

Given two objects a and b in a category C, a **product** of a and b is an object $a{\times}b$ of C together with two morphisms $\pi_1 : a{\times}b \to a$ and $\pi_2 : a{\times}b \to b$ such that for any pair of morphisms $f : x \to a$ and $g : x \to b$ there exists a unique morphism $h : x \to a{\times}b$ such that

$$h\,;\pi_1 = f \quad \text{and} \quad h\,;\pi_2 = g\ .$$

The following diagram represents this situation.

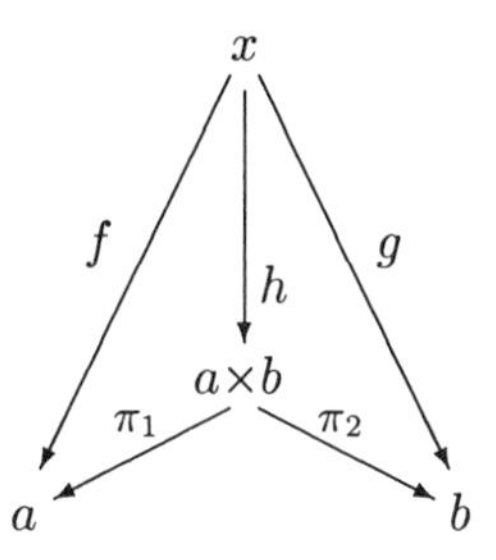

For example, in the category Sets products are given by Cartesian products with the standard projection functions. In the category arising from a partial order $(C, \leq)$, products correspond to greatest lower bounds.

More generally, limits are defined as follows. A **diagram** (in a category C) is a functor $\delta : \mathsf{X} \to \mathsf{C}$, where the category X is called the **shape** of the diagram. A **cone** for a diagram $\delta : \mathsf{X} \to \mathsf{C}$ is an object a of C together with a family of morphisms $\pi_x : a \to \delta x$ for each $x \in |\mathsf{X}|$ such that $\pi_x\,;\delta f = \pi_y$ for each $f : x \to y$ in X, as in the following diagram.

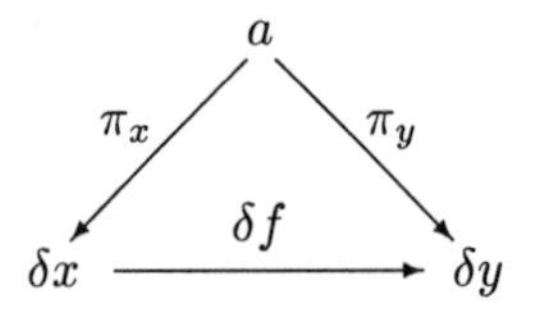

We write $\pi : a \Rightarrow \delta$ to indicate that a and the family of morphisms π_x are a cone for δ.

A **limit** of a diagram $\delta : \mathsf{X} \to \mathsf{C}$ is a cone $\pi : a \Rightarrow \delta$ such that for any other cone $\rho : b \Rightarrow \delta$ there is a unique morphism $h : b \to a$ such that $h\,;\pi_x = \rho_x$ for all $x \in |\mathsf{X}|$.

As an example of this general situation, products in a category C are limits of diagrams $\delta : \mathsf{X} \to \mathsf{C}$, where X is the category with two objects 1 and 2 and no arrows (except identities). The product $a{\times}b$, together with the two projections π_1 and π_2, is a limit of the diagram δ which maps 1 to the object a and 2 to b.

Another important example of limits is the following, will be useful for talking about parallel composition with synchronisation. Suppose we are given two morphisms $l : a \to c$ and $r : b \to c$ in a category C (this is the same as supposing there is a diagram $\delta : \mathsf{Sync} \to \mathsf{C}$, where Sync is the category pictured at the end of Section 10.1.1). A **pullback** (of l and r) is an object $a \times_c b$ of C together with two morphisms $\pi_1 : a \times_c b \to a$ and $\pi_2 : a \times_c b \to b$ such that $\pi_1\,; l = \pi_2\,; r$ (that is, $\pi : a \times_c b \Rightarrow \delta$), with the 'universal property' that for any $f : x \to a$ and $g : x \to b$ such that $f\,; l = g\,; r$ there is a unique $h : x \to a \times_c b$ such that $h\,;\pi_1 = f$ and $h\,;\pi_2 = g$.

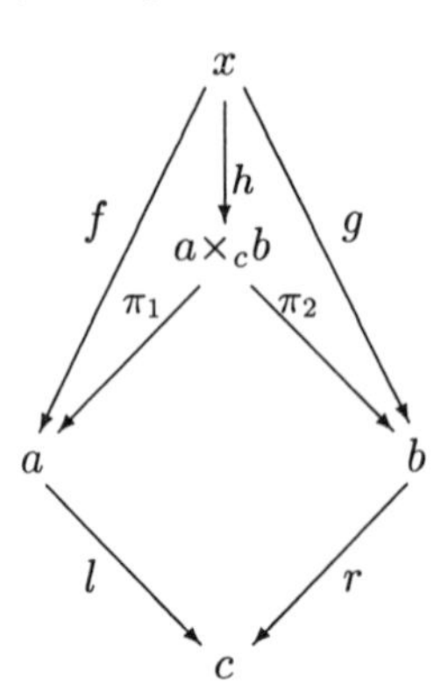

We call $a{\times}_c b$ the **pullback object**. For example, in Sets the pullback object of l and r would be $\{(x, y) \in a{\times}b \mid l(x) = r(y)\}$. Pullbacks, and limits in general, can be thought of as combining the objects in a given diagram in a minimal way such as to make all the projections commute with the arrows of the diagram. In following sections, we use limits to model parallel composition of systems of object specifications. It is useful to know when an arbitrary diagram of such specifications can be composed: we say that a category C is **complete** iff there is a limit in C for every diagram $\delta : \mathsf{X} \to \mathsf{C}$.

10.1.4 Monoids

Objects have state, and the specification of a class of objects fixes the visible attributes of an object's state, as well as the methods by which that state may be changed. An object class is therefore an abstraction of objects in the real world, which may reveal unexpected facets, or be altered by all manner of unforeseen circumstances. Abstracting yet further from reality, we might proscribe the simultaneous effect of two or more methods on an object's state; doing so, we impose a monoid structure on the fixed set of methods proper to an object class. Applying methods one after the other corresponds to multiplication in the monoid, and applying no methods corresponds to the identity of the monoid.

A **monoid** is a set M with an associative binary operation $\bullet_{\mathcal{M}} : M \times M \to M$, usually referred to as 'multiplication', which has an identity element $e_{\mathcal{M}} \in M$. If $\mathcal{M} = (M, \bullet_{\mathcal{M}}, e_{\mathcal{M}})$ is a monoid, we often write just M for $\mathcal{M}$, and e for $e_{\mathcal{M}}$; moreover for $m, m' \in M$, we usually write $m\, m'$ instead of $m \bullet_{\mathcal{M}} m'$.

For example, A^*, the set of lists containing elements of A, together with concatenation $+\!\!+ : A^* \times A^* \to A^*$ and the empty list $[\,] \in A^*$, is a monoid. This example is especially important for the material in later sections.

A **monoid homomorphism** is a structure preserving map between the carriers of two monoids. In other words, $f : (M, \bullet_M, e_M) \to (N, \bullet_N, e_N)$ means that f is a map $M \to N$ which distributes over multiplication and preserves identities. That is,

$$\begin{aligned} f(m \bullet_M m') &= f(m) \bullet_N f(m') \\ f(e_M) &= e_N \end{aligned}$$

for all $m, m' \in M$.

This gives us a category Mon of monoids and monoid homomorphisms. Moreover, Mon is complete, and there is an interesting relationship between limits in Mon and independent sums. For example, consider the following diagram in Mon.

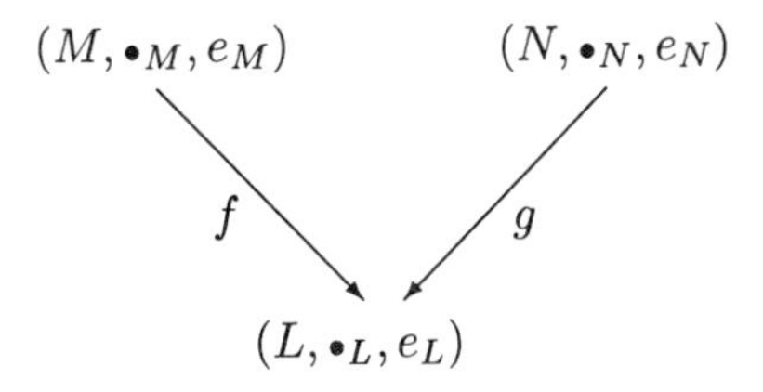

The pullback object of this diagram in Mon has

$$\{(m, n) \in M \times N \mid f(m) = g(n)\}$$

as carrier. Multiplication in the pullback object is defined componentwise

$$(m, n) \bullet (m', n') \;=\; (m \bullet_M m',\, n \bullet_N n')$$

and the identity is (e_M, e_N).

We might think of M and N as being specifications of objects which share a common subobject L, and the morphisms f and g as restricting programs of M and N to programs of L. The pullback object represents the parallel composition of M and N synchronised at L, or their independent sum. An element $m \in M$ gives rise to an element $\widehat{m} = (m, e_N)$ of the pullback object, and similarly for $n \in N$. These elements satisfy the commutativity axioms of independent sums, because

$$\widehat{m}\,\widehat{n} \;=\; (m, e_N)\,(e_M, n) \;=\; (m, n) \;=\; (e_M, n)\,(m, e_N) \;=\; \widehat{n}\,\widehat{m}\ .$$

The program $\widehat{m}\,\widehat{n}$ can be seen as representing the parallel composition of the programs m and n. As with independent sums, the commutativity of their composition expresses the irrelevance of any temporal ordering of their execution.

10.1.5 Right Actions of Monoids

In following sections we simply view the methods of an object class as a monoid; a natural extension of this analogy is to view an object's state as a *right action* of its methods.

A **right action** of a monoid $\mathcal{M} = (M, \bullet, e)$ consists of a set S and a binary operation $\oplus : S \times M \to S$ such that

$$\begin{aligned} s \oplus e \;&=\; s \\ s \oplus (m \bullet m') \;&=\; (s \oplus m) \oplus m' \end{aligned}$$

for all $s \in S$ and $m, m' \in M$.

If $(S, \oplus)$ and $(T, \otimes)$ are right actions of $\mathcal{M}$, a morphism of right actions $f : (S, \oplus) \to (T, \otimes)$ is a map $f : S \to T$ such that

$$f(s \oplus m) \;=\; f(s) \otimes m\ .$$

This gives us a category $\mathsf{RA}(\mathcal{M})$ of right actions of $\mathcal{M}$. This category is also complete. For example, the pullback object of

$$(S, \oplus) \xrightarrow{\;f\;} (T, \otimes) \xleftarrow{\;g\;} (U, \ominus)$$

in $\mathsf{RA}(\mathcal{M})$ is the right action $(P, \odot)$, where

$$P = \{(s, u) \in S \times U \mid f(s) = g(u)\}$$

and $\odot$ is defined by

$$(s, u) \odot m = (s \oplus m,\; u \ominus m)\ .$$

If $\mathcal{M}$ is the monoid of methods in the specification of an object class, and if we ignore the attributes in the specification, it is possible to think of a right action of $\mathcal{M}$ as being a 'model' or 'implementation' of the specification. A useful property of such 'models' is that they move back along monoid morphisms (cf. the same property of models and signature morphisms in the institutions of Goguen and Burstall [Goguen and Burstall, 1992]).

Proposition 1 Every monoid homomorphism $f : \mathcal{M} \to \mathcal{N}$ induces a functor $\mathsf{RA}(f) : \mathsf{RA}(\mathcal{N}) \to \mathsf{RA}(\mathcal{M})$.

Proof. Let $(S, \oplus)$ be a right action of $\mathcal{N}$. We define $\mathsf{RA}(f)(S, \oplus)$ to be $(S, \oplus_f)$, where

$$s \oplus_f m \;=\; s \oplus f(m) \; .$$

It is straightforward to check that this is a right action of $\mathcal{M}$. Moreover for $g : (S, \oplus) \to (T, \otimes)$ a morphism of right actions on $\mathcal{N}$, we define $\mathsf{RA}(f)(g)$ to be just $g : (S, \oplus_f) \to (T, \otimes_f)$. That this is a morphism of right actions follows from the following equalities:

$$g(s \oplus_f m) \;=\; g(s \oplus f(m)) \;=\; g(s) \otimes f(m) \;=\; g(s) \otimes_f m \; .$$

□

10.1.6 Sheaves

Sheaf theory is used in many branches of mathematics, the underlying theme in its various applications being the passage from local to global properties [Gray, 1980]. It provides a formal notion of coherent systems of observations: a number of consistent observations of various aspects of an object can be uniquely pasted together to give an observation over all of those aspects. The passage from local to global properties, and the pasting together of local observations of behaviour allow sheaf theory to be usefully applied in computer science, to give models for concurrent processes [Monteiro and Pereira, 1986, Dubey, 1991, Lilius, 1993] and objects [Goguen, 1992, Ehrich *et al.*, 1991, Wolfram and Goguen, 1992, Cîrstea, 1995]. We give a basic definition of 'sheaf' below; fuller accounts can be found in [Tennison, 1975, Lane and Moerdijk, 1992].

We may consider a sheaf as giving a set of observations of an object's behaviour from a variety of 'locations'. The notion of location is formalised by the following

Definition 2 A **complete Heyting algebra** is a partially ordered set $(C, \leq)$ such that:

- for all $c, d \in C$, there is a greatest lower bound $c \wedge d$
- for all subsets $\{c_i \mid i \in I\}$ of C, there is a least upper bound $\bigvee_{i \in I} c_i$
- greatest lower bounds distribute through least upper bounds:

$$(\bigvee_{i \in I} c_i) \wedge d \;=\; \bigvee_{i \in I} (c_i \wedge d) \; .$$

□

For example, any topological space with the subset inclusion ordering is a complete Heyting algebra. Another important example, the subsystems of a given system, is described in Section 10.4 below.

Definition 3 Let C be a complete Heyting algebra; a **presheaf** F on C is a functor from C^{op} to Sets. That is, for each $c \in C$ there is a set $F(c)$, and for $c, d \in C$ such that $c \leq d$, there is a **restriction function** $F_{c \leq d} : F(d) \to F(c)$, subject to the following conditions:

- $F_{c \leq c} = id_{F(c)}$, the identity on the set $F(c)$; and
- if $c \leq d \leq e$, then $F_{d \leq e} ; F_{c \leq d} = F_{c \leq e}$.

□

Notation 4 For a presheaf F on C, if $c \leq d$ in C and $x \in F(d)$, we often write $x|_c$ for $F_{c \leq d}(x)$. □

Note that the observations given by a presheaf can be *structured* in the sense that presheaves can be defined as functors to some concrete category of structures. For example, we consider presheaves of monoids below, which are functors $C^{\mathrm{op}} \to$ Mon. In this case the restriction functions are required to preserve that structure; thus, for example, the restriction functions of presheaves of monoids are monoid homomorphisms.

A sheaf is a presheaf which allows families of consistent local observations to be pasted together to give a global observation.

Definition 5 A presheaf F is a **sheaf** iff it satisfies the following **pasting condition**:

- if $c = \bigvee_{i \in I} c_i$ and $x_i \in F(c_i)$ is a family of elements for $i \in I$ such that $x_i|_{c_i \wedge c_j} = x_j|_{c_i \wedge c_j}$ for all $i, j \in I$, then there is a unique $x \in F(c)$ such that $x|_{c_i} = x_i$ for all $i \in I$.

□

For example, suppose that a system is just a functor $F : C^{\mathrm{op}} \to$ Sets, where $c \leq c'$ means that c is a subsystem of c', and where $F(c)$ is the set of observations that can be made of teh subsystem c. Then if F is a sheaf, we can paste together, in a unique way, *local* observations at a set of subsystems c_i, to give a *global* observation of the system at $c = \bigvee_{i \in I} c_i$, provided that the local observations do not contradict each other. This idea is further developed in the following sections.

10.2 Process Classes

A class of objects is specified by declaring the attributes of objects in the class, by declaring the methods that operate upon their states, and by defining the effect of the methods on the attributes. If we ignore, for the moment, the attributes of objects, and concentrate on just the methods, then we arrive at a rather simplistic notion of *process* classes. This notion is too simplistic to have any intrinsic interest, but we use it to illustrate the ideas behind our approach to interconnections of objects. Perhaps the simplest way of specifying a class of processes is to state the set of events that the processes of that class can engage in. Thus we might say that a process class specification is just a set. This would correspond to specifying the alphabet of a process in a formalism such as CSP [Hoare, 1985a]. More generally, we might instead say that a process class specification is a monoid, which can be thought of as specifying either the programs that a class of processes can carry out, or the history traces of the events that it engages in. Using monoids rather than alphabets also has the advantage that, for the purposes of comparing classes of processes, we can use monoid morphisms, rather than renaming of events, to compare the actions of processes.

Consider, for example, a clock as a process that can engage in only one event, a 'tick'. The monoid of programs for the class of clocks is the set of lists $\{\surd\}^*$, where $\surd$ denotes a tick of a clock. A slightly more complex example might be a process which can engage in two distinct events, say called 'dot' and 'dash', so that the corresponding monoid is $\{\mathrm{dot}, \mathrm{dash}\}^*$. Suppose we want to conjoin such a process with a clock in such a way that a dot lasts for one tick of the clock, but a dash takes two ticks. This is expressed by the monoid morphism

$$f : \{\mathrm{dot}, \mathrm{dash}\}^* \to \{\surd\}^*$$

where

$$\begin{aligned} f([\,]) &= [\,] \\ f(l\,\mathrm{dot}) &= f(l)\,\surd \\ f(l\,\mathrm{dash}) &= f(l)\,\surd\surd \end{aligned}$$

for all $l \in \{\mathrm{dot}, \mathrm{dash}\}^*$. The morphism f takes a list of dots and dashes and returns the amount of ticks of the clock required. For example, if a history of a 'dot-dash' process is the list 'dot dash dot', then the corresponding history of the clock process is '$\surd\surd\surd\surd$' — four ticks.

This morphism f can be used to synchronise two dot-dash processes which share the same clock process. The result of such a synchronisation should consist of histories of each of the dot-dash processes provided that these take the same number of ticks of the clock. For example, one process might go 'dot dash dot' while the other goes 'dash dash'; this would be acceptable, as both of these histories require four ticks of the clock. Given two dot-dash processes with a common clock, the result of their parallel composition is the pullback object of the following diagram in Mon, where $C = \{\surd\}^*$ and $D = \{\mathrm{dot}, \mathrm{dash}\}^*$.

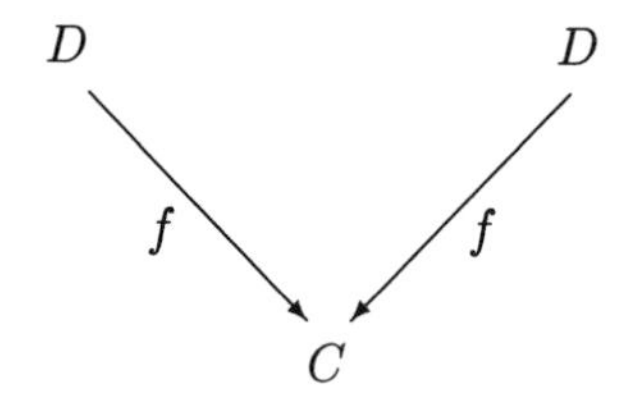

The pullback object of this diagram is $\{(d_1, d_2) \in D \times D \mid f(d_1) = f(d_2)\}$. That is, given two lists d_1 and d_2 of dots and dashes, which represent histories of two processes, the pair (d_1, d_2) is a history of the parallel composition of the two processes provided that d_1 and d_2 both require the same number of ticks of the shared clock.

The point illustrated by this example is that systems can be thought of as diagrams, and that the behaviour of a system is its limit. In developing this for interconnections of systems of objects we are following ideas and results from categorical systems theory (cf. Goguen [Goguen, 1970, Goguen, 1992]).

It is possible to contrive some rather curious examples of the behaviour of composite processes. Let

$$\begin{aligned} A &= \{a\}^* \\ B &= \{b\}^* \\ C &= \{a, b\}^* \\ D &= \{a, b\}^* \end{aligned}$$

be four processes, arranged as in Figure 10.1 where the morphisms between

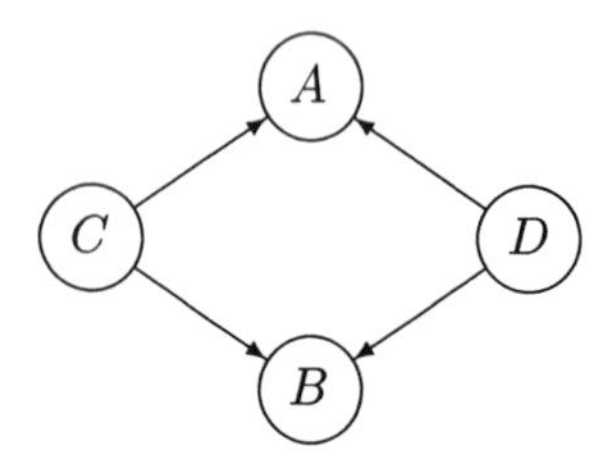

Figure 10.1: C and D observe A and B

the processes are the obvious restriction functions, e.g., the arrow from C to A restricts a list over $\{a, b\}$ to a list over $\{a\}$. The idea is that A is a process which can participate only in the event a, B can participate only in the event b, and C and D are two processes which can either observe a happening in A or b happening in B. The behaviour of this system is the limit in Mon, which has as carrier the set

$$\{(w, x, y, z) \in A \times B \times C \times D \mid \ y|_A = w = z|_A \ \wedge \ y|_B = w = z|_B\} \ .$$

An element of this set is (a, b, ab, ba), because $ab|_A = a = ba|_A$ and $ab|_B = b = ba|_B$. This suggests that C observes a happening in A and then observes b happening in B, whereas D observes first b happening in B and *then* a happening in A. Monteiro and Pereira [Monteiro and Pereira, 1986] refer to such a trace as an 'impossible behaviour', although we see nothing unacceptable with such *relativistic* behaviour, since none of the objects share a clock. Indeed, synchronising the system with a clock which is shared by all objects is the only reasonable way of avoiding such 'impossible behaviours'.

10.3 Object Specification

In the previous section we outlined a relationship between limits in the category of monoids and the independent sum of process specifications. This relied on viewing the set of methods of a class of processes as a monoid. However, an object class specification also declares a number of attributes, which allow the observation of aspects of an object's state. An object class specification should therefore also give a set D representing the data values that the attributes might take. If an object is to have more than one attribute, then the set D can be thought of as the set of tuples of all the attributes. Moreover, there should be some means of specifying the way in which methods affect an object's attributes. A simple way of achieving this is to provide a function that maps methods to attribute values: this function can be thought of as returning the attribute values after evaluating the given method in an object's initial state.

Definition 6 An **object class specification** is a triple (P, ε, D), where P is a monoid, D is a set, and ε is a function from the carrier of P to D. We often write $P \stackrel{\varepsilon}{\to} D$ instead of (P, ε, D). □

For example, the specification of a stack object might be $P \stackrel{\varepsilon}{\to} D$, where:

- P is the set of lists over $\{\text{pop}\} \cup \{\text{push}(n) \mid n \in \omega\}$;
- D is the set $\{\text{error}\} \cup \omega$, representing the value on the top of the stack; and
- ε is defined by

$$\begin{aligned} \varepsilon([\,]) &= \text{error} \\ \varepsilon(p\,\text{push}(n)) &= n \\ \varepsilon(p\,\text{pop}) &= \varepsilon(\mathit{tail}(p)) \end{aligned}$$

  for all $p \in P$, where

$$\begin{aligned} \mathit{tail}([\,]) &= [\,] \\ \mathit{tail}(p\,\text{push}(n)) &= p \\ \mathit{tail}(p\,\text{pop}) &= \mathit{tail}(\mathit{tail}(p)). \end{aligned}$$

Consider two object class specifications (P, ε, D) and (P', ε', D'). A morphism from the one to the other should consist of a translation $f : P \to P'$ of programs and a translation $g : D \to D'$ of state attributes, such that for any program $p \in P$, if we translate p and then evaluate the result in (P', ε', D'), we should get the same result as evaluating p and then translating the resulting state attributes. In other words, we require that $\varepsilon'(f(p)) = g(\varepsilon(p))$, i.e., the following diagram commutes.

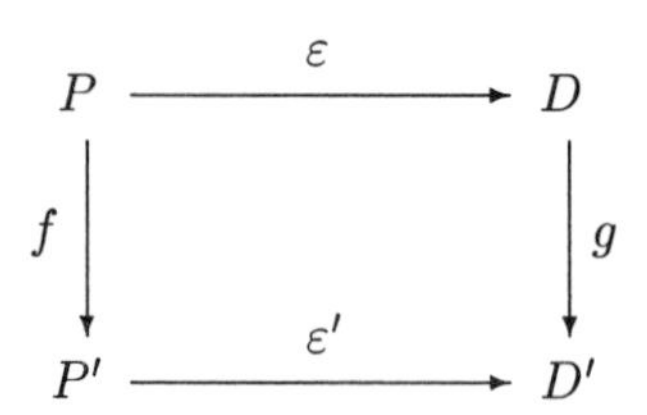

Note that f should be a monoid morphism, so, strictly speaking, we should have $U(f) : U(P) \to U(P')$ as the leftmost arrow in this diagram, where U is the forgetful functor from Mon to Sets; however, omitting U is in keeping with the notation of Section 10.1.4.

Definition 7 **A morphism of object class specifications**

$$(P, \varepsilon, D) \to (P', \varepsilon', D')$$

is a pair (f, g), with $f : P \to P'$ a monoid morphism and $g : D \to D'$ in Sets, such that $\varepsilon\,; g = f\,; \varepsilon'$. This gives rise to a category Obj of object class specifications and morphisms. □

Note that, because f is a monoid morphism rather than just a map between sets of methods, we allow methods to be translated into composite programs.

Inheritance is modelled by inclusions of object class specifications (cf. [Ehrich *et al.*, 1991, Goguen, 1992, Costa *et al.*, 1994], for example). The intuition is that the inheriting specification specialises the inherited specification by adding more methods and more attributes. Thus if (P, ε, D) inherits from (P', ε', D'), there should be an inclusion $P' \hookrightarrow P$, and a restriction $g : D \to D'$ of state attributes.

Definition 8 **A containment of object class specifications** is a morphism $(f, g) : (P, \varepsilon, D) \to (P', \varepsilon', D')$ such that f has a pre-inverse $i_f : P' \to P$ (that is, $i_f\,; f = id_{P'}$). This gives a subcategory $\mathsf{Obj}_{\subseteq}$ of Obj with object class specifications as objects and containments as morphisms. □

This definition implies that $\varepsilon' = i_f\,; \varepsilon\,; g$; that is, the effect of inherited methods on state attributes is not altered.

In the previous section, we suggested that interconnection of objects, at least as far as methods are concerned, is captured by limits on the category of monoids. Moreover, it makes sense to use limits for the state attributes of aggregated objects: the state space of a composite object will be composed of the state spaces of its component parts in a coherent way. Therefore, the following proposition allows us to build specifications of classes of composite objects.

Proposition 9 The category Obj of object class specifications is complete. □

In fact, limits in Obj are constructed by taking limits of methods and of state attributes. The following notation allows us to give an easy characterisation of limits in Obj.

Notation 10 Note that a diagram $\mathsf{X} \to \mathsf{Obj}$ is the same thing as a functor $P : \mathsf{X} \to \mathsf{Mon}$, a functor $D : \mathsf{X} \to \mathsf{Sets}$, and a natural transformation $\varepsilon : P\,; U \to D$, where U is the forgetful functor $\mathsf{Mon} \to \mathsf{Sets}$. Extending the notation of Definition 6, we write diagrams of object class specifications as $P \xrightarrow{\varepsilon} D$. □

Proof of Proposition 9. Suppose $P \xrightarrow{\varepsilon} D$ is a diagram of object class specifications, and let $\pi : P' \Rightarrow P$ be the limit of P in Mon, and let $\varpi : D' \Rightarrow D$ be the limit of D in Sets. Then composing π with ε gives a cone $\pi\,;\varepsilon : P' \Rightarrow D$ of D, which induces a morphism $\varepsilon' : P' \to D'$:

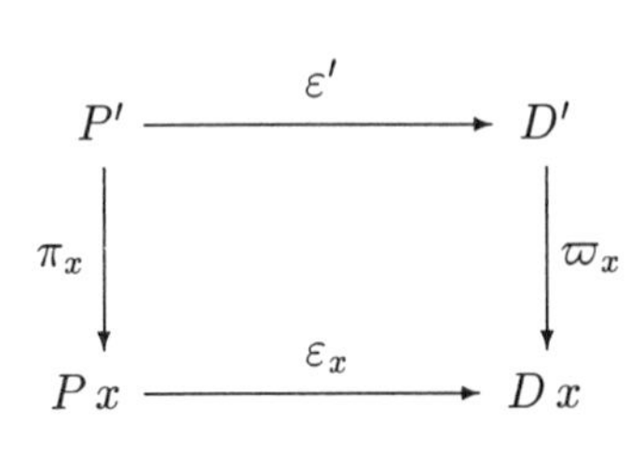

It is clear from this diagram that $(\pi, \varpi) : (P' \xrightarrow{\varepsilon'} D') \Rightarrow (P \xrightarrow{\varepsilon} D)$ is a cone for $(P \xrightarrow{\varepsilon} D)$, and it is straightforward to check that it is a limiting cone. □

This describes how to paste together an arbitrary diagram of object class specifications: parallel composition of objects is given by limits.

It is also possible to paste together *models* of object class specifications, where a model is something that evaluates methods in a way that agrees with the specification.

Definition 11 **A model** for an object class specification $P \xrightarrow{\varepsilon} D$ is a tuple $(s, S, \oplus, \alpha)$, where $(S, \oplus)$ is a right action of P, and $s \in S$ and $\alpha : S \to D$ are such that

$$\alpha(s \oplus p) \;=\; \varepsilon(p)$$

for all $p \in P$. □

Given a diagram of object specifications and models for each of the specifications in the diagram, we can build a model for the limit of the diagram.

Proposition 12 Let $P \overset{\varepsilon}{\to} D : \mathsf{X} \to \mathsf{Obj}$ be a diagram of object class specifications, and suppose that for each $x \in |\mathsf{X}|$ we have a $(Px \overset{\varepsilon_x}{\to} Dx)$-model $(s_x, Sx, \oplus_x, \alpha_x)$, where $S : \mathsf{X} \to \mathsf{Sets}$ is a functor, and $s : 1 \to S$, $\oplus : S{\times}P \to S$ and $\alpha : S \to D$ are natural transformations[1]. Then these models give rise to a model of the limit of $P \overset{\varepsilon}{\to} D$. □

This model is constructed by taking the limit of $S : \mathsf{X} \to \mathsf{Sets}$; the operations required to make this a model of the limit of $P \overset{\varepsilon}{\to} D$ can be constructed by using the universal properties of the limits of S, P and D. The details are left as an exercise for the reader.

10.4 System Specification

In the previous section we presented limits of diagrams of objects as a means of composing object class specifications into specifications of complex object classes. In this section we generalise this by considering systems of objects and interconnections of such systems. As in the previous section, our goal is to examine ways of composing systems of objects to form larger systems.

There are various ways of defining systems of objects. The most general way also seems to provide the richest structure for system specifications. The following definitions are due to Goguen [Goguen, 1970, Goguen, 1972b, Goguen, 1975b].

Definition 13 A **system specification** is a diagram $\delta : \mathsf{X} \to \mathsf{Obj}$ of object class specifications. □

The category X can be thought of as defining the 'shape' or 'topology' of the system.

Morphisms of system specifications should specify how systems are to be composed into larger systems; in particular, such a morphism should identify the shared components of systems which are to be composed. Suppose $\delta : \mathsf{X} \to \mathsf{Obj}$ and $\zeta : \mathsf{Y} \to \mathsf{Obj}$ are system specifications; a morphism from δ to ζ should interpret δ in ζ. That is, there should be a functor $F : \mathsf{X} \to \mathsf{Y}$ which interprets the topology of δ in that of ζ, and also a means of comparing the object class specifications $\delta(x)$ and $\zeta(F(x))$ for any object $x \in |\mathsf{X}|$. In particular, we want a means of encoding the methods and state attributes of $\zeta(F(x))$ as methods and state attributes of $\delta(x)$; this may be achieved by a natural transformation $\eta : F\,;\zeta \to \delta$, as in the following picture.

---

[1] In saying that $s : 1 \to S$ is a natural transformation we mean that $s_x \in Sx$ for each $x \in |\mathsf{X}|$, and for each arrow $f : x \to y$ in X we have $Sf(s_x) = s_y$.

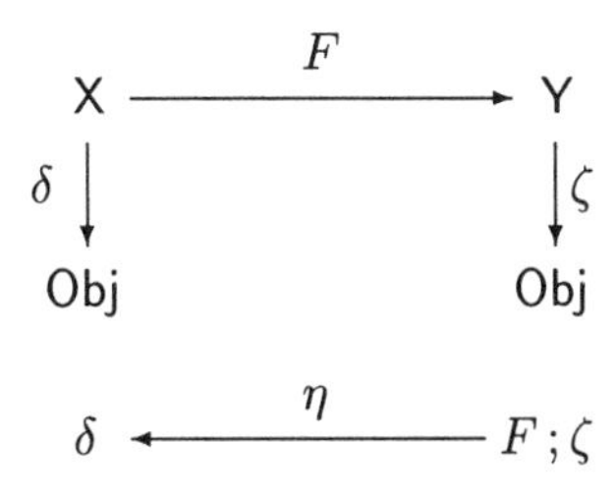

Definition 14 Given system specifications $\delta : \mathsf{X} \to \mathsf{Obj}$ and $\zeta : \mathsf{Y} \to \mathsf{Obj}$, a **morphism of system specifications** $\delta \to \zeta$ is a pair (F, η) where $F : \mathsf{X} \to \mathsf{Y}$ is a functor and $\eta : F\,;\zeta \to \delta$ is a natural transformation. This gives a category Sys of system specifications. □

As we argued in the previous section, a system can be thought of as an object, by taking the limit of that diagram. In accordance with the slogan of categorical systems theory 'Behaviour is given by limits' (cf. Goguen [Goguen, 1992]), this is made explicit in the following

Definition 15 The functor BEH : Sys → Obj takes a system specification to its limit in Obj, called its **behaviour**. □

In fact, objects can be viewed as systems composed of only one object. This gives a functor Obj → Sys, which is right adjoint to BEH (Goguen [Goguen, 1972b]).

Just as objects were composed into complex objects by taking limits of diagrams, so too are systems composed into larger systems.

Definition 16 An **interconnection of system specifications** is a diagram $\delta : \mathsf{X} \to \mathsf{Sys}$. The **result** of an interconnection is its colimit in Sys. □

The following proposition states that the category Sys is cocomplete; arbitrary diagrams of systems can be composed into larger systems by taking colimits. The proposition is a special case of a theorem of Goguen's [Goguen, 1972b], which asserts the result for any complete category of objects.

Proposition 17 Every interconnection of system specifications has a result. □

Moreover, the behaviour of the result of an interconnection $\delta : \mathsf{X} \to \mathsf{Obj}$ can be computed by taking the limit in Obj of the limits of each system $\delta(x)$, that is, the limit of $\delta\,;\mathrm{BEH} : \mathsf{X} \to \mathsf{Obj}$.

Another way of structuring objects into systems is to consider only topologies that have the form of a partial order. In this way, we consider only subobject relationships between components of a system.

Definition 18 A **PO-system** is a functor $S : \mathsf{C}^{\mathsf{op}} \to \mathsf{Obj}$ where C is a partially ordered set $(C, \leq)$. □

PO-systems are special kinds of system specifications, and morphisms of PO-systems are just system specification morphisms. The idea is that a PO-system is one that is constructed hierarchically from smaller systems by means of parallel connection with synchronisation, like the cell $(X\|Y)\|_Y(Y\|Z)$ that we considered at the beginning of this chapter.

Analogously to the above, we can define interconnections of PO-systems as diagrams, whose results are given by colimits, and analogously to Proposition 17 we have

Proposition 19 The category of PO-systems is cocomplete. □

In other words, PO-systems may be constructed hierarchically.

Finally, we can begin to relate these results and the sheaf theoretic model of Monteiro and Pereira [Monteiro and Pereira, 1986], and also the categorical systems slogan 'Objects are Sheaves' [Goguen, 1992], by showing that any PO-system specification gives rise in a canonical way to a sheaf.

Definition 20 A **sheaf of objects** is a PO-system $S = P \xrightarrow{\varepsilon} D : \mathsf{C}^{\mathrm{op}} \to \mathsf{Obj}$ where C is a complete Heyting algebra and P and D are sheaves. A morphism of sheaves of objects is a system specification morphism whose functor part distributes through greatest lower bounds. □

In order to construct a sheaf of objects from a PO-system, we first show that every partial order gives rise to a complete Heyting algebra.

Definition 21 Given a partially ordered set $\mathsf{C} = (C, \leq)$, its **topology** is the complete Heyting algebra $\Omega\mathsf{C} = (\Omega C, \subseteq)$, where ΩC is the collection of downwards-closed subsets of C (i.e, subsets $X \subseteq C$ such that if $x \in X$ and $c \leq x$ then $c \in X$). Moreover, define the functor $H : \mathsf{C}^{\mathrm{op}} \to \Omega\mathsf{C}^{\mathrm{op}}$ by $Hc = \{x \in C \mid x \leq c\}$. □

In fact, Ω is a functor from the category of partial orders to the category of complete Heyting algebras, and is left adjoint to the forgetful functor which views a complete Heyting algebra as a partial order, and H is the unit of this adjunction.

Now we construct sheaves of objects as follows.

Definition 22 Given a PO-system $S = P \xrightarrow{\varepsilon} D : \mathsf{C}^{\mathrm{op}} \to \mathsf{Obj}$ let $S^{\Omega} : \Omega\mathsf{C}^{\mathrm{op}} \to \mathsf{Obj}$ be the functor that takes a downwards-closed subset $X \subseteq C$ to the limit of the diagram $X^{\mathrm{op}} \hookrightarrow C^{\mathrm{op}} \xrightarrow{S} \mathsf{Obj}$. □

Proposition 23 S^{Ω} is a sheaf of objects. □

The proof of this proposition uses basic properties of limits; we omit the details. Because S^{Ω} is defined by limits, for any $c \in C$ there is a projection $\eta_c : S^{\Omega}(Hc) \to Sc$. In fact, we have a natural transformation $\eta : (H\,; S^{\Omega}) \to S$ and therefore a morphism of system specifications $(H, \eta) : S \to S^{\Omega}$. This morphism satisfies a special property which expresses the canonicity of the construction of S^{Ω}.

Theorem 24 Let T be a sheaf of objects. For every system specification morphism $(F, \zeta) : S \to T$ there is a unique morphism of sheaves $(G, \theta) : S^{\Omega} \to T$ such that $(H, \eta) ; (G, \theta) = (F, \zeta)$. □

This theorem states that there is an adjunction between PO-systems and sheaves of objects, so that we can view S^{Ω} as the best possible way of extending S to a sheaf of objects.

10.5 Some Speculations

We have given a simple formalisation of object classes and systems of objects, and studied ways in which systems of objects may be interconnected. Much of the motivation for this account comes from Goguen and Diaconescu's independent sum construction [Goguen and Diaconescu, 1994], which captures parallel composition with synchronisation. As in Monteiro and Pereira's sheaf theoretic model of concurrency [Monteiro and Pereira, 1986], we use limits in the category of monoids to obtain the commutativity properties which express truly concurrent parallel composition. Following Goguen [Goguen, 1992] we model both behaviour and parallel composition by means of limit constructions, and we take a step towards the 'objects are sheaves' paradigm by giving an adjunction between PO-systems and sheaves. We expect that, by using a more general notion of sheaf as a functor on a category with a Grothendieck topology [Lane and Moerdijk, 1992], we can obtain an adjunction between system specifications and sheaves of objects.

Our definition of object class specification is similar to abstract definitions of automata in categories [Adámek, 1981, Arbib and Manes, 1974, Goguen, 1972a, Goguen, 1975a]. Since our definition was motivated by Goguen and Diaconescu's independent sum construction in hidden sorted algebra, it would be interesting to examine the relationship between Nerode equivalence in automata theory and behavioural equivalence in hidden sorted algebra. Mazurkiewicz's 'process monoids' [Mazurkiewicz, 1984] suggests that there is a relation between our work and Nerode equivalence.

An important aspect of object orientation that we have not covered is object creation and deletion. It might be possible to treat this in a simple way by looking at restrictions of systems to subcategories. Since a system is a functor $S : \mathsf{X} \to \mathsf{Obj}$, deleting an object would correspond to restricting S to a subcategory of X; similarly, object creation would correspond to extending S to a supercategory. Such an approach to dynamic systems is a very interesting area for future research.

Another very interesting and exciting area we intend to explore is the relationship between sheaf theoretic models of objects and the logic of toposes. Every topos has an 'internal logic' which is a kind of intuitionistic type theory [Lambek and Scott, 1986, Lane and Moerdijk, 1992]. Since the category of sheaves on a particular category form a topos, it might be possible to use that logic to express and constructively prove useful properties of systems of objects. Moreover, the category of right actions of a monoid is a topos, so it would be interesting to examine its internal logic as a means of reasoning about models of specifications. Our use of monoids to capture the structure of programs is perhaps somewhat arbitrary: the topos of representations of a group would be appropriate to the free group structure of programs presented in [Dijkstra, 1993].

Chapter 11

Refinement of Concurrent Object-Oriented Programs

FOOPS is a concurrent object oriented specification language having an executable subset [Goguen and Meseguer, 1987, Rapanotti and Socorro, 1992]. It includes a functional language derived from OBJ [Goguen and Winkler, 1988], which is a first order, purely functional language providing an algebraic style for the specification, rapid prototyping, and implementation of abstract data types. OBJ also supports the declaration of "loose" theory specifications as well as "tight" executable specifications.

FOOPS extends OBJ by providing a simple "declarative" style for object oriented programming and specification. It supports classes of objects with associated methods and attributes, object identity, dynamic object creation and deletion, overloading, polymorphism, inheritance with overriding, dynamic binding, parameterized programming [Goguen, 1989], and many additional features. Here we consider an extension of FOOPS with features for specifying systems of distributed and autonomous objects. Essentially, this extension allows the definition of non terminating (autonomous) methods, and provides constructors (called method combiners) for expressing concurrency, nondeterminism, and atomic execution.

An operational semantics [Plotkin, 1981] for this extension of FOOPS is given in [Borba and Goguen, 1994, Borba, 1995]; it describes the behaviour of FOOPS *method expressions* (commands). Based on this semantics, here we define a notion of refinement that indicates whether the behaviour of a FOOPS system is simulated by another system, so that we can replace the latter by the former without noticing any difference; this extends to (method) expressions, programs, and specifications as well. Although we focus on FOOPS, our definition of refinement is quite independent of this language; it essentially depends on minimal assumptions about the operational semantics, which are satisfied by the FOOPS operational semantics given in [Borba and Goguen, 1994, Borba, 1995].

We also illustrate the use of refinement for stepwise formal development of programs in FOOPS; we use examples involving memory cells and different implementations of buffers. Furthermore we show how the definition of refinement can be used to derive laws of FOOPS method combiners. We include proofs of refinement in order to give a better insight on our approach, by showing how it works in detail, what aspects it considers, how complex the proofs are, and by analysing to what extent the proofs can be mechanized.

Similarly to [Milner, 1989], [Mosses, 1992], and [Hennessy, 1988], we use operational semantics' framework to define refinement and to reason about specifications. So, in order to develop notions of refinement for FOOPS, we adapt ideas from [Milner, 1989], [Milner, 1971], and [Park, 1981], where a general approach for defining refinement (and equivalence) of transition systems is developed and justified; this approach is quite simple and provides an effective proof technique for verifying refinement. We also consider aspects of data refinement [Hoare, 1972, Jifeng *et al.*, 1986, Goguen and Malcolm, 1994].

This chapter is organized in the following way. First, we give an overview of FOOPS. Then we introduce and justify the notion of refinement of states and its associated proof technique. After that we derive basic properties of refinement of states. Then we introduce the notions of refinement of expressions and specifications, deriving basic properties of those relations as well. Following that, the proof technique is used for verifying refinement of several examples. Based on that, we identify different aspects considered by the definition of refinement, evaluate our approach, and comment on alternative approaches for refinement of concurrent object oriented programs.

11.1 Overview of FOOPS

FOOPS extends OBJ with some concepts from object oriented programming. This motivates two central design decisions (see [Goguen and Wolfram, 1991, Goguen and Meseguer, 1987, Rapanotti and Socorro, 1992, Socorro, 1993] for more details about the design of FOOPS): data elements are not objects, and classes are not modules. In this way, FOOPS provides different constructs for defining abstract data types (ADTs) for values and classes for objects. Consequently, there are two constructs for specifying inheritance.

The clear distinction between data elements and objects divides the language in two parts: the functional level and the object level. At each level, there are two kinds of modules: one is used to define executable code, and is simply called a *module*; the other, called a *theory*, is used to specify properties of the operations of an abstract data type or a class.

11.1.1 Functional Level

The functional level of FOOPS is a syntactical variant of OBJ. At this level it is possible to define abstract data types, where the keywords `sort` and `fn` respectively introduce the name of the set of data elements, and the associated operation (function) symbols. A very simple functional theory is

```
fth TRIV is
  sort Elt .
endfth
```

It introduces the sort `Elt`. The models of `TRIV` are just sets.

As an example of a parameterized functional level module, we consider (piece by piece) a module `LIST`, which defines lists of elements of a given sort. This module is parameterized by the sort of the elements in a list:

```
fmod LIST[E :: TRIV] is
  pr NAT .
```

The following declaration introduces a sort for nonempty lists and another for lists,

```
    sorts NeList List .
    subsorts Elt < NeList < List .
```

where elements are considered singleton (nonempty) lists and nonempty lists are, of course, lists, as indicated by the subsort relationship (<). This is what specifies inheritance at the functional level; for example, as all elements of `Elt` are elements of `List`, all functions associated to `List` can also be used for the elements of `Elt`.

The empty list is represented by the constant `nil`, and `_._` denotes the function that concatenates two lists.

```
    fn nil : -> List .
    fn _._ : List List -> List [assoc id: nil] .
    fn _._ : NeList List -> NeList .
    fn _._ : NeList NeList -> NeList .
```

The underscores in `_._` serve as placeholders for the arguments of this function. As indicated by the attributes, this function is associative and has `nil` as identity.

Some other functions are `head`, which gives the first element of a non empty list; `tail`, which maps a non empty list to one obtained by removing its first element; and `#_`, which gives the number of elements in a list.

```
    fn head : NeList -> Elt .
    fn tail : NeList -> List .
    fn #_ : List -> Nat .
```

The meaning of these functions is given by axioms (equations).

```
    ⋮
    ax tail(E . L) = L .
    ax # nil = 0 .
    ax # (L . E) = # L + 1 .
  endfmod
```

The keyword `ax` precedes an axiom, and `endfmod` indicates the end of a functional module.

11.1.2 Object Level

At the object level, it is possible to define classes, which are collections of (potential) objects with same attributes and methods. FOOPS has a general computational model where objects are naturally distributed and (internally) concurrent. Objects are dynamically created and deleted, and there are special operations for performing these actions. Furthermore, each object has an unique identifier, which is used by other objects for access. In this way, methods and attributes have at least one object identifier as argument, indicating which method is going to execute the corresponding operation.

Let us consider an object module **BUFFER** defining a class of bounded buffers. This module is parameterized by the capacity of buffers (a positive natural number), specified by the functional requirements theory **MAX**:

```
fth MAX is
  pr NAT .
  fn max : -> NzNat .
endfth
```

The sort of positive natural numbers is represented by **NzNat** (it is defined in **NAT**). The module **BUFFER** is also parameterized by the sort of the elements to be stored in buffers:

```
omod BUFFER[E :: TRIV, M :: MAX] is
  pr LIST[E] .
```

Here the elements of a buffer are stored in a list; so, it is necessary to import the functional module `LIST`, instantiating it with the argument module giving the sort of elements.

The class `Buffer` of bounded buffers is introduced by the declaration

```
class Buffer .
```

Attributes are defined as operations from an object identifier to a value that denotes a current property of the related object. Objects of **Buffer** have the attribute **elems**, corresponding to the list of elements in a buffer.

```
at elems_ : Buffer -> List [hidden] .
```

As indicated by the declaration **[hidden]**, the attribute **elems** is only visible inside **BUFFER**; so, clients of the objects of **Buffer** cannot directly look at the elements stored in buffers. In addition to **elems**, two more attributes are associated to **Buffer**:

```
at empty?_ : Buffer -> Bool .
at full?_ : Buffer -> Bool .
```

The attribute **empty?** indicates whether the buffer is empty, whereas **full?** indicates whether the buffer contains the maximum number of elements. Like attributes, methods are defined as operations having an object of its class as parameter. They might also have some extra parameters. Methods either evaluate to a special result or to the identifier of the object that performs it. For objects of `Buffer`, the available methods are the following: `reset`, which removes all elements from a buffer; `get`, which removes the first element of a non-empty buffer and gives it as result; `put`, which inserts an element at the end of a buffer, if it is not full; and `del`, which removes the first element of a non empty buffer.

```
me reset : Buffer -> Buffer .
me get : Buffer -> Elt .
me put : Buffer Elt -> Buffer .
me del : Buffer -> Buffer [hidden] .
```

The last one is hidden because we do not allow clients to remove an element from a buffer unless it is going to be used, which can be done using `get`.

Axioms

Attributes can be classified as *stored* or *derived*. The value of a stored attribute is kept as part of the local state of an object. On the other hand, the value of a derived attribute is not stored by an object, but can be computed from the values of other attributes. Hence, one must specify how this is done; in FOOPS, we use equations for that. If no equation is given for an attribute, it is considered a stored attribute.

For `Buffer`, we define `elems` as a stored attribute. The others are derived; so, we introduce the following equations:

```
⋮
ax empty? B = (elems B) == nil .
ax full? B = #(elems B) == max .
```

This indicates that the buffer is empty if the list of the elements stored in it is empty; also, the buffer is full if the size of its associated list is `max`.

Equations defining attributes can only contain functions, attributes, and object identifiers. This kind of equation is interpreted as left-to-right rewrite rules, but attributes are atomically evaluated, without interference from the execution (evaluation) of methods.

The behavior of methods can be specified by two different kinds of axioms. A *direct method axiom* (DMA) specifies how a stored attribute is updated by a given method. In fact, a DMA is an equation such that its left hand side (LHS) indicates its associated attribute and method, whereas its right hand side (RHS) is an expression specifying the new value for the attribute to be updated. For instance, the behavior of `reset` is given by the DMA

```
ax elems(reset(B)) = nil .
```

which specifies that after the execution of `reset` by an object `B`, the value of `elems`, for `B`, is `nil`.

Further examples of DMAs are

```
cx elems(put(B,E)) = (elems B) . E if not(full? B) .
cx elems(del(B)) = tail(elems B) if not(empty? B) .
```

where the methods are only executed if the (enabling) conditions are satisfied; otherwise, the evaluation is suspended. The new value for the specified attribute is computed in terms of the method arguments and the current attribute values. If there is no axiom specifying the new value for a stored attribute after the execution of a given method, this method does not update that attribute. This is called the *frame assumption*; it avoids writing equations indicating that some attributes are not updated.

The evaluation of methods specified by DMAs is atomic and yields the identifier of the object which executes the method; only this objects is modified, and its attributes are updated as specified. As for attribute equations, the axiom's RHS and condition must be formed by functions, attributes, and object identifiers.

Alternatively to DMAs, *indirect method axioms* (IMAs) may be used for defining methods. IMAs are equations that specify how a method is defined in terms of other operations; this is indicated by a method expression, i.e., an expression formed by methods, attributes, functions, object identifiers, and method combiners (operators on method expressions). For example, the method `get` is specified by the IMA

```
acx get(B) = result head(elems B) ; del(B)
if not(empty? B) .
```

where `result_;_` is a method combiner which evaluates its first argument (from left to right) and then evaluates the second one, yielding the value resulting from the evaluation of the first argument.

Similarly to DMAs, no method symbol or method combiner is allowed in an IMA's condition. IMAs are interpreted as left-to-right rewrite rules. Whereas the evaluation of the IMA's condition is atomic, the evaluation of the IMA's RHS is not atomic and may be interfered by the execution of other methods. However, atomicity can be achieved by using the atomic evaluation operator `[_]`, which atomically executes its argument, without interference from the execution of other methods. We assume that IMAs introduced by the keyword `acx` (or `aax`) have their condition and RHS atomically evaluated. In fact, an IMA like

```
acx m(O) = e if c .
```

stands for the following declarations:

```
me m' : C -> C' .
ax m'(O) = e if c .
ax m(O) = [m'(O)] .
```

where `m'` is any method symbol not used in the specification where the original axiom appears.

Method Combiners

In addition to `result_;_` and `[_]`, FOOPS provides other method combiners: sequential composition, `_;_`; (interleaving) parallel composition, `_||_`; (external) nondeterministic choice, `_[]_`; and conditional, `if_then_else_fi`.

Object Creation and Deletion

Dynamic object creation and deletion are respectively provided in FOOPS by the following operators:

- `new.C() : -> C`,
- `new : C -> C`, and
- `remove : C -> C`,

for each class `C`.

For a given class `C`, the operator `new.C()` creates an object of `C` with a nondeterministically choosen identifier that is not already being used for another object. This identifier is given as the result of the evaluation of the operator.

The operator `new` creates an object of the same class as the object identifier given as argument, if this identifier is not associated to another object (otherwise, the operation cannot be executed). This identifier is used for the created object and yielded by the operator.

The operator `remove` receives an object identifier as argument, removes its associated object and yields this identifier. If this identifier does not correspond to an object in the state, the operation is not evaluated. Contrasting to `new`, the argument of `remove` might be an arbitrary expression, it does not have to be an object identifier; however, it is supposed to yield an identifier.

Objects can also be introduced together with the definition of their associated class, where values for their stored attributes are specified. These are called specified objects and are particularly useful when defining classes of recursive data structures such as stacks and linked lists. Specified objects have the same status as objects created at runtime; this means that they can be modified and removed.

11.2 Refinement for FOOPS

A FOOPS specification (a non-parameterized module) defines a FOOPS system; it describes the structure of the objects that the system may have, and the behaviour of the methods and attributes associated to those objects. Object creation, object deletion, and method execution change the state of a FOOPS system. Such a state consists of information about existing objects in the system (database state), and expressions being concurrently evaluated.

Here we want to investigate whether a state is simulated (refined) by another state. Basically, a state P is simulated by a state Q if whatever can be observed by performing experiments with Q can also be observed by performing the same experiments with P. In this case, if experiments are done with Q, we cannot detect whether Q or P is being used, even thought, the experiments done with P might let us make more observations than if Q were used.

We have to define the notions of experiment and observation that will be of interest for us. The only kind of experiment that we can make with a FOOPS system is to invoke *visible* operations with arbitrary arguments. This results in a new state having the invoked operation as one of the expressions to be evaluated. Here are the aspects that we can observe from a state P:

1. the values of *visible* attributes of the objects in P;

2. the results yielded by the expressions being evaluated in P;

3. whatever can be observed by performing experiments with the states that can be immediately reached from P due to the execution of the expressions in P.

Note that the relation of simulation between states depends only on the notions of experiment and observation. Indeed, one state might simulate another even if they have different (internal) structures, and are associated to systems specified by different specifications. However, the experiments and observations should be meaningful for both systems. In general, given states P and Q respectively related to specifications P and Q, Q may only simulate P if the observations and experiments introduced by P are also introduced by Q. Only in this way we can compare the effects of performing the *same* experiment with P and Q.

Now we make more precise what experiments are of interest for us. First we consider that only the following are **operations**: methods, attributes, and object creation and deletion routines. In addition to *operations*, an user of a FOOPS system can require the evaluation of functions. However, the evaluation of functions does not directly change the system database, nor depends on the system state; that's why they are not considered for experiments. We use $Ops(\mathsf{P})$ to denote the family of *operation* symbols associated to a specification P; this family is indexed by the rank of the operations.

Second, we assume that the arguments of visible operations are object identifiers or elements of the ADTs corresponding to the sorts in the arity of those operations. Although complex expressions formed by identifiers, visible operations, and functions could be used as arguments, this would have the same effect as invoking a sequence of simpler expressions as experiments; so it is redundant to consider that possibility here.

Hereafter the family of sets of **experiments** (expressions formed by the application of a visible operation to object identifiers and elements of ADTs) associated to a specification P is denoted by $\mathcal{E}(\mathsf{P})$. This family is indexed by the sorts or classes associated to experiments; so we use $\mathcal{E}_u(\mathsf{P})$ to denote the element of the family $\mathcal{E}(\mathsf{P})$ corresponding to type u.

Note that $\mathcal{E}(\mathsf{P}) \subseteq \mathcal{E}(\mathsf{Q})$ also indicates that the states of systems specified by Q have the same visible attributes as the states associated to P, since we consider that attributes are operations. Furthermore, it is easy to see that in order to check whether $\mathcal{E}(\mathsf{P}) \subseteq \mathcal{E}(\mathsf{Q})$, it is enough to check whether

1. the visible operations of P are also visible operations of Q, having the same name and rank;

2. the object identifiers associated to P are also associated to Q; and

3. the ADTs of P corresponding to sorts of the results yielded by experiments, or to sorts having elements used to make experiments, should also be ADTs of Q.

The last condition also guarantees that states associated to P may possibly be simulated by states associated to a specification Q. In general, this is not implied by $\mathcal{E}(\mathsf{P}) \subseteq \mathcal{E}(\mathsf{Q})$, since the functional terms used to form experiments of P might have a different meaning in Q.

As can be concluded from the discussion above, we assume that functional specifications included in FOOPS specifications have an initial semantics (i.e., they denote ADTs). In this way, it is easier to concentrate on the issues related to refinement at the object level; that's our main focus here. Also it is only a small extension to consider refinement of FOOPS specifications containing functional specifications with initial and loose semantics.

Condition 3 introduced above can be expressed by the following predicate: $\mathsf{P} \leq_{PS(\mathsf{P})} \mathsf{Q}$, which holds if the ADTs specified by P when restricted to $PS(\mathsf{P})$ are also ADTs of Q, where $PS(\mathsf{P})$ denotes the set of **primary** sorts of P; that is, the set formed by the sorts of P having terms yielded by experiments or used to make experiments.

According to the notion of simulation discussed in this section (see Item 3 from the list of aspects that we can observe from states), if P is simulated by Q then states immediately reached from Q simulate states immediately reached from P. However, a better definition of simulation can be obtained if we relax this strong correspondence between state transitions. In fact, it is only important that states immediately reached from Q simulate states eventually reached from P. This is enough because in this case the observations that we can make from Q are equivalent to observations that we can make from P when we fail to observe some intermediate states. This still implies that the observable behaviour associated to Q is a particular case of the observable behaviour associated to P; that's the essence of simulation.

11.2.1 Simulations

Now we start to formalize the finer of the two notions of simulation discussed so far. In order to represent states and reason about state transitions due to evaluating expressions, we use the framework provided by operational semantics. States are represented by configurations of the operational semantics, that is, pairs formed by an expression and a database state, represented as $\langle e, \mathcal{P} \rangle$, for an expression e and a database state $\mathcal{P}$. The first corresponds to the expressions being evaluated in the state, and the second contains information about the objects in the state. When not confusing, database states are just called "states". We use $Conf(\mathsf{P})$ to denote the set of all configurations associated to a specification P, and $I(\mathsf{P})$ to denote the family of sets of object identifiers associated to P (each set is formed by the symbols that can be used to identify objects of a given class).

State transitions are specified by the transition relation

$$\rightarrow_{\mathsf{P}} \subseteq Conf(\mathsf{P}) \times Conf(\mathsf{P}),$$

which defines the operational semantics associated to P; it indicates how an expression is evaluated in a database state. When not confusing, we write $\rightarrow$ instead of $\rightarrow_{\mathsf{P}}$. Similarly, the relation $\rightarrow^*$ denotes the transitive, reflexive closure of $\rightarrow$; also, we write $P\not\rightarrow$ if there is no P' such that $P \rightarrow P'$ (this means that the expression in P cannot be further evaluated). Lastly, if $P \rightarrow^* P'$ then we say that P' is a **$\rightarrow^*$-derivative** of P; a corresponding terminology is used for $\rightarrow$.

We use the term "fully evaluated expression" to refer to object identifiers and *evaluated* functional expressions (elements of ADTs). The notation $\mathcal{FE}(e)$ indicates that the expression e is fully evaluated, according to a specification P that should be understood from the context where $\mathcal{FE}(e)$ is used; otherwise, we write $\mathcal{FE}_{\mathsf{P}}(e)$. For fully evaluated terms p and q respectively related to specifications P and Q, equality is denoted by $p =_{(\mathsf{P},\mathsf{Q})} q$; it holds if p and q are the same object identifiers and $I(\mathsf{P}) \subseteq I(\mathsf{Q})$, or if p and q are the same element of an ADT that is defined by both P and Q.

The notion of simulation between states is defined as the union of all relations having the following special property: if the pair $(P,\ Q)$ is in the relation then whatever can be observed by performing experiments with Q can also be observed by performing experiments with P. Note that this property is exactly what guarantees that one state simulates another.

Relations having this special property are called simulations. Let us formalize this concept now. Hereafter we use P and Q for configurations, $\mathcal{P}$ and $\mathcal{Q}$ for their respective database states, and p and q for their respective expressions.

Definition 11.2.1 Let P and Q be specifications such that $\mathcal{E}(\mathsf{P}) \subseteq \mathcal{E}(\mathsf{Q})$, $I(\mathsf{P}) \subseteq I(\mathsf{Q})$, and $\mathsf{P} \leq_{PS(\mathsf{P})} \mathsf{Q}$. A relation $\mathcal{S} \subseteq Conf(\mathsf{P}) \times Conf(\mathsf{Q})$ is a (P, Q)-**simulation** if $(P,\ Q) \in \mathcal{S}$ implies the following:

1. Whenever $Q \rightarrow Q'$ then, for some P', $P \rightarrow^* P'$ and $(P',\ Q') \in \mathcal{S}$.
2. If $Q\not\rightarrow$ then, for some P',
   (a) $P \rightarrow^* P'\not\rightarrow$;
   (b) $\mathcal{FE}_{\mathsf{P}}(p') \Leftrightarrow \mathcal{FE}_{\mathsf{Q}}(q)$;
   (c) if $\mathcal{FE}_{\mathsf{P}}(p')$ then $p' =_{(\mathsf{P},\mathsf{Q})} q$; and
   (d) $(P',\ Q) \in \mathcal{S}$.
3. For any experiment $exp \in \mathcal{E}(\mathsf{P})$, whenever $\langle exp,\ \mathcal{Q}\rangle \rightarrow^* Q'\not\rightarrow$ then, for some P',
   (a) $\langle exp,\ \mathcal{P}\rangle \rightarrow^* P'\not\rightarrow$;

(b) $(P', Q') \in \mathcal{S}$; and

(c) $(\langle p, \mathcal{P}' \rangle, \langle q, \mathcal{Q}' \rangle) \in \mathcal{S}$.

□

Those conditions reflect the ideas that we have introduced about simulation of states. For example, Item 1 above says that any state immediately reached from Q is related to some state that might eventually be reached from P.

Condition 2 indicates that if the expression in Q cannot be further evaluated then the expression in P might eventually reach the same situation; when this happens, the resulting states will be related by $\mathcal{S}$ and the results yielded by the evaluation of the expression in Q will also be yielded by the evaluation of the expression in P. In other words, the results of the evaluation of expressions in Q might eventually be observed from P.

Lastly, Item 3 says that performing the same experiment with Q and P leads to states related by $\mathcal{S}$; by condition 2, this implies that the experiments yield the same results when performed in both states. Furthermore, if an experiment cannot be performed with Q (its corresponding operation is not enabled, because it cannot be executed in a particular state), then performing the same experiment with P results in a state related to Q by $\mathcal{S}$; this usually means that the experiment cannot be performed with P as well.

Also, it may seem that the formal definition does not require visible attributes to have the same value in related states. However, this is indirectly done by condition 3, since we consider that attributes are used to form experiments.

General Assumptions

Our notion of simulation is based on some assumptions which have not been made explicit so far. Basically, we assume that operations used as experiments are atomic and terminating; that's why the formal definition of simulation does not consider the intermediate states reached during the evaluation of experiments, and assumes that the evaluation always terminates (see condition 3 of the definition of simulation). However, note that the evaluation of experiments might take more than one transition; the extra transitions correspond to the replacement of methods for their definition, as long as the transitions do not change the database state nor depend on it.

Without those assumptions, it is not possible to analyse the effect of experiments in isolation; it would be necessary to consider the effect of experiments consisting of a group of operations being concurrently executed, including an arbitrary number of instances of the same operations. Indeed, it would be very difficult to reason about that.

Furthermore, note that the assumptions introduced above do not imply that all available methods have to be atomic and terminating; indeed, nonatomic and nonterminating methods can be defined as hidden operations. As discussed in [Borba, 1995], visible methods can initiate nonterminating and nonatomic computations by invoking hidden methods as expressions to be evaluated in the background. As usual, nonatomic computations are subjected to interference from the evaluation of other expressions being concurrently evaluated.

Those assumptions suggest a programming methodology: visible methods should be atomic and terminating (in general, it is simple to check that). In fact, it might be very difficult to reason about and predict the behaviour of programs allowing the parallel execution of many nonatomic methods (it is necessary to reason about all possible interleaved interferences caused by those methods). So we suggest that nonatomic methods should only be used in a controlled way, being invoked via visible atomic operations, as discussed above.

Properties

Simulations have interesting properties; some of them we will describe here. Those properties are particularly useful when using simulations to reason about refinement, as we will confirm later. For instance, the next theorem reveals a special (and obvious) simulation.

Theorem 11.2.1 For a specification P, the identity relation on $\mathit{Conf}(\mathsf{P})$, denoted $\mathcal{ID}_{\mathsf{P}}$, is a (P, P)-simulation.

Proof: It easily follows from the reflexivity of $\subseteq$, $\leq_{PS(\mathsf{P})}$, $\mathcal{ID}_{\mathsf{P}}$, and $=_{(\mathsf{P},\mathsf{P})}$, and by observing that

$$Q \rightarrow Q' \text{ implies } Q \rightarrow^* Q'$$

and

$$Q\not\rightarrow \text{ implies } Q \rightarrow^* Q\not\rightarrow.$$

□

As indicated by the following theorem, union of relations preserve simulations.

Theorem 11.2.2 Let P and Q be specifications. For (P, Q)-simulations $\mathcal{S}$ and $\mathcal{S}'$, $\mathcal{S} \cup \mathcal{S}'$ is a (P, Q)-simulation.

Proof: Let us assume that

$$(P,\ Q) \in \mathcal{S} \cup \mathcal{S}'.$$

Therefore

$$(P,\ Q) \in \mathcal{S} \text{ or } (P,\ Q) \in \mathcal{S}'.$$

So we split the proof in two cases. First assume that $(P,\ Q) \in \mathcal{S}$. Now let $Q \rightarrow Q'$. As $(P,\ Q) \in \mathcal{S}$, for some P', we have

$$P \rightarrow^* P' \text{ and } (P',\ Q') \in \mathcal{S}.$$

This trivially implies that $(P',\ Q') \in \mathcal{S} \cup \mathcal{S}'$.

Now let $Q \not\rightarrow$. As $(P, Q) \in \mathcal{S}$, for some P', we have

$$P \rightarrow^* P' \not\rightarrow, \qquad \mathcal{FE}_{\mathsf{P}}(p') \Leftrightarrow \mathcal{FE}_{\mathsf{Q}}(q),$$
$$\mathcal{FE}_{\mathsf{P}}(p') \Rightarrow p' =_{(\mathsf{P},\mathsf{Q})} q, \qquad \text{and } (P', Q) \in \mathcal{S}.$$

Hence $(P', Q) \in \mathcal{S} \cup \mathcal{S}'$.

Finally let $\langle exp, \mathcal{Q}\rangle \rightarrow^* Q' \not\rightarrow$, for any $exp \in \mathcal{E}(\mathsf{P})$. As $(P, Q) \in \mathcal{S}$, for some P', we have

$$\langle exp, \mathcal{P}\rangle \rightarrow^* P' \not\rightarrow,$$
$$(P', Q') \in \mathcal{S}, \text{ and}$$
$$(\langle p, \mathcal{P}'\rangle, \langle q, \mathcal{Q}'\rangle) \in \mathcal{S}.$$

Therefore $(P', Q') \in \mathcal{S} \cup \mathcal{S}'$ and $(\langle p, \mathcal{P}'\rangle, \langle q, \mathcal{Q}'\rangle) \in \mathcal{S} \cup \mathcal{S}'$.

The proof follows in a similar way for the second case: $(P, Q) \in \mathcal{S}'$. □

Relational composition also preserves simulations. This result is established by the following theorem, where $\mathcal{S} ; \mathcal{S}'$, defined as

$$\{(P, Q) \mid \text{for some } O, (P, O) \in \mathcal{S} \text{ and } (O, Q) \in \mathcal{S}'\},$$

denotes the composition of $\mathcal{S}$ with $\mathcal{S}'$.

Theorem 11.2.3 Let O, P, and Q be specifications. For a (P, O)-simulation $\mathcal{S}$ and a (O, Q)-simulation $\mathcal{S}'$, we have that $\mathcal{S} ; \mathcal{S}'$ is a (P, Q)-simulation. □

The proof of this theorem is as easy as the proof of Theorem 11.2.2; it can be found in [Borba, 1995].

11.2.2 Refinement of States

Indeed, many relations are simulations. For example, the identity relation on *Conf*(P) is a (P, P)-simulation, as shown by Theorem 11.2.1. For fixed specifications satisfying the properties required by the definition of simulation, the least simulation is the empty relation, whereas the largest simulation is obtained as the union of all simulations. We take this relation as our definition of simulation (refinement) of states.

Definition 11.2.2 Let P and Q be specifications. A state $Q \in Conf(\mathsf{Q})$ **simulates** (or **refines**) a state $P \in Conf(\mathsf{P})$, denoted $P \sqsubseteq_{(\mathsf{P},\mathsf{Q})} Q$, if there is a (P, Q)-simulation containing the pair (P, Q). Formally,

$$\sqsubseteq_{(\mathsf{P},\mathsf{Q})} = \bigcup\{\mathcal{S} \mid \mathcal{S} \text{ is a (P, Q)-simulation}\}.$$

We usually write $P \sqsubseteq Q$ for $P \sqsubseteq_{(\mathsf{P},\mathsf{Q})} Q$ when P and Q are clear from the context. □

Observe that $\sqsubseteq_{(\mathsf{P},\mathsf{Q})}$ is the empty relation when there is no associated simulation. From this definition, we conclude that in order to prove that a state P is simulated by a state Q, it is enough to find a simulation containing the pair (P, Q). In fact, this is a very effective proof technique.

Now we present some basic properties of the refinement relation on states; those properties will be extensively used for reasoning about refinement. The first theorem confirms a claim made at the beginning of this Section.

Theorem 11.2.4 Let P and Q be specifications such that $\mathcal{E}(\mathsf{P}) \subseteq \mathcal{E}(\mathsf{Q})$, $I(\mathsf{P}) \subseteq I(\mathsf{Q})$, and $\mathsf{P} \leq_{PS(\mathsf{P})} \mathsf{Q}$. The refinement relation on states $\sqsubseteq_{(\mathsf{P},\mathsf{Q})}$ is the largest (P, Q)-simulation.

Proof: By Theorem 11.2.2, $\sqsubseteq_{(\mathsf{P},\mathsf{Q})}$ is a (P, Q)-simulation and includes any other (P, Q)-simulation. □

The refinement relation on states has some reflexive and transitive properties, as indicated by the following theorem.

Theorem 11.2.5 For specifications O, P, and Q,

1. $P \sqsubseteq_{(\mathsf{P},\mathsf{P})} P$; and
2. $P \sqsubseteq_{(\mathsf{P},\mathsf{O})} O$ and $O \sqsubseteq_{(\mathsf{O},\mathsf{Q})} Q$ imply $P \sqsubseteq_{(\mathsf{P},\mathsf{Q})} Q$.

Proof: By Theorem 11.2.1, $\mathcal{ID}_\mathsf{P}$ is a (P, P)-simulation and $(P, P) \in \mathcal{ID}_\mathsf{P}$. This proves 1. Now we prove 2. Let us assume that

$$P \sqsubseteq_{(\mathsf{P},\mathsf{O})} O \text{ and } O \sqsubseteq_{(\mathsf{O},\mathsf{Q})} Q.$$

Hence there is a (P, O)-simulation $\mathcal{S}$ and a (O, Q)-simulation $\mathcal{S}'$ such that

$$(P, O) \in \mathcal{S} \text{ and } (O, Q) \in \mathcal{S}'.$$

Thus we have

$$(P, Q) \in \mathcal{S} ; \mathcal{S}'.$$

By Theorem 11.2.3, $\mathcal{S} ; \mathcal{S}'$ is a (P, Q)-simulation. □

11.2.3 Refinement of Expressions

Based on the notion of refinement of states, we can define refinement of (method) expressions. First observe that the FOOPS system described by a specification P has an associated **initial (database) state**, denoted $\emptyset_\mathsf{P}$, which contains only objects introduced by P and information about their respective attributes. States reached by performing an arbitrary (possibly empty) sequence of experiments with an initial state are called **reachable (database) states**.

For expressions p and q respectively related to specifications P and Q, we say that p is refined by q when the evaluation of p in any reachable state $\mathcal{P}$ of P is simulated by the evaluation of q in a state $\mathcal{Q}$ reached from $\emptyset_\mathsf{Q}$ by performing the same sequence of experiments used to reach $\mathcal{P}$. (By the definition of simulation, this already considers that the evaluation of the expressions might be interfered by experiments.) Hence, when P and Q are the same specification, p is refined by q if the evaluation of p in any reachable state is simulated by the evaluation of q in the same state.

By analysing the definition of simulation of states, it is easy to check that an alternative way of expressing the ideas above is saying that p is refined by q if the evaluation of q in the initial state of Q simulates the evaluation of p in the initial state of P. This is formalized below.

Definition 11.2.3 Let P and Q be specifications. An expression $q \in \mathcal{T}_{\mathsf{Q}}$ **simulates** (or **refines**) $p \in \mathcal{T}_{\mathsf{P}}$, denoted $p \sqsubseteq_{(\mathsf{P},\mathsf{Q})} q$, if

$$\langle p, \emptyset_{\mathsf{P}} \rangle \sqsubseteq_{(\mathsf{P},\mathsf{Q})} \langle q, \emptyset_{\mathsf{Q}} \rangle,$$

where $\mathcal{T}_{\mathsf{P}}$ denotes the family of terms associated to P. □

Reflecting the notion of refinement of states, the notion of refinement of expressions requires the results yielded by related expressions to be the same. Also, due to the direct relationship with refinement of states, the relation of refinement of expressions has reflexive and transitive properties. This is stated by the following theorem.

Theorem 11.2.6 For specifications O, P, and Q,

1. $p \sqsubseteq_{(\mathsf{P},\mathsf{P})} p$; and
2. $p \sqsubseteq_{(\mathsf{P},\mathsf{O})} o$ and $o \sqsubseteq_{(\mathsf{O},\mathsf{Q})} q$ imply $p \sqsubseteq_{(\mathsf{P},\mathsf{Q})} q$.

□

The proof of this theorem is straightforward; it can be found in [Borba, 1995].

11.2.4 Refinement of Specifications

Based on the notion of refinement of expressions, we can introduce a simple definition of refinement of specifications. Basically, a specification P is refined by a specification Q if any experiment of P is refined by its corresponding experiment of Q. This is formalized below.

Definition 11.2.4 A specification P is **refined** by a specification Q, denoted $\mathsf{P} \sqsubseteq \mathsf{Q}$, if

$$exp \sqsubseteq_{(\mathsf{P},\mathsf{Q})} exp,$$

for any $exp \in \mathcal{E}(\mathsf{P})$. □

Note that in order to prove $\mathsf{P} \sqsubseteq \mathsf{Q}$ we need only to prove $\emptyset_{\mathsf{P}} \sqsubseteq_{(\mathsf{P},\mathsf{Q})} \emptyset_{\mathsf{Q}}$.

Similarly to refinement of states and expressions, refinement of specifications has reflexive and transitive properties. This is stated by the following theorem.

Theorem 11.2.7 For specifications O, P, and Q,

1. $\mathsf{P} \sqsubseteq \mathsf{P}$; and

2. $\mathsf{P} \sqsubseteq \mathsf{O}$ and $\mathsf{O} \sqsubseteq \mathsf{Q}$ imply $\mathsf{P} \sqsubseteq \mathsf{Q}$.

□

We omit the proof of this theorem since it is trivial; the details can be found in [Borba, 1995].

11.3 Proving Refinement

In order to explore our notion of refinement, we introduce some FOOPS specifications giving examples of simulation relations and refinement, together with the necessary proofs.

11.3.1 Memory Cells

Let us consider a specification defining a class of memory cells for storing natural numbers:

```
omod CELL is
   pr NAT .
   class Cell .
   at val_ : Cell -> Nat .
   me _:=_ : Cell Nat -> Cell .
   me _:=_+_ : Cell Cell Nat -> Cell .
   vars C C´ : Cell .
   var N : Nat .
   ax val(C := N) = N .
   ax val(C := C´ + N) = val(C´) + N .
endomod
```

Here objects of `Cell` have the attribute `val` (for *val*ue), which indicates the natural number stored in a cell, and two methods for assigning values to a cell. The method _:=_ assigns a given natural number to a cell, whereas _:=_+_ assigns to a cell the value stored in another cell plus a natural number (the cells and the natural number are given as arguments). Both methods are atomic, since they are defined by DMAs.

In order to keep it simple, we first study the simulation of states related to the same specification. Based on the specification `CELL`, let's examine the relationship between the expressions

```
x := 0 ; x := x + 1
```
(11.1)

and

```
x := 1,
```
(11.2)

where `x` is the identifier of an object of `Cell`.

Note that the expressions above may be seen as programs of an imperative programming language. Indeed, by providing a class of memory cells, we can use our notion of simulation for reasoning about (concurrent) imperative programs as well; they are just a particular case of the object oriented programs that we can write using FOOPS.

Both expressions have the effect of assigning `1` to `x`, independently of the state of the system (as long as `x` exists). In a sequential environment, this is enough to guarantee that they are equivalent; we can just consider the input/output behaviour of programs, since the evaluation of expressions cannot be interfered with.

However, interference usually occurs in a concurrent environment; so in order to assure equivalence, we also have to consider if expressions behave in an equivalent way when subjected to the same interference (that's why our notion of refinement considers that experiments can be made with the intermediate states reached during the evaluation of an expression). Indeed, the expressions (11.1) and (11.2) are not equivalent in a concurrent environment: just consider that the evaluation of (11.1) might be interfered—after the execution of `x := 0` and before `x := x + 1`—whereas (11.2) is atomic. For example, if the expressions (11.1) and (11.2) are interfered with by `x := 1`, then they will behave differently. This can be justified by checking that, independently of the system state, the expression

```
x := 1 || x := 1
```

is deterministic: it assigns `1` to `x`. On the other hand,

```
x := 1 || x := 0 ; x := x + 1
```

is nondeterministic: it can either assign `1` to `x`, or assign `2` to `x`, depending on whether `x := 0` is evaluated after or before `x := 1`.

The discussion above also tells us that (11.1) is not a refinement of (11.2), since the behaviour of the first is not a particular case of the behaviour of the second when both are subjected to the same interference. However, we cannot conclude that (11.2) is not a refinement of (11.1). Let us investigate that here, by checking whether

$$\langle \texttt{x := 0 ; x := x + 1}, \mathcal{P}\rangle \tag{11.3}$$

is refined by

$$\langle \texttt{x := 1}, \mathcal{P}\rangle, \tag{11.4}$$

for any reachable database state $\mathcal{P}$.

As discussed in Section 11.2.2, this can be verified by finding a (`CELL`, `CELL`)-simulation relating the configurations (states) described above. As proved by Theorem 11.2.1, such a simulation may exist because the restrictions on related specifications are trivially valid, since the specifications are the same.

A Simulation

Let us now check whether the relation $\mathcal{S}$, consisting of pairs in the forms

$$(\langle \texttt{x := 0 ; x := x + 1}, \mathcal{P}\rangle \quad , \quad \langle \texttt{x := 1}, \mathcal{P}\rangle) \tag{11.5}$$
$$(\langle e, \mathcal{P}\rangle \quad , \quad \langle e, \mathcal{P}\rangle), \tag{11.6}$$

for any reachable database state $\mathcal{P}$ and any expression e, is a simulation. Pairs of the first form should be in $\mathcal{S}$ because we are interested in proving that (11.3) is refined by (11.4), whereas pairs of the second form represent states reached from the states in the first pair, and states resulting from the execution of experiments.

In order to check the conditions in the definition of simulation, we must have some knowledge about the transition relation $\rightarrow$. Although we have not defined it here, the informal description of FOOPS concepts in Section 11.1 gives an informal overview of FOOPS operational semantics and justifies the state transitions discussed in this chapter.

By Definition 11.2.1, $\mathcal{S}$ is a simulation if its pairs satisfy some properties. Now we consider whether those properties are valid for pairs of $\mathcal{S}$ in the first form. First we assume that `x` is in $\mathcal{P}$. So the unique possible transition from configuration 11.4 is

$$\langle \texttt{x := 1}, \mathcal{P}\rangle \rightarrow \langle \texttt{x}, \mathcal{P} \oplus \texttt{val x = 1}\rangle, \tag{11.7}$$

where $\mathcal{P} \oplus$ `val x = 1` denotes the database state derived from $\mathcal{P}$ by updating the attribute `val x` with the value `1`; that's the expected effect of an assignment operation.

From configuration 11.3, the unique possible sequence of transitions is

$$\begin{array}{l} \langle \texttt{x := 0 ; x := x + 1}, \mathcal{P}\rangle \\ \quad \rightarrow \langle \texttt{x ; x := x + 1}, \mathcal{P} \oplus \texttt{val x = 0}\rangle \\ \quad \rightarrow \langle \texttt{x := x + 1}, \mathcal{P} \oplus \texttt{val x = 0}\rangle \\ \quad \rightarrow \langle \texttt{x}, \mathcal{P} \oplus \texttt{val x = 0} \oplus \texttt{val x = 1}\rangle. \end{array}$$

Note that `x := x + 1` is atomically executed (since the behaviour of `_:=_+_` is specified by a DMA), and the second transition is necessary for eliminating the result yielded by `x := 0`. Also, observe that

$$\mathcal{P} \oplus \texttt{val x = 0} \oplus \texttt{val x = 1}$$

is equal to

$$\mathcal{P} \oplus \texttt{val x = 1},$$

since the update specified by

`val x = 0`

is overwritten by the update indicated by `val x = 1`. From this result and the sequence of transitions introduced above, we can say that

$$\langle \texttt{x := 0 ; x := x + 1}, \mathcal{P}\rangle \rightarrow^* \langle \texttt{x}, \mathcal{P} \oplus \texttt{val x = 1}\rangle. \qquad (11.8)$$

After analysing the possible transitions from states, it is straightforward to verify that condition 1 (of Definition 11.2.1) is valid: just consider the unique transitions (11.7) and (11.8), and that the pair

$$(\langle \texttt{x}, \mathcal{P} \oplus \texttt{val x = 1}\rangle, \langle \texttt{x}, \mathcal{P} \oplus \texttt{val x = 1}\rangle)$$

is a particular instance of the pairs of $\mathcal{S}$ in the second form.

Condition 2 is valid for pairs of $\mathcal{S}$ in the first form, since

$$\langle \texttt{x := 1}, \mathcal{P}\rangle \not\rightarrow$$

does not hold, as shown by transition (11.7).

In general, it is trivial to check condition 3 when the database states in the related configurations are the same and the relation in investigation contains the identity relation; that's certainly the case of $\mathcal{S}$. In fact, we do not even have to consider what are the experiments associated to `CELL`: let's assume an arbitrary experiment *exp*. So if

$$\langle exp, \mathcal{P}\rangle \rightarrow^* Q' \not\rightarrow$$

for some state Q', then condition 3a holds by taking P' as Q', and considering the transition above; that's possible because only one specification is being considered, so we are also dealing with only one transition relation: $\rightarrow_{\texttt{CELL}}$. Also, condition 3b is valid, since (Q', Q') is clearly in $\mathcal{S}$ ($\mathcal{S}$ contains the identity). Lastly, we can verify that condition 3c is valid by checking that

$$(\langle \texttt{x := 0 ; x := x + 1}, \mathcal{Q}'\rangle, \langle \texttt{x := 1}, \mathcal{Q}'\rangle)$$

is a particular instance of the pairs of $\mathcal{S}$ in the first form.

This concludes the proof that, when `x` is in $\mathcal{P}$, the pairs of $\mathcal{S}$ in the first form satisfy the properties of pairs in a simulation relation. Now let's assume that `x` is not in $\mathcal{P}$. Condition 1 becomes vacuously true since

$$\langle \texttt{x := 1}, \mathcal{P}\rangle \not\rightarrow \qquad (11.9)$$

because `x` is not in $\mathcal{P}$ (an attribute or a method specified by DMAs can only be executed if its associated object exists). For the same reason,

$$\langle \texttt{x := 0 ; x := x + 1}, \mathcal{P}\rangle \not\rightarrow . \qquad (11.10)$$

The sequential composition cannot be executed because the first expression cannot be executed.

Based on (11.9) and (11.10), we can verify condition 2a, by observing that

$$\langle \texttt{x := 0 ; x := x + 1}, \mathcal{P}\rangle \rightarrow^* \langle \texttt{x := 0 ; x := x + 1}, \mathcal{P}\rangle \not\rightarrow$$

since $\rightarrow^*$ is reflexive. As the expressions in the states of (11.9) and (11.10) are not fully evaluated, conditions 2b and 2c are valid. Lastly, condition 2d is satisfied because the resulting pairs are instances of the pairs of $\mathcal{S}$ in the first form.

It remains only to verify condition 3. However, this follows exactly in the same way as when `x` is in $\mathcal{P}$.

Verifying that pairs of $\mathcal{S}$ in the second form satisfy the conditions in Definition 11.2.1 is trivial; in fact, this is a particular case of what was done to prove Theorem 11.2.1.

Finally, we can concluded that $\mathcal{S}$ is a simulation. From this and Definition 11.2.2, it follows that

$$\langle \texttt{x := 0 ; x := x + 1}, \mathcal{P}\rangle \sqsubseteq \langle \texttt{x := 1}, \mathcal{P}\rangle,$$

for any reachable database state $\mathcal{P}$. In fact, this is enough to guarantee that

```
x := 0 ; x := x + 1
```

is refined by

```
x := 1.
```

Failing to be a Simulation

In the beginning of this section we claimed that

```
x := 1,
```

is not refined by

```
x := 0 ; x := x + 1;
```

We have informally shown that the behaviour of the last is not in general a particular case of the behaviour of the former when both are subjected to the same interference. For further exploring our approach, we investigate what our definition of simulation says about that.

Similarly to what has been done before, let's check if

$$\langle \texttt{x := 1}, \mathcal{P}\rangle \tag{11.11}$$

is refined by

$$\langle \texttt{x := 0 ; x := x + 1}, \mathcal{P}\rangle, \tag{11.12}$$

for any reachable database state $\mathcal{P}$. Basically, we should find a (`CELL`, `CELL`)-simulation containing pairs in the form

$$(\langle \texttt{x := 1}, \mathcal{P}\rangle, \langle \texttt{x := 0 ; x := x + 1}, \mathcal{P}\rangle),$$

for any reachable database state $\mathcal{P}$. By condition 1 from the definition of simulation, and observing that

$$\langle \texttt{x := 0 ; x := x + 1}, \mathcal{P}\rangle \rightarrow \langle \texttt{x ; x := x + 1}, \mathcal{P} \oplus \texttt{val x = 0}\rangle,$$

a simulation containing the pair above must relate the state

$$\langle \texttt{x ; x := x + 1}, \mathcal{P} \oplus \texttt{val x = 0} \rangle \tag{11.13}$$

to some $\rightarrow^*$-derivative of

$$\langle \texttt{x := 1}, \mathcal{P} \rangle.$$

The only $\rightarrow^*$-derivatives of this state are itself and

$$\langle \texttt{x}, \mathcal{P} \oplus \texttt{val x = 1} \rangle \tag{11.14}$$

(see transition (11.7)).

The first $\rightarrow^*$-derivative cannot be related to state (11.13) by any simulation: just consider performing the experiment `val(x)` with both states. The experiment performed with (11.13) yields the value `0`, whereas the experiment performed with $\langle$`x := 1`, $\mathcal{P}\rangle$ does not necessarily yield the same value (only if the value associated to `x` in $\mathcal{P}$ is `0`), and might even be suspended (if `val x` is not defined in $\mathcal{P}$). Hence, condition 3 in the definition of simulation would not be valid.

Also, trying to relate (11.14) and (11.13) does not work, for a similar reason as discussed above. Again, this can be checked by trying the experiment `val(x)` with both states. Essentially, the problem is with the intermediate state (11.13) reached from (11.12), since it is not a simulation of any state reached from (11.11); this is why the behaviour of (11.12) is not a particular case of the behaviour of (11.11). As there is no simulation relating the pairs of states analysed here, we confirm the claim made at the beginning of the section.

Further Examples of Simulation

Following the same line as discussed in the previous sections, we can prove that

$$\texttt{x := x + 1 ; x := x - 1} \tag{11.15}$$

is refined by

$$\texttt{skip(x)}, \tag{11.16}$$

where

```
me skip : Cell -> Cell .
ax skip(C) = C .
```

(note that the method **skip** does not change the state of objects). In fact, a simulation relation very similar to $\mathcal{S}$ can be chosen as the basis for this proof. As in Section 11.3.1, we can also verify that (11.16) is not refined by (11.15).

Perhaps more interesting is verifying that

```
skip(x) ; x := 0
```

is refined by

```
x := 0,
```

and vice versa. We just have to check that the relations denoted by

$$(\langle \texttt{skip(x) ; x} := 0, \mathcal{P}\rangle \quad , \quad \langle \texttt{x := 0}, \mathcal{P}\rangle) \tag{11.17}$$
$$(\langle e, \mathcal{P}\rangle \quad , \quad \langle e, \mathcal{P}\rangle), \tag{11.18}$$

and

$$(\langle \texttt{x := 0}, \mathcal{P}\rangle \quad , \quad \langle \texttt{skip(x) ; x := 0}, \mathcal{P}\rangle) \tag{11.19}$$
$$(\langle \texttt{x := 0}, \mathcal{P}\rangle \quad , \quad \langle \texttt{x ; x := 0}, \mathcal{P}\rangle) \tag{11.20}$$
$$(\langle e, \mathcal{P}\rangle \quad , \quad \langle e, \mathcal{P}\rangle),$$

for any reachable database state $\mathcal{P}$ and any expression e, are (**CELL**, **CELL**)-simulations. We omit the details here.

Also, we can prove that the nondeterministic choice

```
x := 0 [] x := 1
```

is refined by the assignment

```
x := 1,
```

by proving that the relation formed by pairs in the form

$$(\langle \texttt{x := 0 [] x := 1}, \mathcal{P}\rangle \quad , \quad \langle \texttt{x := 1}, \mathcal{P}\rangle) \tag{11.21}$$
$$(\langle e, \mathcal{P}\rangle \quad , \quad \langle e, \mathcal{P}\rangle), \tag{11.22}$$

for any reachable database state $\mathcal{P}$ and any expression e, is a simulation. It is also easy to check that the nondeterministic choice is not a refinement of the assignment.

11.3.2 Buffers

Here we exemplify the use of the simulation technique for proving refinement of states related to different specifications. We consider the specification (implementation) of buffers represented by lists, introduced in Section 11.1.2, and a simplified specification of buffers represented by array and pointers.

First, we introduce the following functional level module defining arrays of elements of an arbitrary sort having a constant (denoted by `err`, as specified by `TRIVE`, which is an extension of `TRIV` with the constant `err`) that is used as the value yielded by the array indexing operation when the array is not initialized for a particular index. We assume that arrays are indexed by natural numbers.

```
fmod ARRAY[E :: TRIVE] is
  pr NAT .
  sort Array .
  fn null : -> Array .
  fn _[_] := _ : Array Nat Elt -> Array .
  fn _[_] : Array Nat -> Elt? .
```

```
    var A : Array .
    vars I J : Nat .
    var E : Elt .
    ax null[I] = err .
    ax (A[I] := E)[I] = E .
    cx (A[I] := E)[J] = A[J] if I =/= J .
  endfmod
```

The empty array is represented by `null`; the expression `a[i] := e` denotes an array identical to `a` except that `e` is stored in position `i`; and `a[i]` is the element of `a` stored in position `i` (if there is no element in this position, `a[i]` yields `err`).

The implementation of buffers using arrays is given by the object level module `BUFFERA`; similarly to `ARRAY`, the sort of the elements to be stored in buffers must have a constant:

```
  omod BUFFERA[E :: TRIVE, M :: MAX] is
    pr ARRAY[E] .
    class Buffer .
    at array_ : Buffer -> Array [hidden] .
    at fst_ : Buffer -> Nat [hidden] .
    at next_ : Buffer -> Nat [hidden] .
    at empty?_ : Buffer -> Bool .
    at full?_ : Buffer -> Bool .
```

This module introduces the same visible attributes as `BUFFER`, but has different hidden attributes: `array`, which corresponds to the array containing the elements of a buffer; and the pointers `fst` and `next`, which contain indexes of the array respectively indicating where the first element of the buffer is stored, and where the next element will be inserted. This module also has the same methods as `BUFFER`:

```
    me reset : Buffer -> Buffer .
    me get : Buffer -> Elt .
    me put : Buffer Elt -> Buffer .
    me del : Buffer -> Buffer [hidden] .
```

As the elements of the array between the two indexes (including `fst`, but not `next`) are supposed to be the elements of the buffer, we can say that the buffer is empty when the pointers point at the same position; it is full when the difference between the pointers `next` and `fst` is `max`:

```
    ⋮
    ax empty? B = (fst B) == (next B) .
    ax full? B = (next B) - (fst B) == max .
```

where - denotes the operation of subtraction on natural numbers. The operation `reset` is implemented by setting both pointers to `0`, and the array to `null`:

```
    ax next(reset(B)) = 0 .
    ax fst(reset(B)) = 0 .
    ax array(reset(B)) = null .
```

Deleting the first element of a buffer is implemented by incrementing the **fst** pointer by one; the operation **get** is specified as in **BUFFER**, considering that the first element in a buffer **B** is denoted by **(array B)[fst B]**. Note that **del** and **get** can only be performed if the buffer is not empty:

```
    ax fst(del(B)) = (fst B) + 1 if not(empty? B) .
    acx get(B) = result (array B)[fst B] ; del(B)
    if not(empty? B) .
```

Lastly, inserting an element in a buffer involves storing it in the next available position of the array, and increasing the **next** pointer by one:

```
    cx next(put(B,E)) = (next B) + 1 if not(full? B) .
    cx array(put(B,E)) = (array B)[next B] := E
    if not(full? B) .
  endomod
```

However, this can only be done if the buffer is not full. Note that all visible operations of **BUFFER** and **BUFFERA** are atomic. This enable us to use our refinement technique.

Refinement of Parameterized Modules

In order to further explore our approach, we should now try to prove that the operations specified in **BUFFERA** simulate the corresponding operations specified in **BUFFER**. As we have seen in previous sections, this can be done by verifying simulation of states. However, our notion of simulation assumes that states are associated to specifications (unparameterized FOOPS modules), whereas **BUFFER** and **BUFFERA** are parameterized specifications (parameterized FOOPS modules). Using our notion, we can only verify simulation of states associated to particular instantiations of parameterized modules.

It would be desirable if we could check whether states associated to an arbitrary instantiation of **BUFFER** with arguments satisfying **TRIVE** and **MAX** are simulated by states associated to the instantiation of **BUFFERA** with the same arguments. This is necessary for ensuring that if we instantiate both modules in the same way (as long as the arguments satisfy **TRIVE** and **MAX**) then the operations resulting from the instantiation of the first are simulated by the operations resulting from the instantiation of the second.

In fact, we can use our definition of simulation in order to achieve the general result mentioned above. We just have to consider the (unparameterized) specifications obtained from **BUFFER** and **BUFFERA** by assuming the minimum information about their instantiation. This information is given by the requirements theories **TRIVE** and **MAX**. Intuitively, if we can prove refinement for the "minimum instantiation" of the parameterized modules, we should be able to prove refinement for any other instantiation.

However, the "minimum instantiation" of the parameterized modules gives a minimum family $\mathcal{E}$ of experiments, since comparing with other instantiations, it has the least number of functions. Fortunately, we need only a small extension of our definition of simulation to consider the general form of (expressions denoting) experiments, rather than the experiments themselves. Contrasting with $\mathcal{E}$, the general form of experiments does not depend on a particular instantiation. So instead of analysing the effects of particular experiments (see condition 3 of Definition 11.2.1), we should analyse the effect of the invocation of visible operations without assuming any information about their arguments (except that they are identifiers or elements of ADTs). That's the approach that we are going to use now, in order to reason about the modules defining buffers. Hereafter we also use the names `BUFFER` and `BUFFERA` when we want to refer to the "minimum instantiation" of the corresponding parameterized modules.

First, we have to check whether the specifications of buffers can be related by simulations; according to Definition 11.2.1, $\mathcal{E}(\texttt{BUFFER})$ should be contained in $\mathcal{E}(\texttt{BUFFERA})$, and

$$\texttt{BUFFER} \leq_{PS(\texttt{BUFFER})} \texttt{BUFFERA} \tag{11.23}$$

should hold. The set of primary sorts of `BUFFER` is formed by `Elt` and `Bool`. In both specifications, the definition of the functions associated to the primary sorts is basically given by `TRIVE` and the standard definition of an ADT of booleans (included in any specification). Also, some extra functions are introduced: `BUFFER` introduces `head`, whereas `BUFFERA` introduces `_[_]`. However, the equations defining those functions indicate that they yield an existing element of `Elt`; they do not introduce any extra element to `Elt`. Hence, we can conclude that (11.23) is valid, since the difference between the two specifications does not affect their meaning; the denoted ADTs are the same.

Also note that both specifications introduce the same visible operations; this and (11.23) guarantees that the experiments we can make with states of `BUFFER` can also be made with states of `BUFFERA`, if we assume that both specifications have the same family of object identifiers (in practice, we can assume that an element of the family I is formed by names prefixed by the name of the class associated to this element).

From what discussed above, we can also conclude that the requirements on specifications are satisfied for any instantiation of the parameterized modules defining buffers, assuming that they are instantiated with the same arguments satisfying the minimum requirements. Just observe that

- all instantiations of a module have the same family of visible operations, if the arguments are functional modules (they do not define methods or attributes);
- the ADTs of a particular instantiation of `BUFFER` and `BUFFERA` are the same, since they are essentially defined by their arguments, which are the same as well; and
- we assume that all instantiations of a module have the same related set of object identifiers.

A Simulation

Now we can verify the simulation of states associated to the specifications of buffers that we have previously introduced. The definition of `BUFFERA` is based on our intuitions that a buffer can be represented by an array `a`, and indexes `i` and `j`, considering that the elements of the buffer are the elements of `a` from index `i` to index `j`, denoted by `i < a > j`, where

```
vars I J : Nat .
var A : Array .
fn _ < _ > _ : Nat Array Nat -> List .
ax I < A > J = if I >= J then nil
                else A[I] . (I + 1) < A > J fi .
```

So let's check whether such a buffer represented by array and pointers simulates a buffer represented by a list `l` whenever `i <= j` and `l` is equal to `i < a > j`, with respect to the functional theory `EB` obtained by extending `LIST` with `ARRAY` and the definition above.

Following Section 11.2.2, we can check if our intuitions are correct by trying to prove that a relation $\mathcal{S} \subseteq \mathit{Conf}(\texttt{BUFFER}) \times \mathit{Conf}(\texttt{BUFFERA})$ containing states in the following form is a simulation:

$$(\langle e, \mathcal{P}\rangle \quad , \quad \langle e, \mathcal{Q}\rangle), \tag{11.24}$$

where e is either a fully evaluated expression of a primary sort of `BUFFER` or an experiment or object identifier of `BUFFER`, and $\mathcal{P}$ and $\mathcal{Q}$ are reachable database states, respectively associated to `BUFFER` and `BUFFERA`, such that

- the objects in $\mathcal{P}$ and $\mathcal{Q}$ are the same;
- either `array b`, `fst b`, and `next b` are defined in $\mathcal{Q}$, or none is defined; also, the attribute `elems b` has a defined value in $\mathcal{P}$ iff the attributes `array b`, `fst b`, and `next b` have defined values in $\mathcal{Q}$;
- the condition `fst b <= next b` is satisfied in $\mathcal{Q}$; and
- the value denoted by `elems b` in $\mathcal{P}$ is equal, with respect to `EB`, to the value denoted by

  ```
  (fst b) < (array b) > (next b)
  ```

  in $\mathcal{Q}$,

for any buffer `b` in $\mathcal{P}$.

Before verifying that $\mathcal{S}$ is a simulation, note that the experiments associated to `BUFFER` have one of the following general forms: `empty?(b)`, `full?(b)`, `reset(b)`, `put(b,e)`, `get(b)`, `new.Buffer()`, `new(b)`, and `remove(b)`, for an arbitrary buffer `b` and an element `e` of sort `Elt`.

Let us now check whether the pairs of $\mathcal{S}$ satisfy the conditions in Definition 11.2.1. We first assume that e has the general form `reset(b´)`, for an arbitrary buffer `b´`, and that `b´` is in $\mathcal{P}$ (this implies that `b´` is also in $\mathcal{Q}$, since both states are related by $\mathcal{S}$). Thus, the unique transition from

$$\langle \texttt{reset(b´)}, \mathcal{Q}\rangle \tag{11.25}$$

leads to

$$\langle \texttt{b´}, \mathcal{Q} \oplus \texttt{fst b´ = 0} \oplus \texttt{next b´ = 0} \oplus \texttt{array b´ = null}\rangle, \tag{11.26}$$

whereas the unique transition from

$$\langle \texttt{reset(b´)}, \mathcal{P}\rangle \tag{11.27}$$

leads to

$$\langle \texttt{b´}, \mathcal{P} \oplus \texttt{elems b´ = nil}\rangle. \tag{11.28}$$

This clearly implies that

$$\langle \texttt{reset(b´)}, \mathcal{P}\rangle \rightarrow^* \langle \texttt{b´}, \mathcal{P} \oplus \texttt{elems b´ = nil}\rangle.$$

So in order to check condition 1, we just have to prove that the states (11.28) and (11.26) are related by $\mathcal{S}$. First, note that those resulting states have the same fully evaluated expression (`b´`), as required by $\mathcal{S}$. Second, let $\mathcal{P}'$ and $\mathcal{Q}'$ be the (reachable) database states respectively in (11.28) and (11.26). Thus observe the following:

- the objects in $\mathcal{P}'$ and $\mathcal{Q}'$ are the same, since $\mathcal{P}$ and $\mathcal{Q}$ have the same objects, and the execution of `reset(b´)` does not create any objects;
- the attribute `elems b´` has a defined value in $\mathcal{P}'$, and `array b´`, `fst b´`, and `next b´` also have defined values in $\mathcal{Q}'$;
- in $\mathcal{Q}'$, the condition `(fst b´) <= (next b´)` is equal to `0 <= 0`, which is trivially valid; and
- the value of `elems b´` in $\mathcal{P}'$ is `nil`, whereas the value of

  `(fst b´) < (array b´) > (next b´)`

  in $\mathcal{Q}'$ is `0 < null > 0`, which is the same as `nil`, by the definition of `_ < _ > _`.

Instead of considering the attributes of all objects in $\mathcal{P}'$ and $\mathcal{Q}'$, we consider only properties of the stored attributes of `b´`, since those are the only attributes that change due to the execution of `reset(b´)`; we can conclude that the corresponding properties for the other attributes are valid from what is assumed about $\mathcal{P}$ and $\mathcal{Q}$ (they are related by $\mathcal{S}$).

Condition 2 is vacuously valid, since there is a transition from state (11.25).

Now it remains to check condition 3. Considering experiments in the form `empty?(b)`, we observe that the unique transition from

$$\langle \texttt{empty?(b)}, \mathcal{Q}\rangle \tag{11.29}$$

leads to

$$\langle [\![\texttt{i == j}]\!]_{\texttt{BA}}, \mathcal{Q} \rangle,$$

where `BA` is an abbreviation for FE(`BUFFERA`) (the set of functional equations of `BUFFERA`, and `i` and `j` are respectively the values of `fst b` and `next b` in $\mathcal{Q}$ (this implies that `array b` has value `a` in $\mathcal{Q}$, for some array `a`). If those attributes are not defined in $\mathcal{Q}$, no transition is possible from (11.29). Similarly, no transition is possible from

$$\langle \texttt{empty?(b)}, \mathcal{P} \rangle, \tag{11.30}$$

if `elems b` is not defined in $\mathcal{P}$. However, if this is not the case, there is a unique transition from (11.30); it leads to

$$\langle [\![\texttt{l == nil}]\!]_{\texttt{B}}, \mathcal{P} \rangle,$$

where `B` is an abbreviation for FE(`BUFFER`) and `l` is the value of `elems b` in $\mathcal{P}$.

As $\mathcal{P}$ and $\mathcal{Q}$ are related by $\mathcal{S}$, we can deduce that `l` is equal, with respect to `EB`, to `i < a > j`. So, using `EB`, by the definition of `_ < _ > _`, we can prove that `l == nil` is equal to `i == j`. It is also easy to see that `EB` is a conservative (protecting) extension of the functional theory of `LIST`. Hence we can assume that $[\![e]\!]_{\texttt{B}}$ is the same as $[\![e]\!]_{\texttt{EB}}$ for any e of `BUFFER`. Therefore we have that $[\![\texttt{l == nil}]\!]_{\texttt{B}}$ is the same as $[\![\texttt{i == j}]\!]_{\texttt{B}}$. It is then easy to check that the resulting states are related by $\mathcal{S}$, since `i == j` is a term of `Bool`, a primary sort of `BUFFER`. We have omitted the details about the functional level reasoning above because here we concentrate on the object level of FOOPS (see [Goguen, 1994] for the details and justification).

Also, as $\mathcal{P}$ and $\mathcal{Q}$ are related by $\mathcal{S}$, the attributes of `b` are defined either in both states or in none; hence, either both transitions are possible or none is possible. Thus, it follows that condition 3 is valid considering experiments in the form `empty?(b)`.

Similarly, we can check that condition 3 is valid for experiments in the form `full?(b)`. Also, the proof that the condition is valid for experiments in the form `reset(b)` follows the same lines of our proof that pairs of $\mathcal{S}$ having `reset(b´)` as e satisfy condition 1. We omit the details here.

For experiments in the form `put(b,e)`, let's first assume that the stored attributes of `b` are defined in $\mathcal{Q}$ (this implies that `elems b` is defined as `l` in $\mathcal{P}$, for some list `l`), and that the value of `full? b` in $\mathcal{Q}$ is `false`. So by the definition of `full?`, `j - i == max` is equal to `false`, where `i` and `j` are respectively the values of `fst b` and `next b` in $\mathcal{Q}$. Hence, as `l` is equal to `i < a > j`, we can infer that `# l == max` is equal (with respect to `EB`) to `false`; that is, the value of `full? b` in $\mathcal{P}$ is `false` as well.

In this case, the unique transitions caused by `put(b,e)` are

$$\langle \texttt{put(b,e)}, \mathcal{Q} \rangle \rightarrow$$
$$\langle \texttt{b}, \mathcal{Q} \oplus \texttt{next b = j + 1} \oplus \texttt{array b = a[j] := e} \rangle, \tag{11.31}$$

where `a` is the value of `array b` in $\mathcal{Q}$, and

$$\langle \texttt{put(b,e)}, \mathcal{P}\rangle \rightarrow \langle \texttt{b}, \mathcal{P} \oplus \texttt{elems b = l . e}\rangle. \qquad (11.32)$$

Observe that these resulting states are related by $\mathcal{S}$, because the resulting expressions are the same fully evaluated term; the database states $\mathcal{P}$ and $\mathcal{Q}$ are related by $\mathcal{S}$; no new objects were created; the updated attributes were already defined before the transitions; the condition `i <= j + 1` trivially follows from `i <= j`; and we can prove that `l . e` is equal, with respect to `EB`, to

```
i < a[j] := e > j + 1,
```

since `l` is equal to `i < a > j`.

Now we verify that condition 3 also follows when the stored attributes of `b` are not defined in $\mathcal{Q}$, or when the value of `full? b` in $\mathcal{Q}$ is `true`. In those cases, we can prove that either `elems b` is not defined in $\mathcal{P}$, or that and the value of `full? b` in $\mathcal{P}$ is `true`. Hence, no transition is possible; the experiments cannot be performed with $\mathcal{P}$ and $\mathcal{Q}$. This implies condition 3.

As proved when considering experiments in the form `empty? b`, the value of `empty? b` in $\mathcal{Q}$ is the same as its value in $\mathcal{P}$. So when analysing the effect of experiments in the form `get(b)`, we assume that the value of `empty? b` in both $\mathcal{P}$ and $\mathcal{Q}$ is `false`. Otherwise, the experiment cannot be performed neither in $\mathcal{P}$ nor in $\mathcal{Q}$, and it is then trivial to check condition 3.

Assuming that the buffers are not empty, the unique transitions are the following:

$$\langle \texttt{get(b)}, \mathcal{Q}\rangle \rightarrow \langle [\![\texttt{a[i]}]\!]_{\texttt{BA}}, \mathcal{Q} \oplus \texttt{fst b = i + 1}\rangle, \qquad (11.33)$$

and

$$\langle \texttt{get(b)}, \mathcal{P}\rangle \rightarrow \langle [\![\texttt{head(l)}]\!]_{\texttt{B}}, \mathcal{P} \oplus \texttt{elems b = tail(l)}\rangle. \qquad (11.34)$$

Note that the evaluation of `get` is atomic, and that `a[i]` is equal (with respect to `EB`) to `head(l)`, since `l` is equal to `i < a > j`, and `i < j` (the buffer is not empty); so it also follows that `i <= j + 1`. It is then easy to check that the resulting states are related by $\mathcal{S}$, and that condition 3 is valid. We omit the details here.

The object creation operations `new.Buffer()` and `new(b)` just add an object to the database state; they do not initialize attributes. In fact, the behaviour of those operations just depends on the family of identifiers associated to a specification, and the objects in the state; the first determines what identifiers may be chosen by `new.Buffer()`, whereas the other determines when `new(_)` can be evaluated (it might be suspended, if its argument is the identifier of an object in the state). As `BUFFER` and `BUFFERA` have the same associated family of identifiers, and the objects in $\mathcal{P}$ and $\mathcal{Q}$ are the same, the creation operations have the same effect in $\mathcal{P}$ and $\mathcal{Q}$. It follows that condition 3 holds for this kind of experiment. We omit the details again.

Similarly to `new(b)`, the experiment `remove(b)` only depends on what objects are in the state. So it will behave in the same way when performed with

$\mathcal{P}$ and $\mathcal{Q}$. If `b` is an object in the state, this operation removes all its stored attributes from the state. Hence, it follows that the resulting states are related by $\mathcal{S}$, and that condition 3 is satisfied. This concludes the proof that pairs of $\mathcal{S}$ relating `reset(b´)` satisfy the conditions of a simulation relation whenever `b´` is in $\mathcal{P}$.

Now we assume that `b´` is not in $\mathcal{P}$ (this implies that `b´` is not in $\mathcal{Q}$ as well). Observe that checking the effect of experiments for this case is the same as checking the effect of experiments when `b´` is in $\mathcal{P}$, as previously discussed. Hence, we just need to verify conditions (1) and (2). The first is vacuously true; there is no transition from

$$\langle \texttt{reset(b´)}, \mathcal{Q}\rangle,$$

since `b´` is not in $\mathcal{Q}$. The second condition can be verified by checking that `reset(b´)` cannot be evaluated in $\mathcal{P}$ as well, since `b´` is not in $\mathcal{P}$; so we may conclude that

$$\langle \texttt{reset(b´)}, \mathcal{P}\rangle \rightarrow^* \langle \texttt{reset(b´)}, \mathcal{P}\rangle \not\rightarrow,$$

the resulting states are related by $\mathcal{S}$, and the expression `reset(b´)` is not fully evaluated.

In order to complete the proof that $\mathcal{S}$ is a (BUFFER, BUFFERA)-simulation, we should analyse the pairs of $\mathcal{S}$ having expressions in other forms different from `reset(b´)`. However, that's essentially what we have proved when considering the effects of performing experiments with the database states related by $\mathcal{S}$; we have just used `b` instead of `b´`.

11.3.3 Laws of FOOPS Method Combiners

The notion of refinement of expressions introduced here can also be used for proving general properties of FOOPS method combiners, if those properties are expressed as refinement inequations. Besides providing a better insight on the semantics of method combiners, those properties may be used for proving refinement, or deriving correct implementations from specifications.

For example, we can assert that the sequential composition method combiner is associative by verifying that

$$p \; ; \; (q \; ; \; o) \sqsubseteq_{(\mathsf{P},\mathsf{P})} (p \; ; \; q) \; ; \; o \tag{11.35}$$

and

$$(p \; ; \; q) \; ; \; o \sqsubseteq_{(\mathsf{P},\mathsf{P})} p \; ; \; (q \; ; \; o), \tag{11.36}$$

for any specification P and method expressions $p, q, o \in \mathcal{T}_{\mathsf{P}}$.

In order to verify 11.35, we have to prove that

$$\langle p \; ; \; (q \; ; \; o), \emptyset_{\mathsf{P}}\rangle \sqsubseteq_{(\mathsf{P},\mathsf{P})} \langle (p \; ; \; q) \; ; \; o, \emptyset_{\mathsf{P}}\rangle.$$

This can be directly verified by checking that the relation containing $\mathcal{ID}_{\mathsf{P}}$ and all pairs in the form

$$(\langle p \; ; \; (q \; ; \; o), \mathcal{D}\rangle \quad , \quad \langle (p \; ; \; q) \; ; \; o, \mathcal{D}\rangle), \tag{11.37}$$

for any $\mathcal{D} \in Db(-\mathsf{P})$, p, q, and o, is a (P, P)-simulation. This is straightforward, by using Theorem 11.2.1 and the rules of the operational semantics for sequential composition:

$$\langle e \ ; \ f, \mathcal{D}\rangle \rightarrow \langle e' \ ; \ f, \mathcal{D}'\rangle \ \textit{if} \ \langle e, \mathcal{D}\rangle \rightarrow \langle e', \mathcal{D}'\rangle \tag{11.38}$$

and

$$\langle v \ ; \ e, \mathcal{D}\rangle \rightarrow \langle e, \mathcal{D}\rangle \tag{11.39}$$

where v is a fully evaluated expression. Basically, we just have to observe the following facts. First let p be a fully evaluated expression. By Rule 11.39, we have

$$\langle p \ ; \ q, \mathcal{D}\rangle \rightarrow \langle q, \mathcal{D}\rangle.$$

Therefore, by Rule 11.38, we have

$$\langle (p \ ; \ q) \ ; \ o, \mathcal{D}\rangle \rightarrow \langle q \ ; \ o, \mathcal{D}\rangle.$$

But, by Rule 11.39, we also have

$$\langle p \ ; \ (q \ ; \ o), \mathcal{D}\rangle \rightarrow \langle q \ ; \ o, \mathcal{D}\rangle.$$

Second let's assume that p is not a fully evaluated expression and $\langle p, \mathcal{D}\rangle\not\rightarrow$. Hence

$$\langle p \ ; \ q, \mathcal{D}\rangle \not\rightarrow .$$

Therefore

$$\langle (p \ ; \ q) \ ; \ o, \mathcal{D}\rangle \not\rightarrow,$$

since $p \ ; \ q$ is clearly not fully evaluated. But we also have

$$\langle p \ ; \ (q \ ; \ o), \mathcal{D}\rangle \not\rightarrow .$$

Finally assume that p is not a fully evaluated expression and $\langle p, \mathcal{D}\rangle \rightarrow \langle p', \mathcal{P}\rangle$, for some p' and $\mathcal{P}$. By Rule 11.38, we have

$$\langle p \ ; \ q, \mathcal{D}\rangle \rightarrow \langle p' \ ; \ q, \mathcal{P}\rangle.$$

Hence, by Rule 11.38, we have

$$\langle (p \ ; \ q) \ ; \ o, \mathcal{D}\rangle \rightarrow \langle (p' \ ; \ q) \ ; \ o, \mathcal{P}\rangle.$$

But then we also have

$$\langle p \ ; \ (q \ ; \ o), \mathcal{D}\rangle \rightarrow \langle p' \ ; \ (q \ ; \ o), \mathcal{P}\rangle.$$

The Inequation 11.36 and many other properties of method combiners can be proved in a similar way.

11.4 Aspects Considered by Refinement

Now we analyse the aspects that are considered by our definition of refinement. We discuss what $P \sqsubseteq Q$ tell us about the states P and Q, in addition to the fact that whatever can be observed by performing experiments with Q can also be observed doing the same experiments with P. This clarifies the aspects that have to be considered when defining refinement for a concurrent object oriented programming language. Hereafter let's assume that $P \sqsubseteq Q$.

It is well known in other frameworks that refinement can be achieved by reducing nondeterminism [Milner, 1989, Hoare, 1985, Morgan, 1994]. Here we follow the same approach. At most, Q should be as nondeterministic as P; in other words, Q should be more predictable than P. This is what guarantees that the behaviour of Q is a particular case of the behaviour of P, as expressed by condition 1 from the definition of simulation, and exemplified in Section 11.3.1, where we check that a nondeterministic choice is refined by an assignment.

Data refinement [Hoare, 1972, Jifeng *et al.*, 1986] is also guaranteed by our definition of refinement. In fact, we abstract from the particular representation of object states, as long as they are observation equivalent (i.e., whatever can be observed by means of experiments from one can also be observed from the other). In the definition of simulation, this is reflected by condition 3; in the examples, this is illustrated in Section 11.3.2, where we prove refinement of programs that use different state structures.

Although we have not given any example of refinement of parallel expressions, concurrency is considered by our definition of refinement. In particular, we consider the effects of interference on the intermediate states reached during the execution of an expression. As illustrated in Section 11.3.1, the expression

```
x := 0 ; x := x + 1
```

is not a refinement of `x := 1` exactly because we consider possible interaction.

In fact, refinement can also be achieved by reducing the possibilities of interferences (from the execution of other operations) during the execution of an expression. This guarantees that the execution of Q in parallel with other expressions is at least as deterministic as the execution of P in parallel with the same expressions. In general, this means that the behaviour we can observe from Q can also be observed from P.

One obvious way of reducing the possibilities for interferences is to reduce the number of intermediate states reached during the evaluation of an expression. For instance, this can be done by providing an atomic implementation of a nonatomic expression, as illustrated in Section 11.3.1, where we proved that

```
x := 0 ; x := x + 1
```

is refined by `x := 1`. In the definition of simulation, this is reflected by condition 1, where a $\twoheadrightarrow$-transition from Q can be matched by many transitions from P.

However, note that Q can lead to even more intermediate states than P, as long as those extra states are simulations of neighbouring states, which in turn should be simulations of corresponding intermediate states reached from P. In this way, no extra interference is possible in Q. In other words, refinement can also be obtained by introducing and removing *stuttering* steps (i.e., idle steps modulo observation equivalence of states). In the examples, this was illustrated by the equivalence (defined as refinement in both ways) of

```
skip(x) ; x := 0
```

and `x := 0`. In the definition of simulation, this is also reflected by condition 1, where a $\twoheadrightarrow$-transition from Q can be matched by none transition from P.

This is what supports the refinement of atomic operations, since an atomic operation may be refined by a nonatomic operation as long as the additional transitions in the evaluation of the latter correspond to stuttering steps. In this way the sequence of intermediate states reached during the execution of the nonatomic operation can be split in two parts such that all the states in one part are equivalent to the state prior to the evaluation of the operation, and all states in the other part are equivalent to the state after the execution of the operation. Hence this kind of refinement increases the level of parallelism without increasing the level of nondeterminism.

11.5 Conclusions

We introduced a notion of refinement for FOOPS together with an effective proof technique. This establishes a basis for formal development of software in FOOPS. We described definitions of refinement for FOOPS system states, expressions, and specifications. We also derived basic properties of those relations. In [Borba, 1995], it is proved that the relation of refinement of expressions introduced here is a congruence with respect to the following FOOPS method combiners: sequential composition, parallel composition, result combiner (**result**_;_), conditional, and the *internal* choice method combiner (which can be defined in terms of the *external* choice combiner _[]_). Choosing the parallel composition combiner as example, it is proved that

$$p \sqsubseteq_{(\mathsf{P},\mathsf{Q})} q$$

implies

$$p \mid\mid o \sqsubseteq_{(\mathsf{P},\mathsf{Q})} q \mid\mid o,$$

for any expression o formed by experiments of P and the method combiners mentioned above. In [Borba, 1995], it is also derived a weaker compositionality result for the atomic evaluation operator.

We concentrated on the object level of FOOPS and only considered refinement of "flat" specifications. Definitions of refinement and equivalence for the functional level are described in [Goguen, 1994]. Although we focused on FOOPS, our definition of refinement only depends on minimal assumptions about FOOPS and its operational semantics; those assumptions were described here. This suggests that our approach could be adapted for other object oriented concurrent languages having remote procedure call as a mechanism for interaction and synchronization.

In order to illustrate the use of our notion of refinement for stepwise formal development of programs in FOOPS, we considered examples involving memory cells and different implementations of buffers. We also explained how to derive laws of FOOPS method combiners from the definition of refinement. By analysing those examples we observed which aspects are relevant for refinement of concurrent object oriented programs: data refinement (including dynamic data structures), nondeterminism, concurrency, interference, and refinement of atomic operations. We also concluded that the definition of refinement presented here is divergence insensitive; that is, it may relate an expression not having a divergent behavior to an expression having a divergent behavior. Just observe that, by the definition of simulation, a $\rightarrow$-transition from a state can be matched by no transitions from a related state.

We included proofs of refinement together with detailed explanation in order to give a better insight on our approach, by showing how it works in detail. In particular, we can conclude that proofs of refinement using the simulation technique are not complicated. The major task involved is to find a candidate simulation relation. This requires some knowledge about the behaviour of expressions, and the relation between different representations for object states. Checking the conditions from the definition of simulation is a routine task which requires only the application of the transition rules of the operational semantics and checking whether the resulting states are related. All those points show that the framework of operational semantics and simulations can be effectively used for defining refinement of concurrent object oriented programs.

Well established approaches for refinement and equivalence of sequential programs [Morgan, 1994, Hoare *et al.*, 1987, Bauer *et al.*, 1989] are of limited use for refinement of concurrent systems, since they are based on semantic frameworks that consider only the relational (input/output) behaviour of programs. By contrast, in order to define a notion of refinement for FOOPS, we used the frameworks of structural operational semantics [Plotkin, 1981] and bisimulations [Park, 1981]. Based on those frameworks, notions of equivalence have also been developed for process algebras [Milner, 1989] and a general notation for giving the semantics of programming languages [Mosses, 1992].

An approach for refinement of object *based* concurrent programs is described in [Jones, 1992], where many examples of formal development of programs are presented. Most examples are simple and elegant, and use a few refinement preserving transformation rules. Assertional "Hoare style" inference rules for reasoning about rely and guarantee conditions [Jones, 1983] are also used for proving refinement. However, no general definition of refinement is proposed; although the semantics of the language presented in [Jones, 1992] is given in terms of a process algebra [Milner *et al.*, 1989], the notions of equivalence and refinement of the algebra cannot be directly used to define the notion of refinement of the language [Jones, 1993].

Another approach for refinement of object oriented concurrent programs is described in [Lano and Goldsack, 1994], which gives an overview of a general notion of refinement for a concurrent object oriented specification language. However, no properties of refinement or associated proof techniques are discussed.

Comparing with those alternative approaches for the refinement of concurrent object oriented programs, our approach is unique in the sense of providing both a general definition of refinement and an effective proof technique.

Acknowledgements

Thanks to Steve Schneider and Grant Malcolm for giving many suggestions which helped to simplify the work described here. The first author also had the opportunity to discuss in detail most of the aspects presented here with Grit Denker and Prof. Hans Dieter Ehrich, during an enjoyable visit to Braunschweig, supported by the IS-CORE project.

Chapter 12

Static Typing for Object-Oriented Languages

12.1 Introduction

One of the major features of object oriented languages is inheritance, and the capability to bind methods dynamically, at runtime, depending on the class of the receiver. This capability allows for a high deegree of reuse and customization of code. In some languages (*e.g.* Simula, C++) this capability has been combined with static type checking, whereas others, of which Smalltalk is the most widespread example, have no static type system. Smalltalk follows dynamic typing: every object has its immutable class, it reacts to message according to the class it belongs to and causes a runtime error ("`object does not understand message`") if it is sent a message for which its class has no corresponding method.

Adding a type system to a dynamically typed language would offer the benefits of type safety (ie the runtime error "`object does not understand message`" would never occur) and of considerable performance enhancement. For instance, [Johnson *et al.*, 1988] report an up to tenfold speedup of program execution due to the possibility of determining statically the method to be executed depending on the classes of the receiver and arguments.

Type systems for Smalltalk have been developed since the early 80's, [Suzuki, 1989, Johnson, 1993, Graver, 1989, Borning and Ingalls, 1982, Bracha and Griswold, 1993]. This retrospective fitting of a type system opens challenging questions. In particular, the notions of subtype and subclass are similar but different, a problem which has been tackled in several different ways. Also, there exist dependencies between receiver, or argument types and the result type (consider the case where the message returns its second argument), which require a form of dependent product types. Furthermore, container classes (*e.g.* Set, Collection) require the idea of parameterized classes or types.

We suggest ST&T, a novel type system, in which:

- The subtype relationship is an extension of the subclass relationship.
- A method may have several signatures; the type of a message expression (or message sent) is determined by the signatures of the message which "fit best" the types of the receiver and arguments.
- Method signatures can express a functional relation between the result type and any of the argument or receiver types (which is a generalization of the type Self).
- Parameterized types express container types.

As a consequence of these features:

- Type-checking a method does not require type checking methods in the subclasses of the class containing the method body.
- More programs are type correct in our system than in previous type systems.

The soundness of this approach has been demonstrated in [Drossopoulou and Yang, 1995].

The remainder of this paper is organized as follows: In section 12.2 we introduce the problem of the difference between subtypes and subclasses, summarize various solutions suggested so far, and outline the ST&T approach. In sections 12.3 and 12.4 we introduce the type system suggested for ST&T: we describe the possible types for variables and methods and explain in terms of type inference rules the types for message expressions. In section 12.5 we discuss the implications of our approach, outline the proof of soundness and describe the ST&T implementation. Finally, in section 12.6 we draw some conclusions, and discuss further possible work.

12.2 Subclases and Subtypes

Subclasses in object oriented languages inherit and possibly extend the capabilities of their superclass. Thus one would expect that subclasses are subtypes - in the sense that every object from a subclass may appear in every context which expects an object from a superclass. Astonishingly, it turns out, that in general -unless special care has been taken when designing the programming language- this is *not* the case. This fact has already been observed by [Cook and Palsberg, 1989] and various solutions have been adopted.

In order to explain this, we first clarify the notion of type and class.

12.2.1 Type vs. Class

The *type* traditionally describes the behaviour of an entity, its interface to other entities. In the case of object oriented programming, the type of an object defines all the messages the object understands, *i.e.* the number and type of arguments, and the type of the result (in other words, the signature of the message). For instance, an object of type Point understands the message x, and returns an integer (its x-coordinate).

The *class* describes the way this behaviour is achieved, *i.e.* the implementation of the object's interface. For instance, an object of class Point2D, as shown in figure 12.1 (cartesian representation of points), is represented by a tuple containing its x- and y- coordinates, and the method x returns the first value from the tuple. Or, an object of class PointPolar (polar representation of points) is represented by a tuple containing the angle and radius and the method x returns the cosine of the angle multiplied by the radius.

Thus we see that the concepts of type and class are tightly related, but not identical. A class *implements* a type. For instance, the classes Point2D and PointPolar implement the type Point. Moreover, we see that it is possible for two objects to belong to the same type, but to different classes.

Thus, it makes sense to distinguish between classes and types. Indeed, such a distinction has been sugeested, among others, by [Canning *et al.*, 1989, Cook and Palsberg, 1989, Bruce, 1992, Bruce, 1991].

On the other hand, a class determines a type. For instance, the methods that appear in the class Point2D determine the messages that the objects will understand, and the type of the arguments and results. For this reason, in most programming languages the distinction between class and type has been abandoned and logically there exist two views of a class: One is the interface (public members in C++, message protocol in Smalltalk) and it plays the role of the type. The other is the implementation (the function bodies and all private members in C++, the method bodies in Smalltalk).

12.2.2 Subtype vs. Subclass

However, as soon as inheritance is introduced to a language the need for distinguishing classes and types becomes stronger:

Traditionally, a type TypeA is a *subtype* of another type TypeB, if all objects of type TypeA satisfy the interface described by TypeB; in other words, if the objects of TypeA understand all the messages understood by TypeB objects, and react in a compatible manner. This means, that in every context where an object of type TypeB is expected, it is legal for an object of TypeA to appear. This is usually expressed by the *subsumption* rule:

$$\frac{\text{exp satisfies TypeA} \qquad \text{TypeA is a subtype of TypeB}}{\text{exp satisfies TypeB}}$$

Subtypes express *interface inheritance.*

On the other hand, a *subclass* expresses *implementation inheritance.* A class ClassA inherits from its superclass ClassB all the state variables and methods defined in ClassB.

The example from figure 12.1 demonstrates why subclasses are not necessarily subtypes: Point3D is a subclass of (ie inherits implementation from)

```
class name Point2D
instance variables xValue yValue
instance methods
    x
        ↑xValue
    addX: anX
        xValue ← anX + xValue
    add: aPoint
        aPoint addX: (self x).
        ↑aPoint
    invertedAdd: aPoint
        ↑aPoint add: self

class name Point3D
superclass name Point2D
instance variable zValue
instance methods
    addZ: aZ
        zValue ←zValue + aZ
    add: aPoint
        aPoint addX: (self x).
        aPoint addZ: zValue.
        ↑aPoint
```

Figure 12.1: Subclasses are not subtypes - Smalltalk example

the expression	returns a
`p2 add: p2`	Point2D object
`p2 add: p3`	Point3D object
`p3 add: p2`	run time error
`p3 add: p3`	Point3D object
`p2 invertedAdd: p2`	Point2D object
`p2 invertedAdd: p3`	runtime error
`p3 invertedAdd: p2`	Point3D object
`p2 invertedAdd: p2`	Point3D object

Figure 12.2: Class of Objects returned, where p2 is a Point2D, p3 is a Point3D

Point2D, but is not a subtype of (ie does not fully inherit specification from) Point2D.

The figure 12.2 shows the class of the obje3cts obtained by evaluating some expressions, where p2 is an object of class Point2D, and p3 an object of class Point3D:

Thus, in an environment in which an object of class Point2D executes normally, an object of class Point3D causes errors. This happens, because the method add: in class Point2D expects an argument of class Point2D, whereas the method add: in class Point3D expects an argument of class Point3D. That is, the "type" of add: from class Point3D is more restricted than that of add: in class Point2D, and thus, the type represented by Point3D is not a subtype of the type represented by Point2D.

In general, subclasses inherit behaviour, but do not inherit specification. So far, to our knowledge, the following approaches have been tried:

1. A fully dynamic system, where there is no notion of typing or static type correctness. This is the approach taken by Smalltalk.

2. Explicit and distinguished notions of type and class, subclass and subtype Thus a subclass inherits behaviour, but not the interface. A class satisfies a type and all its supertypes. A subclass inherits from its superclass, but does not necessarily satisfy the type of its superclass. Such an approach is suggested in [Bruce *et al.*, 1993].

3. The notion of type and class coincide, *i.e.* the class defines the type, and all subclasses have a type that is a subtype of the superclass's type. In terms of our example, the declaration of the method add: in the class Point3D would be forbidden.

   Such an approach is taken by C++ and Eiffel. Also, in [Bracha and Griswold, 1993] a type system for Smalltalk is suggested which follows this approach.

4. An explicit notion of class, and an implicit notion of type and fully decouple subclasses from subtypes [Johnson, 1993], [Graver, 1989]. That is, the programmer will only think in terms of classes, but the type checking system will also have the notion of type. In terms of our example, the method add: in the class Point3D would be allowed, and the body of invertedAdd:, although declared only once, needs to be type checked in class Point2D *and* in class Point3D.

5. Consider classes as polymorphic, recursive records, like in [Remy, 1993], [Ghelli, 1991]. Then the types of our examples are represented by

   Point2D=$\mu\alpha$.[xValue : SmallInteger, add :$\alpha \rightarrow \alpha, \ldots$]

   and

   Point3D=$\mu\alpha$.[xValue : SmallInteger, zValue : SmallInteger, add :$\alpha \rightarrow \alpha, ...$].

   Subtypes are obtained by replacing type variables (*e.g.* α) by other types.

   In this approach, the redefinition of add: in Point3D would be legal, and the expressions p2 add: p2 and p3 add: p3 would be type correct, whereas the expressions p3 add: p2 and p2 add: p3 would be type incorrect, although the latter does not cause errors when executed.

6. In our system, [Drossopoulou and Karathanos, 1993, Drossopoulou and Karathanos, 1994, Drossopoulou and Yang, 1995, Drossopoulou *et al.*, December 1995 to appear], types are unions of classes or parameterized classes, subclasses are subtypes, and a method may have several signatures. We give an informal overview of our approach in the next section.

12.2.3 ST&T: An Example

In figure 12.3 we show the ST&T types that would be given for the previous example. Every variable is followed by its type in comments, and the methods contain their signatures in comments.

The type of a variables indicates the classes to whose objects the variables may point to at run time. A type ⟨ClassA⟩ indicates the set of objects of class ClassA, and all *its* subclasses. This intepretation of object types is not different than the usual, as eg [Suzuki, 1989, Johnson, 1993, Graver, 1989, Borning and Ingalls, 1982, Bracha and Griswold, 1993]. Furthermore, the type [ClassA] indicates the objects of Class, but *not* of its subclasses.

For example, the instance variable xValue of the class Point2D has type ⟨Number⟩. This means, that at runtime xValue may point to objects of class Number, or any subclass of Number. *i.e.* it may poinjt to objects of class SmallInteger, Float *etc* . A variable of type ⟨Point2D⟩ may point at run time only to objects of class Point2D.

We assume that we have variables x2 and x3, with x2 :⟨Point2D⟩, and x3 :⟨Point3D⟩, which means, that x2 may point at run run time to objects of class Point3D, or of class Point2D (or of any further subclass of Point2D, if any are

```
class name Point2D
instance variables xValue yValue
        "xValue, yValue: Number "
instance methods
    x
        "≪∀X ≤ Point2D.X ⟶ Number≫ "
        ↑xValue
    addX: anX
        "≪∀X ≤ Point2D.∀Y ≤ Number.X × Y ⟶ X≫ "
        xValue ← anX + xValue
    add: aPoint
        "≪∀X ≤ Point2D.∀Y ≤ Point2D.X × Y ⟶ Y≫ "
        aPoint addX: (self x).
        ↑aPoint
    invertedAdd: aPoint
        "≪∀X ≤ Point2D.∀Y ≤ Point2D.X × Y ⟶ X≫ "
        "≪∀X ≤ Point3D.∀Y ≤ Point3D.X × Y ⟶ X≫ "
        ↑aPoint add: self

class name Point3D
superclass name Point2D
instance variable zValue
        "zValue: Number "
instance methods
    addZ: aZ
        "≪∀X ≤ Point3D.∀Y ≤ Number.X × Y ⟶ X≫ "
        zValue ←zValue + aZ
    add: aPoint
        "≪∀X ≤ Point3D.∀Y ≤ Point3D.X × Y ⟶ Y≫ "
        aPoint addX: (self x).
        aPoint addZ: zValue.
        ↑aPoint
```

Figure 12.3: Subclasses are subtypes in ST&T

defined later on), and x3 may point to objects of Point3D, (or of any further subclass of Point3D, if any are defined later on).

The interpretation of method types is more interesting. Notice, that the types of the message pattern add: and invAdd: consist of two different signatures, and that the latter has only one method body.

Signatures express a functional relationship between the receiver/argument types and the result. For example, the signature of addX: is
$\ll \forall X \leq \text{Point2D}.\forall Y \leq \text{Number}.X \times Y \longrightarrow X \gg$
which expresses that the receiver must be of a subclass of Point2D, the argument must be of a subclass of Number, and that the result will be of the same class as the receiver. (This is another way of speaking about the receiver type, *i.e.* SELF).

Therefore, x2 addX: 3 has type ⟨Point2D⟩, whereas x3 addX: 3 has type ⟨Point3D⟩.

On the other hand, the signature has to "cover" the types of the argument and the receiver, thus, the signature of addZ: does not cover the types ⟨Point2D⟩, ⟨SmallInteger⟩, and therefore the type of x2 addZ: 3 is ⟨TypeError⟩, which indicates that the exception object does not understand message may occur when evaluating this expression..

The signature of add: in class Point2D is
$\ll \forall X \leq \text{Point2D}.\forall Y \leq \text{Point2D}.X \times Y \longrightarrow Y \gg$.
This means that the receiver and argument must be of a subclass of Point2D, and the result will be of the same class as the argument. The signature of add: in class Point3D is $\ll \forall X \leq \text{Point3D}.\forall Y \leq \text{Point3D}.X \times Y \longrightarrow Y \gg$ and this means that the receiver and argument must be of a subclass of Point3D, and the result will be of the same class as the argument. Here we see two interesting features of ST&T: firstly, that it is possible to express a functional relationship between result and *argument* type, and secondly, that when a message pattern has many signatures a kind of best fit matching with the receiver and argument types will take place. The method to be executed at run time is determined by the class of the receiver. Thus, only the signatures whose receiver type fit the type of the receiver object the closest will be considered.

If the receiver of add: has type ⟨Point3D⟩, then the second signature, *i.e.* $\ll \forall X \leq \text{Point3D}.\forall Y \leq \text{Point3D}.X \times Y \longrightarrow Y \gg$, fits better than the first, and therefore we consider only this one. This signature requires the argument to be of type ⟨Point3D⟩, and the result type will be the same as the argument type. Thus, x3 add: x3 has type ⟨Point3D⟩. However, the signature does not cover the types ⟨Point3D⟩, ⟨Point2D⟩, and therefore x3 add: x2 has type ⟨TypeError⟩.

On the other hand, if the receiver of add: is of type ⟨Point2D⟩, then this means that it may be an object of class Point2D or of any of Point2D's subclasses - which includes Point3D. Therefore both signatures fit equally well, and therefore both signatures have to be considered. If the argument has type Point3D, the first signature covers the result and argument types, and yields the result ⟨Point3D⟩, and similarly, the second signature yields the result ⟨Point3D⟩. If the argument has type Point2D, the first signature covers the result and argument

The expression ...	has type ...
x2 add: x2	⟨TypeError⟩
x2 add: x3	⟨Point3D⟩
x3 add: x2	⟨TypeError⟩
x3 add: x3	⟨Point3D⟩
y2 add: y2	[Point2D]
y2 add: y3	[Point3D]
y3 add: y2	⟨TypeError⟩
y3 add: y3	[Point3D]
x2 invertedAdd: x2	⟨TypeError⟩
x2 invertedAdd: x3	⟨TypeError⟩
x3 invertedAdd: x2	⟨Point3D⟩
x3 invertedAdd: x3	⟨Point3D⟩

Figure 12.4: Types of message expressions, where x2:⟨Point2D⟩, ⟨x3⟩:⟨Point3D⟩ and y2:[Point2D], x3:[Point3D]

types, and yields the result ⟨Point2D⟩, but the second signature does not cover the type of the argument, and therefore it yields the result ⟨TypeError⟩.

In figure 12.2.3 we show the types of some more message expressions. Although the evaluation of p2 add: p2 returns an object of class Point2D, the expression x2 add: x2 has type ⟨TypeError⟩. This is so, because in the expression x2 add: x2 we only know the types but not the actual classes of the receiver and arguments. One can say that the expression x2 add: x2 stands for the four cases: p2 add: p2, p2 add: p3, p3 add: p2 and p3 add: p3.

12.3 Types in ST&T

Types in programming languages express the possible set of values an entity could denote at runtime. In Smalltalk we have the following entities: objects, expressions (which consist of variables, literals, pseudovariables and messages), methods and classes.
Thus, in ST&T we have two kinds of types:

- *Object Types* describe the set of possible objects to which a variable may point, or an expression may evaluate to. Object types are unions of open (ie class and all subclasses) or closed (*i.e.* only the class itself) classes. The types of self and super need not be explicitly described.

- *Method Types* are sets of method signatures. A method signature describes the types of the receiver, the arguments and the result. In a method signature the argument and result types are described by object types, but in addition, the result may depend on the receiver or one of the argument types (this caters for self).

Types are defined in terms of classes. A type is enclosed in angle or square brackets, *e.g.* ⟨True⟩ and [True] denote types, whereas True denotes a class.

12.3.1 Some Subclass Relationships in Smaltalk

The subtype relationship in ST&T is based on the subclass relationship of Smalltalk. Therefore, in order to understand some of the following examples, we shall need to remember some subclass relationships.

For example, in the Objectworks\Smalltalk system, False $\sqsubseteq$Boolean, True $\sqsubseteq$Boolean and Boolean $\sqsubseteq$Object. (The symbol $\sqsubseteq$denotes the subclass relationship) Furthermore, SmallInteger $\sqsubseteq$Integer, Integer $\sqsubseteq$Number, and also Number $\sqsubseteq$ArithmeticValue, ArithemticValue $\sqsubseteq$Magnitude, Character $\sqsubseteq$Magnitude and Magnitude $\sqsubseteq$Object.

12.3.2 Object Types

Object types describe instance variables, class variables, local variables and expressions, and they determine the possible classes to which the objects denoted by variables (or by expression evaluation) may belong.

ObjectType	::=	*SingleType* (∨ *SingleType*)* .
SingleType	::=	⟨*ClassName* ⟩| [*ClassName*]
		| ⟨TypeError ⟩| *ParType* | *ApplType*
ParType	::=	*Quant+* . *ParClName* (*VarName* (, *VarName*)*)
ApplType	::=	*ParClName* (*SingleType* (, *SingleType*)*)
Quant	::=	∀ *VarIdent* ≤ *ObjectType* .
ClassName	::=	*Identifier*
ParClName	::=	*Identifier*
VarName	::=	*Identifier*
VarIdent	::=	*Identifier*

Object types can be

- Open or Closed types. A variable with type ⟨ClassA ∨ ClassB⟩ ∨ [ClassC] may denote at runtime an object of class ClassA, or an object of class ClassB, or of a subclass of ClassA, or of a subclass of ClassB, or of class ClassC. Literals and the pseudovariables true and false have an immutable object type, *e.g.* 3 has type [SmallInteger], and true has type [True]. classes in angle brackets indicating that the variable may point to an object of this class or any of its subclasses

- ⟨TypeError⟩ indicates that at run time the exception object does not understand message may be raised.

- parameterized types depending on universally quantified variables; the variable may point to an object of this class with its parameter replaced by a subclass of the bound.

- applications of parameterized types to types, yielding more specific types
- unions of variables types, denoting the set of all objects that may belong to either of the constituing types

For the sake of simplicity we abbreviate object type expressions ⟨ClassA⟩ ∨ ⟨ClassB⟩ to ⟨ClassA ∨ ClassB⟩, and we drop the ⟨-⟩ brachets whenever there is no doubt from the context.

Thus, according to the above syntax, the following are possible object types:

- [ArithmeticValue] ∨ ⟨Integer⟩: a variable of this type may point to objects of class ArithmeticVAlue, SmallInteger, Integer, but NOT Number.
- ⟨Set(Boolean)⟩: denotes such objects that belong to the class Set and store boolean objects in the set.
- ≪∀X ≤ Magnitude.Set(X)≫ : A variable of this type will point at runtime to objects of type ⟨Set(Z)⟩ where Z is a subtype of ⟨Magnitude⟩. It may *e.g.* point to an object of class ⟨Set(Float)⟩ in which case the method select returns a floating point value, or, it may point to an object of class ⟨Set(SmallInteger)⟩ in which case the method select returns a ⟨SmallInteger⟩.

Notice that the bounds of a variable have to be closed, *e.g.* it is not possible to have a object type like ≪∀X ≤ Point.∀Y ≤ X....≫ . In other words, we cannot have dependent sum types as, *e.g.* in [MacQueen, 1986].

12.3.3 The Subtype Relationship

Object types in ST&T are ordered by the *subtype* relationship. A object type is a subtype of another type, if the set of possible objects described by the first is a subset of the set of possible objects described by the latter.[1]

Notice, that our system has no subsumption rule and this will make the subtype relationship simpler to compute. Notice also, that we do not define a subtype relationship for method types. [2]

The subtype relationship is the minimal relationship defined by the following inference rules:

The subtype relationship (≤) is primarily based on the subclass relationship (⊑):

$$\frac{\text{ClassA} \sqsubseteq \text{ClassB}}{\begin{array}{l}\langle\text{ClassA}\rangle \leq \langle\text{ClassB}\rangle\\ [\text{ClassA}] \leq \langle\text{ClassB}\rangle\end{array}}$$

[1]This is a weaker definition than the usual definition of subtype, because it does not require that the objects of a subtype satisfy the behaviour required for the supertype, *i.e.* the subsumption rule is not required

[2]Methods are not first class citizens, they cannot be assigned or passed as parameters. Furthermore, our system does not attempt to check whether a subtype has the subsumption property. For these reasons it is not necessary to define the subtype relationship for method signatures.

This means, that if ClassA is a subclass of ClassB, then the type ⟨ClassA⟩ is a subtype of ⟨ClassB⟩. For instance, SmallInteger is a subclass of Number, therefore ⟨SmallInteger⟩ is a subtype of ⟨Number⟩.
Every type is a subtype of itself and of the error type:

$$\frac{}{\begin{array}{l}\mathsf{TypeA} \leq \langle\mathsf{TypeError}\rangle \\ \mathsf{TypeA} \leq \mathsf{TypeA}\end{array}}$$

The union of a collection of types is smaller than the union of a collection of larger types. In other words, the ∨ operation preserves the ≤ relationship:

$$\frac{\text{for all } i \in 1 \ldots n \text{ there exists a } j \in 1 \ldots m \text{ such that } \mathsf{TypeA_i} \leq \mathsf{TypeB_j}}{\langle \mathsf{TypeA_1} \vee \ldots \vee \mathsf{TypeA_n} \rangle \leq \langle \mathsf{TypeB_1} \vee \ldots \vee \mathsf{TypeB_m} \rangle}$$

For instance, ⟨True ∨ False ∨ Point⟩ is a subtype of ⟨Boolean ∨ Boolean ∨ Magnitude⟩ = ⟨Boolean ∨ Magnitude⟩.
For parameterized types, if the bound of the first is smaller than the bound of the second, then the first is smaller than the second [3]

$$\frac{\mathsf{TypeA} \leq \mathsf{TypeB}}{\ll \forall \mathsf{X} \leq \mathsf{TypeA.ParamClass} \gg \; \leq \; \ll \forall \mathsf{X} \leq \mathsf{TypeB.ParamClass} \gg}$$

Thus, $\ll \forall \mathsf{X} \leq \mathsf{Number.Set(X)} \gg$ is a subtype of $\ll \forall \mathsf{X} \leq \mathsf{Object.Set(X)} \gg$. [4]
It is easy to prove that the ≤ relationship is a partial order. All rules enjoy the subformula property; thus the ≤ relationship is deterministically computable.

12.3.4 Method Signatures

Signature describes constraints on the type of the receiver, the arguments and the result.

MethodSignature	::=	*Quant+∗ Params* ⟶ *ResultType*
Params	::=	*VarName* (× *VarName*)*
ResultType	::=	*ObjectType* | *VarName* | *Function*(*VarName*)
Function	::=	*InstanceOf* | *ClassOf*

The result type may be any object type, or it may depend on the receiver or the argument via the *Function*al dependencies.

- X indicates that the result will be of the same class as the receiver (or thje i-th argument), if the receiver (or the i-th argument) was called X in the signature

[3]the relation is ***covariant*** in the argument, as opposed to most other systems, where it is contravariant, *e.g.* [Cardelli and Mitchell, 1989]. The reason is that our interpretation of subtype is different than ususal; especially our system has no subsumption rule.

[4]Notice, however, that it is not possible to conclude that $\ll \forall \mathsf{Y} \leq \mathsf{Number.Set(Y)} \gg$ is a subtype of $\ll \forall \mathsf{Y} \leq \mathsf{Number.Collection(Y)} \gg$.

- *InstanceOf*(X) indicates that the type of the result will be an instance of the receiver, or the argument which was called X; this function is only allowed if the bound of X is an instance of class Metaclass.

- *ClassOf*(X) indicates that the result will be the class of the receiver or parameter X.

For example:

- $\ll\forall X \leq$ Point2D.X $\longrightarrow$ Number$\gg$: a method with this signature is applicable to objects of class (or subclass of) Point2D, it does not expect any arguments, and it returns a ⟨Number⟩. The method x from figure 12.3 has this signature.

- $\ll\forall X \leq$ ClassA.$\forall Y \leq$ ClassB.X $\times$ Y $\longrightarrow$ X$\gg$: a method with this signature is applicable to objects of class (or subclass of) ClassA, expects arguments of class (or subclass of) ClassB, and returns objects of the same class as the receiver. Thus, if the receiver is of class ClassC, where ClassC is a subclass of ClassA, then the result will be of class ClassC as well.

- $\ll\forall X \leq$ ClassA.X $\longrightarrow$ *ClassOf*(X)$\gg$: a method with this signature is applicable to objects of class (or the subclasses of) ⟨ClassA⟩ and returns the class of the receiver. For instance, a method myClass defined in class ClassA has this signature:

```
myClass
    "≪∀X ≤ ClassA.X ⟶ ClassOf(X)≫ "
    ↑self class
```

Obviously, renaming of variables does not affect the meaning of signatures. For example the signature $\ll\forall X \leq$ Boolean.$\forall Y \leq$ Magnitude.X $\times$ Y $\longrightarrow$ Y$\gg$ is equivalent to the signature $\ll\forall Y \leq$ Magnitude.$\forall X \leq$ Boolean.X $\times$ Y $\longrightarrow$ X$\gg$.

Notice the difference between the signatures $\ll\forall X \leq$ ClassA.X $\longrightarrow$ ClassA$\gg$ and $\ll\forall X \leq$ ClassA.X $\longrightarrow$ X$\gg$: If a message m1 has the first signature, and a message m2 has the latter signature, and ClassB $\sqsubseteq$ClassA, and expr has type ⟨ClassB⟩, then expr m1 has type ⟨classA⟩, whereas expr m2 has type ⟨ClassB⟩.

12.3.5 Receiver Type, Argument Type, Application of a Signature

For a signature Sig we need to speak about the *receiver type*, $[\text{Sig}]^0$, and the *i-th argument type*, $[\text{Sig}]^i$.

For example, in SigA = $\ll\forall X \leq$ Boolean.$\forall Y \leq$ Magnitude.X $\times$ Y $\longrightarrow$ Y$\gg$, the receiver type, $[\text{SigA}]^0$, is ⟨Boolean⟩, and the first argument type is $[\text{SigA}]^1$ = ⟨Magnitude⟩.

In a more interesting example, for the signature SigB $=\ll\forall$X $\leq$ Boolean.$\forall$Y $\leq$ Number.$\forall$Z $\leq$ Array(X).Z $\times$ Y $\longrightarrow$ Set(Y)$\gg$ we have that $[\text{SigB}]^0=\ll\forall$X $\leq$ Boolean.Array(X)$\gg$, and $[\text{SigB}]^1=\langle$Number$\rangle$.

As we discussed earlier, the signature expresses a functional relation between receiver or argument type, and result type. In the case where the result type is closed, as *e.g.* in the signature $\ll\forall$X $\leq$ Boolean.X $\longrightarrow$ SmallInteger$\gg$ the functional relation is trivial, but in the case where the result type is open, it depends on the type of the arguments and/or receiver. For example, in SigB the result is the parameterized class Set applied to the argument.

As we shall argue later, the type of a message expression depends on the types of the receiver and/or arguments. We therefore need the concept of *application* of a signature to a list of types.

For example, the application of SigA to the types the types $\langle$Boolean$\rangle$ and $\langle$SmallInteger$\rangle$, returns the result type SigA($\langle$Boolean$\rangle$,$\langle$SmallInteger$\rangle$) = $\langle$SmallInteger$\rangle$. As a further example, the application SigB($\langle$Array(True)$\rangle$,$\langle$Float$\rangle$) returns the result type $\langle$Set(Float)$\rangle$.

For the sake of brevity, we shall not give here a complete definition of the above terms.

12.3.6 Method Types

$$\textit{MethodType} \quad ::= \quad \textit{MethodSignature}^*$$

Every method body has one or more signatures. The type of a message pattern contains all the signatures from the methods with this pattern.

For example, the method body add: from class Point2D has one signature, namely $\ll\forall$X $\leq$ Point2D.$\forall$Y $\leq$ Point2D.X $\times$ Y $\longrightarrow$ Y$\gg$; the method body add: from class Point3D has another signature, namely $\ll\forall$X $\leq$ Point3D.$\forall$Y $\leq$ Point3D.X $\times$ Y $\longrightarrow$ Y$\gg$. Therefore, the message pattern add: has these two signatures in its method type.

An example of *one* method with *several* signatures is invertedAdd: from figure 12.3.

12.4 Type Rules in ST&T

12.4.1 The Environment

The type of an expression depends on the types of the variables, literals and/or message patterns it consists of. The type of a literal is uniquely determined by the class it belongs to. The types of variables and message patterns are supplied by the user in appropriate comments [5] and they are stored in the *environment.*

The environment, which we represent as Γ, can be understood as a symbol table; it is a sequence of *declarations* of the form x : T, or X $\leq$ T, or of the form $\mathsf{m}_1 : \ldots \mathsf{m}_n ::$ SigB, where T is an object type, x is a variable declared to be of

[5] or they could be automatically inferred in a type inference system

type T, X is a type variabel bound by T, and $m_1 : \ldots m_n :$ is a message selector for which there exists a method with this selector, and which was defined to have SigB.

We shall denote with Γ_x the unique type of x in Γ and with $\Gamma_{m_1:\ldots m_n:}$ the set of all signatures given for $m_1 : \ldots m_n :$ in Γ.

An assertion of the form $\Gamma \vdash$ expr : TypeA signifies that under the types described in the environment Γ, the expression expr has type TypeA. Especially,

$$\frac{\text{x is a variable}}{\Gamma \vdash x : \Gamma_x}$$

For instance, a local variable aPoint declared in Γ to be of type ⟨Point⟩ has type ⟨Point⟩; in an environment which contains the corresponding declarations, the message selector asPoint has the following types: $\ll\forall X \leq \text{Point}.X \longrightarrow \text{Point}\gg$, $\ll\forall X \leq \text{LineSegment}.X \longrightarrow \text{Point}\gg$ and $\ll\forall X \leq \text{Number}.X \longrightarrow \text{Point}\gg$.

The environment contains the method types for all methods (which are provided by the user in form of special comments, or by using the special browser described in section 12.5).

12.4.2 Type Checking Classes

A class is type checked by type checking all its methods, in an environment which contains the types of all methods in the system, and the types of the instance and class variables, which, again, are provided by the user in special comments.

A method may, in general, have several signatures. A method of a class ClassA is checked by checking it for all the signatures which have a recreiver type that is a subclass of ClassA.

When type ckecking a method with message pattern $m_1 : x_1 \ldots m_n : x_n$ for a signature $\ll\forall X_0 \leq \text{type}_0.\forall X_1 \leq \text{type}_1 \ldots \forall X_n \leq \text{type}_n.X_1 \times \ldots \times X_n \longrightarrow \text{type}_{n+1}\gg$ defined in class ClassA, then in addition to the object types for all instance or class variables, the environment will also contain the types of the parameters as inferred by the signature, *i.e.* self:X_0, x_1:X_1, ... x_n:X_n, $X_0 \leq \text{type}_0$, $X_1 \leq \text{type}_1$, ... $X_n \leq \text{type}_n$, ResultType $= \text{type}_{n+1}$[6]. The environment will also

[6]Some readers may wonder why we do not have the simpler extension of the environment by self:type_0, x_1:type_1, ... x_n:type_n, ResultType $= \text{type}_{n+1}$. The reason is that the simpler extension does not carry enough information. Consider, namely the following two methods:

```
class name ClassA
instance methods
    m1
        "≪∀X ≤ ClassA.X ⟶ X≫ "
        ↑self
    m2
        "≪∀X ≤ ClassA.X ⟶ X≫ "
        ↑ClassA new
```

The method m1 is type correct for its signature, but the method m2 is not. In order to make this distinction, we *need* the environment where self:X, X ≤ ⟨ClassA⟩.

contain he types of the temporary variables, again as specified by the user in the appropriate comments.

A method body is type checked by type checking all the expressions it consists of.

12.4.3 Types of Assignments

An assignment is legal when the type of the left hand side is larger than the type of the right hand side. Furthermore, an assignment has a type, namely the type of the right hand side.

$$\frac{\begin{array}{l}\Gamma \vdash \mathsf{x} : \mathsf{TypeA} \\ \Gamma \vdash \mathsf{expr} : \mathsf{TypeB} \\ \mathsf{TypeB} \leq \mathsf{TypeA}\end{array}}{\Gamma \vdash \mathsf{x} \leftarrow \mathsf{expr} : \mathsf{TypeB}}$$

For instance, given a variable x of type ⟨Point ∨ LineSegment⟩ and an expression expr of type ⟨Point⟩ the assignment x ←expr is legal, and has the type ⟨Point⟩.

The fact that the assignment has the smaller type is not arbitrary; for example, if y has type ⟨Point ∨ Number⟩, then although the type of x is not a subtype of the type of y, the assignment y ←x ←expr is legal.

12.4.4 Types of Return Expressions

A return expression is legal, if the type of the expression being returned is a subtype of the reult type (stored in Γ) of the current method and signature.

$$\frac{\begin{array}{l}\Gamma \vdash \mathsf{expr} : \mathsf{TypeA} \\ \Gamma \vdash \mathsf{TypeA} \leq \mathsf{ResultType}\end{array}}{\Gamma \vdash \uparrow \mathsf{expr} : \mathsf{TypeA}}$$

12.4.5 Types of Message Expressions

When considering the type of a message expression, we need to consider the type of the receiver, the type of the arguments, and all possible types for the message selector.

The type of the receiver and arguments may be unions of several single types, *e.g.* ⟨Number ∨ LineSegment⟩. We consider each combination of single types for the receiver and arguments separately, and then build the union of all result types thus calculated. For example, if anExpr has type ⟨Number ∨ LineSegment⟩, and the message pattern aMess has the signature ≪∀X ≤ Object.X ⟶ X≫ , then the type of the message expression anExpr aMess has type ⟨Number ∨ LineSegment⟩.

For each such combination of single types, the type of the message expression is calculated according to the next four steps:

1. Distinguish from all possible signatures of the message, those that are relevant to the actual receiver and arguments. A signature is *relevant* for a message expression, if the types of the receiver expression and arguments expressions are in the subtype *or* supertype relationship with the corresponding receiver or argument types of the signature.

   If the set of relevant signatures is empty, then the expression has type ⟨TypeError⟩.

2. The methods from a subclass override the methods from a superclass. Thus, out of the relevant signatures we only need to consider those whose receiver and argument types "are nearest", or "match best" the receiver and argument.

   A signature *matches* a message expression *better* than another signature, if the receiver type of the first signature is a supertype of the type of the receiver expression, and a subtype but different from the receiver type of the second signature. Also, a signature matches a message expression better than another signature, if they have the same receiver type which is a supertype of the type of the receiver expression, and the argument types of the first signature are supertypes of the types of the corresponding argument expressions, and the argument types of the first signature are subtypes of the corresponding argument types of second signature, and at least one argument type of the first is different from the corresponding argument type in the second signature.

3. Now, check whether the minimal signatures calculated in the previous step "cover" the message expression.

   A set of signatures *covers* a message expression if at least one of the signatures has receiver type and argument types that are supertypes of the corresponding types of the receiver expression and argument expressions, or if, for any signature in the message type, whose receiver type is a subtype of the type of the receiver expression, there exists a signature[7] with the same receiver type, and whose argument types are supertypes of the corresponding types of the argument expressions.

   If the set of relevant signatures does not cover the message expression, then the expression has type ⟨TypeError⟩.

4. The type of the message expression is calculated as the union of all signatures from the minimal set applied to the types of receiver and arguments.

In the remainder of this section we shall give full definitions of the above introduced concepts.

The type of $\mathsf{expr}_0\mathsf{m}_1 : \mathsf{expr}_1 \ldots \mathsf{m}_n : \mathsf{expr}_n$ is the union of all types obtained, when considering the message $\mathsf{m}_1 : \ldots \mathsf{m}_n :$ to be sent to y_0 with arguments y_1

[7] not necessarily a different one

... y_n, and where each of the y_i has one of the single types from the type of $expr_i$.

$$\frac{\begin{array}{l}\Gamma \vdash expr_i : Type_{i,1} \vee Type_{i,2} \ldots \vee Type_{i,n_i} \text{ for all } i \in 0 \ldots n \\ y_i \text{ does not appear in } \Gamma \text{ for all } i \in 0 \ldots n \\ \Gamma, y_0 : Type_{0,j_0}, \ldots y_n : Type_{n,j_n} \vdash y_0 m_1 : y_1 m_2 : \ldots m_n : y_n : T_{j_1,\ldots j_n}\end{array}}{\Gamma \vdash expr_0 m_1 : expr_1 \ldots m_n : expr_n : \bigcup_{j_i \in 1 \ldots n_i} T_{j_1,\ldots j_n}}$$

Thus, from now we shall assume that the types of the receiver and argument, $Type_0, \ldots, Type_n$, are single types.

Relevant signatures

A message pattern may have several signatures. We consider a signature to be relevant for a message expression, if the types of the receiver and arguments of the expression are in the subtype or supertype relationship with the corresponding receiver or argument types of the signature.

$$\frac{[Sig]^i \leq Type_i \text{ or } Type_i \leq [Sig]^i \text{ for all } i \in 0, \ldots n}{Sig \text{ is relevant signature for } Type_0, \ldots, Type_n}$$

For example, $\ll \forall X \leq Magnitude.\forall Y \leq Boolean.X \times Y \longrightarrow \ldots \gg$, and $\ll \forall X \leq ArithmeticValue.\forall Y \leq Object.X \times Y \longrightarrow \ldots \gg$, and $\ll \forall X \leq SmallInteger.\forall Y \leq False.X \times Y \longrightarrow \ldots \gg$, are relevant for $\langle Number \rangle$, $\langle Boolean \rangle$.

On the other hand, the signatures $\ll \forall X \leq False.\forall Y \leq True.X \times Y \longrightarrow \ldots \gg$, $\ll \forall X \leq SmallInteger.\forall Y \leq Number.X \times Y \longrightarrow \ldots \gg$, are not relevant for $\langle Number \rangle$, $\langle Boolean \rangle$.

If the set of relevant signature is empty, then the message expression is type incorrect:

$$\frac{\begin{array}{l}\Gamma \vdash expr_i : Type_i \text{ for all } i \in 0 \ldots n \\ \{Sig | Sig \in \Gamma_{m_1:m_2:\ldots m_n:} \text{ and } Sig \text{ is relevant for } Type_0, \ldots, Type_n\} = \emptyset\end{array}}{\Gamma \vdash expr_0 m_1 : expr_1 \ldots m_n : expr_n : \langle TypeError \rangle}$$

Covering signatures

However, in some cases the signatures do not cover all possibilities. A message expression is illegal (has type $\langle TypeError \rangle$), if the minimla signature set for the message pattern "covers " the types of receiver and arguments.

A signature set *covers* the types $Type_0, \ldots, Type_n$, if ther excists a signature whose receiver type is a supertype of $Type_0$, and the i-th argument type is a supertype of $Type_i$, and if for all subtypes of $Type_0$ if there exists a signature in the set, then then there exists another signature, so, that the i-th argument type is a supertype of $Type_i$.

$$\frac{\begin{array}{l}Sig \text{ is signature} \\ Type_i \leq [Sig]^i \text{ for all } i \in 0 \ldots n\end{array}}{Sig \text{ strongly covers } Type_0, \ldots, Type_n}$$

$$\frac{\begin{array}{l}\text{Sig is signature}\\ \mathsf{Type_i} \leq [\mathsf{Sig}]^i \text{ for all } i \in 1 \ldots n\end{array}}{\mathsf{Sig} \text{ weakly covers } \mathsf{Type_0}, \ldots, \mathsf{Type_n}}$$

$$\frac{\begin{array}{l}S \text{ is a set of signatures}\\ \exists \mathsf{Sig} \in S \;\; \mathsf{Sig}\text{covers strongly}\mathsf{Type_0}, \ldots, \mathsf{Type_n}\\ \forall \mathsf{T} < \mathsf{Type_0}, \mathsf{Sig} \in S, [\mathsf{Sig}]^0 = \mathsf{T} \;\; \exists \mathsf{Sig}' \in \mathsf{S} \; [\mathsf{Sig}']^0 = \mathsf{T},\\ \qquad\qquad\qquad\qquad \mathsf{Sig}'\text{covers weakly } \mathsf{Type_0}, \ldots, \mathsf{Type_n}\end{array}}{\mathsf{S} \text{ covers } \mathsf{Type_0}, \ldots, \mathsf{Type_n}}$$

The signature Sig4=$\ll\forall$X $\leq$ Number.$\forall$Y $\leq$ Boolean.X $\times$ Y $\longrightarrow$ SmallInteger$\gg$ covers strongly the types $\langle$SmallInteger$\rangle$,$\langle$Boolean$\rangle$; but it does not cover strongly the types $\langle$ArithmeticValue$\rangle$,$\langle$False$\rangle$ nor does it cover strongly the types $\langle$Integer$\rangle$, $\langle$Object$\rangle$. Sig4 covers weakly the types $\langle$SmallInteger$\rangle$,$\langle$Boolean$\rangle$, and the types $\langle$ArithmeticValue$\rangle$,$\langle$False$\rangle$ but it does not cover weakly the types $\langle$Integer$\rangle$, $\langle$Object$\rangle$. Strong covering is a stronger requirement than weak covering. Also, any signature with no arguments weakly covers any type.

For a message expression to be legal, *at least one* minimal signature must strongly cover the types of the receiver and arguments, and for each receiver subtype *at least one* minimal signature must weakly cover the types of the arguments. Otherwise the type of the method expression is $\langle$TypeError$\rangle$.

The best firring sinatures, $\mathcal{M}(R, \mathsf{Type_0}, \ldots, \mathsf{Type_n})$, are calcuated according to the definitions in the next section, 12.4.5.

$$\frac{\begin{array}{l}\Gamma \vdash \mathsf{expr_i} : \mathsf{Type_i} \text{ for all } i \in 0 \ldots n\\ R = \{\mathsf{Sig} | \mathsf{Sig} \in \Gamma_{\mathsf{m_1:m_2:\ldots m_n:}} \text{ and Sig relevant for } \mathsf{Type_0}, \ldots, \mathsf{Type_n}\}\\ M = \mathcal{M}(R, \mathsf{Type_0}, \ldots, \mathsf{Type_n}) \text{ does not cover } \mathsf{Type_0}, \ldots, \mathsf{Type_n}\end{array}}{\Gamma \vdash \mathsf{expr_0 m_1 : expr_1 \ldots m_n : expr_n} : \langle\mathsf{TypeError}\rangle}$$

Best fitting signatures

Clearly, for given types of receiver and arguments, several signatures can be relevant. For example, in the expression v3 add: v3 the type of receiver and argument are $\langle$Point3D$\rangle$ and $\langle$Point3D$\rangle$. The message pattern add: has two signatures, $\ll\forall$X $\leq$ Point2D.$\forall$Y $\leq$ Point2D.X $\times$ Y $\longrightarrow$ Y$\gg$, and $\ll\forall$X $\leq$ Point3D.$\forall$Y $\leq$ Point3D.X $\times$ Y $\longrightarrow$ Y$\gg$. Both signatures are relevant for the types $\langle$Point3D$\rangle$ and $\langle$Point3D$\rangle$. However, because the method to be executed depends on the class of the receiver, we only need to consider the second signature; we say that the second signature "fits better" than the first.

In general, if two relevant signatures prescribe different receiver types which are supertypes of the type of the receiver of the message, then, we only want to consider the signature with the closest fitting receiver type, *i.e.* if TypeOfRec is the type of the receiver, and TypeOfRec $\leq$ [SigA]^0 and [SigA]^0 $\leq$ [SigB]^0, then SigA "matches better" than SigB, and only SigA will be considered.

The matches better relationship is an ordering relationship, which depends on the type of the receiver, $\mathsf{Type_0}$, and the types of the arguments $\mathsf{Type_1}$, ... $\mathsf{Type_n}$. An assertion SigA $\ll_{\mathsf{Type_0}, \ldots, \mathsf{Type_n}}$ SigB means that the expression where the receiver has type $\mathsf{Type_0}$, and the arguments have types $\mathsf{Type_1}$, ... $\mathsf{Type_n}$ is better matched by the signature SigA than by the signature SigB.

A signature matches better than another signature if its receiver type is a supertype of the type of the receiver, and a subtype and different than the receiver type of the second signature.

$$\frac{\begin{array}{l}\mathsf{SigA}, \mathsf{SigB} \text{ are signatures}\\ \mathsf{Type_0} \leq [\mathsf{SigA}]^0\\ [\mathsf{SigA}]^0 \leq [\mathsf{SigB}]^0\\ [\mathsf{SigA}]^0 \neq [\mathsf{SigB}]^0\end{array}}{\mathsf{SigA} \ll_{\mathsf{Type_0},\ldots,\mathsf{Type_n}} \mathsf{SigB}}$$

For example, given $\mathsf{s1}{=}\ll \forall \mathsf{X} \leq \mathsf{Point3D}.\forall \mathsf{Y} \leq \mathsf{Point2D}.\mathsf{X} \times \mathsf{Y} \longrightarrow \ldots \gg$, $\mathsf{s2}{=} \ll \forall \mathsf{X} \leq \mathsf{Point2D}.\forall \mathsf{Y} \leq \mathsf{Point3D}.\mathsf{X} \times \mathsf{Y} \longrightarrow \ldots \gg$, $\mathsf{s1} \ll_{\mathsf{Point3D},\mathsf{Point3D}}$ s2 Notice, that only the receiver types are taken into consideration for the above rule, and the fact that the argument type of the second signature is a subclass of the argument type of the second signature did not affect the matches better $\ll_{\mathsf{Point3D},\mathsf{Point3D}}$-relationship. Also, notice, that the type of the receiver, $\mathsf{Type_0}$ is important, and, for example, it is NOT the case that $\ll \forall \mathsf{X} \leq \mathsf{Point3D}.\forall \mathsf{Y} \leq \mathsf{Point2D}.\mathsf{X} \times \mathsf{Y} \longrightarrow \ldots \gg \ \ll_{\mathsf{Point2D},\mathsf{Point3D}} \ll \forall \mathsf{X} \leq \mathsf{Point2D}.\forall \mathsf{Y} \leq \mathsf{Point2D}.\mathsf{X} \times \mathsf{Y} \longrightarrow \ldots \gg$.

Notice that for different receiver types the concept of better fitting signature only depends on the type of the receiver, and not on the types of the arguments. This is so, because in Smalltalk (as in most object oriented programming languages) the method to be executed at run time is selected depending on the class of the receiver, and is independent of the class of the arguments. On the other hand, where the receiver type of the signature is a subtype of the receiver type in the expression there is no concept of better fitting signature.

Notice also, that if two relevant signatures prescribe different receiver types which are supertypes of the type of the receiver of the message, then one of the two signatures will be fitting better than the other.

On the other hand, if two signatures for a message selector have the same receiver type, and all the argument types of the first are smaller than the argument types of the second, and the argument types are smaller than the types of the corresponding arguments in the expression, then the first signature matches better than the latter.

$$\frac{\begin{array}{l}\mathsf{SigA}, \mathsf{SigB} \text{ are signatures}\\ \mathsf{Type_0} \leq [\mathsf{SigA}]^0 = [\mathsf{SigB}]^0\\ \mathsf{Type_i} \leq [\mathsf{SigA}]^i \leq [\mathsf{SigB}]^i \text{ for all } i \in 1\ldots n\\ [\mathsf{SigA}]^i \neq [\mathsf{SigB}]^i \text{ for one } i \in 1\ldots n \text{ at least}\end{array}}{\mathsf{SigA} \ll_{\mathsf{Type_0},\ldots,\mathsf{Type_n}} \mathsf{SigB}}$$

For example, given $\mathsf{sig1}{=} \ll \forall \mathsf{X} \leq \mathsf{Point3D}.\forall \mathsf{Y} \leq \mathsf{Point2D}.\mathsf{X} \times \mathsf{Y} \longrightarrow \ldots \gg$ $\mathsf{sig2}{=}\ll \forall \mathsf{X} \leq \mathsf{Point3D}.\forall \mathsf{Y} \leq \mathsf{Object}.\mathsf{X} \times \mathsf{Y} \longrightarrow \ldots \gg$. $\mathsf{sig1} \ll_{\mathsf{Point3D},\mathsf{Point3D}}$ sig2. Notice, that the $\ll_{\ldots}$ relationship depends on the types considered. For instance, it is not the case that $\mathsf{sig1} \ll_{\mathsf{Point3D},\mathsf{Object}} \mathsf{sig2}$ The $\ll_{\mathsf{Type_0},\ldots,\mathsf{Type_n}}$-relationship is a partial order.

And now, for a message expression, we only consider the relevant signatures that best match the types of the receiver and arguments.

The *minimal signatures* set, $\mathcal{M}(S, \mathsf{Type}_0, \ldots, \mathsf{Type}_n)$, is the uniquely defined set that contains all minimal signatures in the sense of the $\ll_{\mathsf{Type}_0,\ldots,\mathsf{Type}_n}$-ordering. That is, $\mathcal{M}(S, \mathsf{Type}_0, \ldots, \mathsf{Type}_n) \subseteq S$ and for any pair of signatures SigA, SigB with SigA, SigB $\in S$, SigA $\ll_{\mathsf{Type}_0,\ldots,\mathsf{Type}_n}$ SigB implies that either there exists a further SigC such that SigC $\ll_{\mathsf{Type}_0,\ldots,\mathsf{Type}_n}$ SigA, or that SigA $\in \mathcal{M}(S, \mathsf{Type}_0, \ldots, \mathsf{Type}_n)$. [8]

Type of a message expression

For the set of minimal signatures, we consider the application of the signatures to the receiver and argument types:

$$\frac{\Gamma \vdash \mathsf{expr_i} : \mathsf{Type_i} \text{ for all } i \in 0, \ldots n \qquad M \text{ is the minimal signature set, and covers } \mathsf{Type}_0, \ldots, \mathsf{Type}_n}{\Gamma \vdash \mathsf{expr}_0 \mathsf{m}_1 : \ldots \mathsf{m}_n : \mathsf{expr}_n : M \circ \mathsf{Type}_0, \ldots, \mathsf{Type}_n}$$

According to the above, the expression v3 invertedAdd: v3 has type ⟨Point3D⟩.

An example

In figure 12.5 we show how the steps from above determine the types of some message expressions. Remember, that x2 :⟨Point2D⟩, x3 :[Point3D], y2 :[Point2D], and y3 :[Point3D]. Also we call $\ll \forall X \leq \mathsf{Point2D}.X \times [\mathsf{Point2D}] \longrightarrow X \gg$ by $S_1$and $\ll \forall X \leq \mathsf{Point3D}.\forall Y \leq \mathsf{Point2D}.X \times Y \longrightarrow X \gg$ by S_2.

12.5 Discussion, Implementation, Soundness

In [Ghelli, 1991] it is also suggested that a message pattern may have several signatures, and the best fitting signature is selected. However, every method body has just one signature, and also, the signatures of the methods in the subclass are either subtypes or supertypes of the signatures in the superclass, thus allowing for both covariant and covariant extensions of the signature types.

In ST&T several signatures are legal for one method body. For example, in the absence of the subclass Point3D, for the method add: from 12.1, both SigA=$\ll \forall X \leq \mathsf{Point2D}.\forall Y \leq \mathsf{Point2D}.X \times Y \longrightarrow Y \gg$ and SigB= $\ll \forall X \leq \mathsf{Point2D}.\forall Y \leq \mathsf{Point2D}.X \times Y \longrightarrow \mathsf{Point2D} \gg$ would be valid signatures. In this case, the first signature is "better" than the second, in the sense that it gives a more accurate description of the result type.[9] This raises the question of existence of principal signatures; this question is still open.

In our system any signatures are allowed for the method in a subclass whose superclass has a method of the same message pattern.

[8] Notice, that because $\ll_{\ldots}$ is only a partial order, the set of minimal signatures may contain more than one signature; examples for this are discussed in section 12.4.5.

[9] Consider, for example, the expression v2 add: v3. According to the first signature it is of type ⟨Point3D⟩, whereas according to the second it has type ⟨Point2D⟩

the expression	Step 1: relevant $\mathcal{R}$	Step 2: cover?	Step 3a: match better $\ll$	Step 3b: minimal $\mathcal{M}$	Step4 the type
x2 invAdd: y3	S_2	NO	-	S_2	⟨TypeError⟩
x2 invAdd: y2	S_2, S_1	YES	-	S_2, S_1	⟨Point2D⟩
x2 invAdd: x3	S_2	NO	-	S_2	⟨TypeError⟩
x2 invAdd: x2	S_2, S_1	NO	-	S_2, S_1	⟨TypeError⟩
x3 invAdd: y3	S_2	YES	-	S_2	⟨Point3D⟩
x3 invAdd: y2	S_2, S_1	YES	$S_2 \ll S_1$	S_2	⟨Point3D⟩
x3 invAdd: x3	S_2	YES	-	S_2	⟨Point3D⟩
x3 invAdd: x2	S_2, S_1	YES	$S_2 \ll S_1$	S_2	⟨Point3D⟩
y2 invAdd: y3	-	NO	-	-	⟨TypeError⟩
y2 invAdd: y2	S_1	YES	-	S_1	[Point2D]
y2 invAdd: x3	-	NO	-	-	⟨TypeError⟩
y2 invAdd: x2	S_1	NO	-	S_1	⟨TypeError⟩
y3 invAdd: y3	S_2	YES	-	S_2	[Point3D]
y3 invAdd: y2	S_2, S_1	YES	$S_2 \ll S_1$	S_2	[Point3D]
y3 invAdd: x3	S_2	YES	-	S_2	[Point3D]
y3 invAdd: x2	S_2, S_1	YES	$S_2 \ll S_1$	S_2	[Point3D]

Figure 12.5: Calculation of the types of some message sends

This has the consequence, that when a new method is added to a subclass, under circumstances, it can "invalidate" the original method in the superclass. For example, the method add: from Point3D is type incorrect when the receiver has type ⟨Point2D⟩. The only type correct use of the message add: is one where the type of the receiver is [Point2], or, of course ⟨Point3D⟩.

On the other hand, if the signature of a method in a subclass requires its argument types to be the same or supertypes of the argument types of the signature of the superclass (*i.e.* is a subtype in the sense of, say, [Cardelli and Mitchell, 1989] of the signature in the superclass), then it holds that the method of the subclass does not "invalidate" the method from the superclass. In that sense, the antimonotonicity rule applies in ST&T too.

Type checking must take place *after* the complete development of a program. Namely, if a Smalltalk program is checked and found type correct, and then a new method is added into the program, then other parts of this program that were found type correct would have been redendered type incorrect. This can be tackled by restricting the modifications allowed in a module after it has declared its export list.

ST&T has been implemented in a prototype version [Karathanos, 1993], where the programmer is requested to give the types of all instance variables, class variables, temporary variables, and methods, and points to a method and starts the type checker, which then checks the method body.

Our type checking browser is an extension of the original system browser of the Visualworks Parcplace ST80 development system, a view of it is shown in figure 6 in the appendix. It was developed by creating appropriate subclasses of the class Browser for the new browser, and of all classes representing the nodes of the abstract syntax tree, *e.g.* MethodNode, AssignmentNode *etc* . Additional classes for the representation of types had to be developed. The existing system is about 3000 lines of Smalltak code, and was used to check several example programs, and student programs. We were surprised that we sometimes had given wrong types for programs we had developed and fully tested. Further work is necessary to automate testing of larger pieces of code.

In [Drossopoulou and Yang, 1995] we demonstrate the soundness of our approach. We very much follow the approach from [Castagna *et al.*, 1992, Castagna, 1994]. We define $\lambda^{\&,S}$, a concise language of expressions which models the most representative issues in dynamically bound object oriented languages: objects, classes, methods, a sublcass relationship, message sends that consist of one or more message bodies sent to a receiver expression and possibly to several argument expressiogs.

We give the operational semantics for $\lambda^{\&,S}$-expressions in terms of a rewrite relation $\longrightarrow\!\!\triangleright$. Every object has its immutable class, which constitutes its *dynamic* type. Message sends are resolved by consulting the dynamic type of the receiver, and choosing the method body whose receiver type matches best the type of the receiver value. The application of a method body to expressions is described by substitution, and by β-reduction.

We develop a type inference system, which describes the *static* types of expressions, We then demonstrate the soundness of the type system, by proving that the operational semantics preserves the static types. In other words, for any expressions E, E', and type T with $\Gamma \vdash \mathsf{E} : \mathsf{T}$, and $\mathsf{E} \longrightarrow\!\!\rhd \mathsf{E}'$ it also holds that $\Gamma \vdash \mathsf{E}' : \mathsf{T}'$ and $\Gamma \vdash \mathsf{T} \leq \mathsf{T}'$.

12.6 Conclusions and Further Work

A static type system for Smalltalk was developed. We were able to adopt the intuitive appraoch, whereby subclasses are subtypes, by adopting a modified interpretation for the subtype relationship.

A method may have several signatures, and this introduces the need to introduce the notions of best-fitting signatures, and covering signatures. This facts also abolishes the need to re-type check methods in the subclasses.

The representation of the types of arguments and the receiver by a variable which may appear in the result type, leads to an elegant solution to the problem of representing the type of self. Also, it allows for a richer language of types, and a more accurate description of the dynamic behaviour of Smalltalk. For instance, our type checker will recognize v2 add: v3 to be of class Point3D, and thus the expression v3 invertedAdd: (v2 add: v3) will be accepted as type correct, whereas it would be rejected in most other type systems.

We plan to work further in a better instrumentation of our implementation and checking of larger programs, and extensions to allow multiple inheritance, and dependent sum types, invetigate the possibility of type inference.

Acknowledgements

We are grateful to Ralph Johnson for a lot of information and advice at the initial stages of our work, to Gilad Braha, for ready communication and discussion about his approach, to Ross Paterson for pointing out important parallels to our work, and to Theoni Pittoura for her careful reading and helpful comments on this paper.

Appendix: The Extended Type Checking Browser

A view of the ST&Tbrowser can be seen in figure 6.

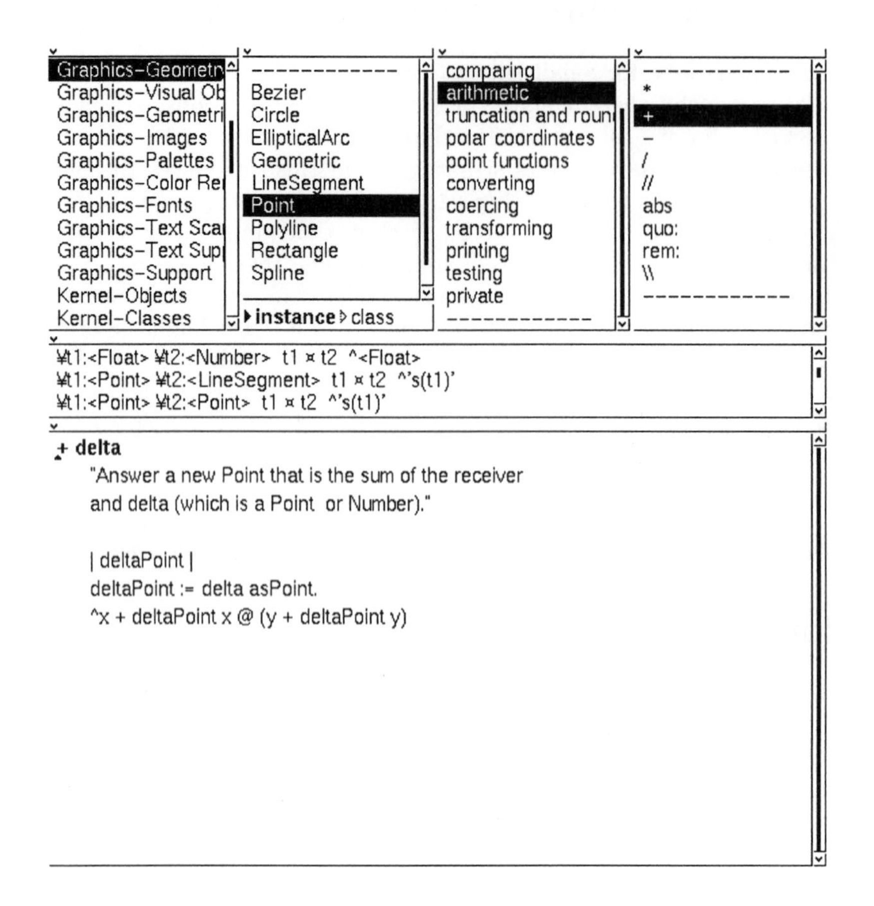

Figure 12.6: The Type checking Browser

Chapter 13

A Note on the Semantics of Inclusion Polymorphism

13.1 The Concept of Inclusion Polymorphism

Inclusion polymorphism is one of the fundamental innovations of object technology (others are the adoption of the object as the unit of encapsulation and the use of inheritance for incremental programming), and is widely touted as enhancing the extensibility and maintainability of object oriented programs compared with structured (e.g. Pascal) or object-based (e.g. Ada-83) programs.

In object oriented programming (OOP), an entity (variable) can denote an object of its static (declared) type, T, or any of T's *subtypes*.

The dynamic (actual) type of the object attached to the entity is determined only at run-time. When a message is sent to an entity, the action taken is determined by the object that the entity is attached to (late binding) i.e. the object's identity, state and class (the entity's dynamic type). This is subtype/*inclusion polymorphism* [Cardelli and Wegner, 1985]:

- Polymorphism - an entity can denote an instance of many distinct types and thus the same message can be interpreted in many distinct ways
- Inclusion - the type of any object attached to an entity is a subtype of the entity's static type.

The fundamental notion here is that of a *subtype*. Various synonyms are used for *subtyping* in the literature, including *type compatibility*, *type conformance*,

substitutivity, *is-a* and *substitutability* . There are several possible interpretations of subtype ranging from subclass (or derived or child class) via signature subtype [Cardelli and Wegner, 1985] to behavioural subtype [America, 1991]. Given types S and T, S is a behavioural subtype of T iff every instance of S

is behaviourally substitutable for (i.e. behaves compatibly with) some instance of T.

We focus on the behavioural interpretation of subtyping in this chapter, and call the resulting kind of subtype polymorphism *semantic inclusion polymorphism*, since all subtypes are required to specialize the semantic properties of their supertypes. Our aim is to give a definitive formalization of semantic inclusion polymorphism using the very general framework of structured logical theories that can serve as a semantic basis for both specification [Kent, 1995] and programming languages [Goldblatt, 1982].

13.2 Formalization of Inclusion Polymorphism

We use many sorted first order deterministic dynamic logic, which is appropriate for expressing properties of sequential object-oriented systems. However, we depart from the conventional literature (where many sorts are used - one for each class) in assuming a single sort of object identities. For each class, a predicate for class membership is defined on this object identity (ID) domain. An identity satisfies the class predicate if it denotes an instance of that class. Static and dynamic subclasses and roles [Wieringa *et al.*, 1994] can all be modelled uniformly in this way. Static subclasses e.g. S can be specified by the axiom:

$$\forall a : ACT\ \forall x : ID\ (S(x) \rightarrow [a]S(x)),$$

where ACT is the sort of actions, where an action is a primitive state-changing operation/event. Here, a is an action, and [a]S(x) is a *modal* formula expressing that x is a member of S after the occurence of a. Note that conventional formulations where each class has its own sort of identities and subsort relations to denote subtyping can be translated into our framework by very well established techniques (see Section 4.3 of [Enderton, 1972]).

13.2.1 Object Types, Classes and Theories

An object type is a theory. It is interpreted as the set of all sentences that are true of any instance of the type.

Example 1 *We assume standard presentations for the theory of strings, integers, booleans, characters etc. Here is a specification (finite presentation) of the theory of unbounded stacks of characters:*

theory *CharStack*
features
$state : STRING$
$top : CHAR$
$size : INTEGER$
$pop : ACT$
$push(c : CHAR) : ACT$

$isempty : BOOLEAN$
axioms
$\neg isempty \rightarrow state = concatenate(string(top), [pop]state)$
$\forall c : CHAR([push(c)]state = concatenate(string(c), state))$
$isempty = (size = 0)$
$size = len(state)$
end theory *CharStack*

The type is given by the consequential closure of these axioms (including those imported from the presentations of $STRING$, $CHAR$, $INTEGER$ *etc.). This type induces a theory of the class of CharStack objects. These have identities that are a subset of* ID*, the universal set of object identities:* $\forall x(CharStack(x) \rightarrow ID(x))$*. The class theory of CharStack is specified as follows:*

class *CharStack*
signature
$\_.state : ID \rightarrow STRING$
$\_.top : ID \rightarrow CHAR$
$\_.size : ID \rightarrow INTEGER$
$\_.pop : ID \rightarrow ACT$
$\_.push(\_) : ID \times CHAR \rightarrow ACT$
$\_.isempty : ID \rightarrow BOOLEAN$
$new : STRING \rightarrow ID$
axioms
$\forall x : ID(CharStack(x) \rightarrow$
$(\neg x.isempty \rightarrow x.state = concatenate(string(x.top), [x.pop]x.state))$
$\wedge \forall c : CHAR([x.push(c)]state = concatenate(string(c), x.state))$
$\wedge\ x.isempty = (x.size = 0)$
$\wedge\ x.size = len(x.state))$
$\forall s : STRING(new(s).state = s)$
end class *CharStack*

There is a very simple relationship between the instance theory (type) of CharStack and its class theory, as illustrated above. An extra argument of the type's identity sort (in this case, CharStack) is added to each feature, and the axioms are universally quantified over this first argument. In addition, the class theory includes axioms about creation routines for new instances of the class.

The class theory of a class, C includes theorems of the form:

$$\forall x : C\ (\phi(x)) \text{ i.e. } \forall x : ID\ (C(x) \rightarrow \phi(x))$$

but in general does not contain axioms of the form:

$$\forall x : ID\ (\phi(x) \rightarrow C(x))$$

where ϕ is a predicate (unless C has a feature with result type, C). Thus the class specification gives only necessary (but not sufficient) conditions for membership of the class, C. Note that an instance of class, C need not be created using class C's creation procedure. However, the intent of the class specification (presentation) is to define both necessary and sufficient conditions for membership of the class, C.

If the class theory of C were finitely (as opposed to just effectively) presentable, say by axioms $\theta_1, \ldots, \theta_n$, then we could replace the presentation with a single sentence:

$$\forall x : ID(\theta_1(x) \wedge \ldots \wedge \theta_n(x) \leftrightarrow C(x))$$

which gives necessary and sufficient conditions for membership in C. Unfortunately this is not possible in general since most theories are not finitely presentable but instead require (finitely many) infinite axiom schemata to specify them. We must either use second-order logic (quantification over predicates) or express the sufficiency of the class theory in our (informal) metalanguage. We adopt the latter approach.

13.2.2 Behavioural Subtyping - Simple Cases

Consider first a double ended queue (deque) for which characters can be inserted and removed at both ends. Any deque provides all the behaviour of some stack i.e. has all the same properties, as well as some additional ones. The particular stack that a given deque simulates depends upon its state. We say that the class of deques is a behavioural subtype of the class of stacks. This intuition is formalized below.

Example 2 *Double Ended Queues of Characters*

theory *CharDeque*
features
$state : STRING$
$top : CHAR$
$size : INTEGER$
$pop : ACT$
$push(c : CHAR) : ACT$
$bottom : CHAR$
$pull : ACT$
$append(c : CHAR) : ACT$
$isempty : BOOLEAN$
axioms
$\neg isempty \rightarrow state = concatenate(string(top), [pop]state)$
$\neg isempty \rightarrow state = concatenate([pull]state, string(bottom))$
$\forall c : CHAR([push(c)]state = concatenate(string(c), state))$
$\forall c : CHAR([append(c)]state = concatenate(state, string(c)))$

$isempty = (size = 0)$
$size = len(state)$
end theory *CharDeque*

It is obvious that:

$$InstanceTheory(CharStack) \subseteq InstanceTheory(CharDeque)$$

since the presentation of CharDeque contains that of CharStack and logical consequence is monotonic. Thus we can safely assume that:

$$\forall x : ID \ (CharDeque(x) \rightarrow CharStack(x))$$

since being an instance of $CharDeque$ is sufficient to guarantee all the properties required for membership of class $CharStack$. If x is an object identity and x is an instance of $CharDeque$ then the object denoted by x satisfies the instance theory of $CharDeque$, but since this contains the instance theory of $CharStack$, this ensures that the object denoted by x satisfies the instance theory of $CharStack$, and so by the sufficiency of the instance theory, x must actually be an instance of $CharStack$ also.

Note that in the above example, every stack is simulated by some deque. However, it is not necesary that every instance of a class is simulated by an instance of its behavioural subtype. This is illustrated by the following example of squares and rectangles with horizontal and vertical sides.

Example 3 *Squares and Rectangles*

theory *Rectangle*
features
$move(x, y : INTEGER) : ACT$
$top, left, right, bottom : INTEGER$
axioms
$top \geq 0 \wedge left \geq 0 \wedge bottom \geq top \wedge right \geq left$
$\forall x, y : INTEGER((left + x \geq 0 \wedge top + y \geq 0 \rightarrow$
$\quad ([move(x, y)]bottom) = bottom + y \wedge ([move(x, y)]left) = left + x \wedge$
$\quad ([move(x, y)]top) = top + y) \wedge ([move(x, y)]right) = right + x))$
end theory *Rectangle*

theory *Square*
features
$move(x, y : INTEGER) : ACT$
$top, left, side, right, bottom : INTEGER$
axioms
$top \geq 0 \wedge left \geq 0 \wedge side \geq 0 \wedge bottom = top + side \wedge right = left + side$
$\forall x, y : INTEGER((left + x \geq 0 \wedge top + y \geq 0 \rightarrow$

$([move(x,y)]side) = side \wedge$
$(([move(x,y)]left) = left + x \wedge ([move(x,y)]top) = top + y))$
end theory *Square*

Now, InstanceTheory(Rectangle) ⊆ InstanceTheory(Square). This holds even though not every Rectangle instance can be thought of as a Square instance i.e. Square is not an adequate reification of Rectangle since it has a stronger class invariant. For example:
$top = left = 0 \wedge right = 10 \wedge bottom = 20 \rightarrow$
$([move(20,10)]top = 10 \wedge [move(20,10)]right = 30)$ *is a theorem of the object type, Rectangle. Since there is no value of side such that:*

$$20 = 0 + side \text{ and } 10 = 0 + side$$

it follows that the antecedent of the above implication is equivalent to false in the theory of type, Square and hence this is also a theorem of Square. Thus we can safely assume that:

$$\forall x : ID\ (Square(x) \rightarrow Rectangle(x)).$$

13.2.3 Behavioural Subtyping - Contravariance

The examples of the previous section have identical feature signatures in the class and its behavioural subtype class. However, this is not a necessary condition for behavioural subtyping. Contravariance [Harris, 1991] is the relationship between the signatures of a feature, say f, of two classes, say A and B, whereby the argument types of $A.f$ are generalizations of those of $B.f$ and the result type of $A.f$ is a specialization of that of $B.f$. This is widely considered to be a necessary (though not sufficient) condition for A to be a behavioural subtype of B. Illustrative examples are given and formalized below. The reader will notice that from now on the examples are contrived to demonstrate the full range of possibilities. It must be said that most (if not all) of the important cases of behavioural subtyping are covered by the simple case definition of the previous subsection.

Example 4 *Two contrived classes:*

theory *MyClass*
features
$equal(a : MyClass) : BOOLEAN$
$val : INTEGER$
axioms
$\forall a : MyClass(equal(a) \leftrightarrow same\_int(val, a.val))$
end theory *MyClass*

theory *MyNewClass*
features

$equal(a : ID) : BOOLEAN$
$val : INTEGER$
axioms
$\forall a : ID(MyClass(a) \rightarrow (equal(a) \leftrightarrow same\_int(val, a.val)))$
$\forall a : ID(\neg MyClass(a) \rightarrow (equal(a) = false))$
end theory *MyNewClass*

Note that the axioms of the class theory presentation of MyClass must be imported into both presentations in order to compute their theories. This is because MyClass is a supplier of both classes. Since MyClass is a subsort of ID, there is a contravariance relationship between the argument sorts of MyNewClass and MyClass.
Note that $InstanceTheory(MyClass) \subseteq InstanceTheory(MyNewClass)$.

Example 5 *Two more contrived classes:*

theory *AnotherClass*
features
$combine(a : ID) : ID$
$val : INTEGER$
axioms
$\forall a : ID(YetAnotherClass(a) \wedge combine(a) \rightarrow$
$(AnotherClass(combine(a)) \wedge combine(a).val = (val + a.val)))$
end theory *AnotherClass*

theory *YetAnotherClass*
features
$combine(a : ID) : ID$
$state : INTEGER$
$val : INTEGER$
axioms
$\forall a : ID(AnotherClass(a) \wedge combine(a) \rightarrow$
$(YetAnotherClass(combine(a)) \wedge$
$combine(a).state = state \wedge combine(a).val = (val + a.val)))$
end theory *YetAnotherClass*

Intuitively, YetAnotherClass is a behavioural subtype of AnotherClass but the rigorous demonstration is not immediately obvious since it is not possible to prove that:

$$InstanceTheory(AnotherClass) \subseteq InstanceTheory(YetAnotherClass)$$

without assuming that: $\forall x : ID(YetAnotherClass(x) \rightarrow AnotherClass(x))$.

So it is not sufficient to say that S is a subtype of T iff:

$$InstanceTheory(T) \subseteq InstanceTheory(S).$$

Sometimes (as illustrated above) we must assert the relationship $\forall x(S(x) \rightarrow T(x))$ and use that to establish the theory inclusion relationship.

So it is more accurate to describe subtype relations as consistent or not with the instance theories of types, rather than derivable from them:

Definition 1 *Given some collection, T, of object types and their instance and class theories, we say that the subtype relation, R on T is* **consistent** *if for every* $(A, B) \in R$,

$$Th(InstanceTheory(B) \cup Sub) \subseteq Th(InstanceTheory(A) \cup Sub).$$

where Sub is the collection of axioms $\{\forall x : ID(A(x) \rightarrow B(x)) \mid (A, B) \in R\}$, *and:*

Definition 2 $Th(A)$ *is the consequential closure of the set of sentences (axioms), A.*

Although there is no consensus [Maung, 1993b] about the meaning of behavioural subtyping, the above definition seems to correspond to the most general in the published literature. However, there are valid subtype relations that do not satisfy this definition (see examples below).

13.2.4 Behavioural Subtyping - General Case

The examples in this section demonstrate why contravariance is not a necessary condition for behavioural subtyping. We formalize each example and then outline a general definition which covers these cases.

Example 6 *Consider three object types, MyString, MyInt and Union where Union is essentially a disjoint union of MyString and MyInt with tag field, kind. It is intuitively obvious that both MyInt and MyString are behavioural subtypes of Union. However, as seen by the signatures of the types given below, the feature signatures of MyInt and MyString are not in a contravariant relationship with those of Union. In particular, the signature of the plus feature, which gives concatenation for my_strings and addition for my_ints, has a specialized rather then generalized argument type. Furthermore, Union has features that do not appear in the signature of MyString. This apparent paradox is resolved by the formalization below.*

Assume the mathematical sorts INTEGER, BOOLEAN and STRING with their usual theories. The sort INTSTR is defined as the union of INTEGER and STRING i.e. $\forall x(INTEGER(x) \vee STRING(x) \leftrightarrow INTSTR(x))$. *ID is the sort of object identities and MyString, MyInt and Union are predicates over the domain of ID. We define the instance theories of the three object types, MyString, MyInt and Union as follows:*

theory *MyString*

features
$kind : INTEGER$
$length : INTEGER$
$val : STRING$
$plus(a : ID) : ID$
$equal(a : ID) : BOOLEAN$
axioms
$kind = 0$
$length = len(val)$
$\forall a : ID\ (MyString(a) \rightarrow$
$\quad MyString(plus(a)) \wedge plus(a).val = concatenate(val, a.val))$
$\forall a : ID(MyString(a) \rightarrow equal(a) = same\_string(val, a.val))$
end theory *MyString*

theory *MyInt*
features
$kind : INTEGER$
$val : INTEGER$
$plus(a : ID) : ID$
$equal(a : ID) : BOOLEAN$
$square : BOOLEAN$
$sqrt : INTEGER$
axioms
$kind = 1$
$\forall a : ID(MyInt(a) \rightarrow MyInt(plus(a)) \wedge plus(a).val = val + a.val)$
$\forall a : ID(MyInt(a) \rightarrow equal(a) = same\_int(val, a.val))$
$square = is\_square(val)$
$square \rightarrow sqrt = square\_root(val)$
end theory *MyInt*

theory *Union*
features
$kind : INTEGER$
$length : INTEGER$
$val : INTSTR$
$plus(a : ID) : ID$
$equal(a : ID) : BOOLEAN$
$square : BOOLEAN$
$sqrt : INTEGER$
axioms
$kind = 0 \leftrightarrow STRING(val)$
$kind = 1 \leftrightarrow INTEGER(val)$
$kind = 0 \rightarrow length = len(val)$
$\forall a : ID(Union(a) \wedge a.kind = 0 \wedge kind = 0 \rightarrow$
$\quad Union(plus(a)) \wedge a.kind = 0 \wedge plus(a).val = concatenate(val, a.val))$
$\forall a : ID(Union(a) \wedge a.kind = 1 \wedge kind = 1 \rightarrow$

$\quad\quad Union(plus(a)) \wedge a.kind = 1 \wedge plus(a).val = val + a.val)$
$\forall a : ID(Union(a) \wedge a.kind = 0 \wedge kind = 0 \rightarrow$
$\quad\quad equal(a) = same\_string(val, a.val))$
$\forall a : ID(Union(a) \wedge a.kind = 1 \wedge kind = 1 \rightarrow equal(a) = same\_int(val, a.val))$
$kind = 1 \rightarrow square = is\_square(val)$
$kind = 1 \wedge square \rightarrow sqrt = square\_root(val)$
end theory *Union*

Assuming that $\forall x : ID(MyString(x) \leftrightarrow (x.kind = 0 \wedge Union(x)))$ [1], *it can be shown that:*

$$Th(InstanceTheory(Union) \cup \{kind = 0\}) = InstanceTheory(MyString).^2$$

Similarly, assuming $\forall x : ID(MyInt(x) \leftrightarrow (x.kind = 1 \wedge Union(x)))$, *it follows that:*

$$Th(InstanceTheory(Union) \cup \{kind = 1\}) = InstanceTheory(MyInt).$$

Intuitively, the predicate and the theory relationship are expressing the same information in different ways (under our assumption that satisfying a class's instance theory is sufficient for class membership), and indeed the above results are sufficient to establish the validity of the predicates.

Example 7 *We now introduce another class, NewString which is similar to MyString except that newstrings can be plussed with my_ints, the result being the sum of the my_int argument and the length of the newstring receiver. Clearly, NewString is a behavioural subtype of MyString. However, although the argument type of the feature plus of NewString is a generalization of that of MyString, it's result type is also a generalization, contradicting the contravariance rule.*

Assume the sorts of the previous example and NewString is a predicate over the domain of ID. *We define the instance theory of the object type, NewString as follows:*

theory *NewString*
features
$kind : INTEGER$
$length : INTEGER$
$val : STRING$
$plus(a : ID) : ID$

[1] Note that $\forall x : ID\ (MyString(x) \rightarrow Union(x))$ can be derived from this.

[2] We assume the theory generated by the language with signature given by the union of the signatures of *MyString* and *Union*

$equal(a : ID) : BOOLEAN$
axioms
$kind = 0$
$length = len(val)$
$\forall a : ID(MyString(a) \rightarrow$
$\quad MyString(plus(a)) \wedge plus(a).val = concatenate(val, a.val))$
$\forall a : ID(MyInt(a) \rightarrow MyInt(plus(a)) \wedge (plus(a).val = len(val) + a.val)$
$\forall a : ID(MyString(a) \rightarrow equal(a) = same\_string(a.val, val))$
end theory *NewString*

Assuming that $\forall x : ID(MyString(x) \vee MyInt(x) \leftrightarrow Union(x))$ *(which follows from the previous example), it can be shown that:*

$$InstanceTheory(MyString) \subseteq InstanceTheory(NewString).$$

Example 8 *Prime is an object type consisting of those my_ints that happen to be prime numbers. Unlike MyInt, it has no sqrt feature since no prime number has an integer square root. However, Prime is clearly a behavioural subtype of MyInt. Assume the mathematical sorts of the previous examples and that Prime is a predicate over the domain of* ID*. We define the instance theory of the object types, Prime as follows:*

theory *Prime*
features
$kind : INTEGER$
$val : INTEGER$
$plus(a : ID) : ID$
$equal(a : ID) : BOOLEAN$
$square : BOOLEAN$
axioms
$kind = 1$
$is\_prime(val)$
$\forall a : ID(MyInt(a) \rightarrow MyInt(plus(a)) \wedge plus(a).val = val + a.val)$
$\forall a : ID(MyInt(a) \rightarrow equal(a) = same\_int(val, a.val))$
$square = false$
end theory *Prime*

It can be shown that:

$$InstanceTheory(MyInt) \subseteq InstanceTheory(Prime).$$

Definition 3 *Given a collection of object types, O and a (putative) subtype relation, R over them, this relation is valid precisely if there is a collection of sentences, S which imply the subtype relation i.e.* $Sent(R) = \bigcup_{(A,B)\in R}\{\forall x : ID\ (A(x) \rightarrow B(x))\}$ *and which when unioned with the union of class theories of the types implies the translation of the collection of sentences,* $trans(S)$ *for the instance theories of the types i.e.* $S \cup \bigcup\{ClassTheory(T) \mid T \in O\} \Rightarrow trans(S)$ *and* $S \Rightarrow Sent(R)$ *are both derivable from first order deterministic dynamic logic.*

Note that the validity of $trans(S)$ *under the assumption of S justifies the assumption of S (since an instance theory gives necessary and sufficient conditions for membership of a class).*

Example 9 *Sentence as theory relationship:*
$\forall x : ID(A(x) \wedge \theta(x) \leftrightarrow B(x) \wedge \phi(x))$ *is translated as:*

$$Th(InstanceTheory(A) \cup \{\theta\}) = Th(InstanceTheory(B) \cup \{\phi\})$$

where A and B are class membership predicates and θ *and* ϕ *are just formulae.*

13.3 Inclusion Polymorphism in Practice

In the previous section, we developed a definition of behavioural subtyping that is theoretically satisfactory. Next, we briefly investigate the practical application of this new definition. C++ [Stroustrup, 1991] is the most popular OOPL for commercial software development and there are established communities of developers using each of the following OOPLs: Smalltalk-80 [Goldberg and Robson, 1983], Objective-C [Cox, 1986] and Eiffel [Meyer, 1992].

13.3.1 C++

C++ behavioural subtyping is trivially covered by the simple case (section 2.2) since the signatures of virtual member functions cannot be redefined in derived classes. This is clearly less flexible than the other OOPLs mentioned but appears to suffice for practical purposes.

13.3.2 Eiffel

The safe, static type system of Eiffel has been the subject of much debate and controversy. Eiffel is also interesting from our perspective because it is a commercial OOPL that combines object technology with formal methods. Eiffel has been designed to support the *programming by contract* method, where the behaviour of objects is specified by *contracts* (assertions that are included in the class code). Assertions are categorized as preconditions, postconditions, class invariants or loop invariants and correspond to their namesakes in formal methods. Assertions can be checked at run-time to assist in debugging code, and can also be extracted from source code using tools of the Eiffel programming environment yielding formal documentation. More abstractly, contracts document the partitioning of responsibilities between different classes of a system and control the use of inheritance by Eiffel's assertion redefinition rule. This rule is intended to ensure that child classes have stronger class invariants, weaker feature preconditions and stronger feature postconditions. It can be subverted by the determined hacker but it seems to be intended to ensure that child classes implement behavioural subtypes of their parents. However, the Eiffel static type system, which allows child classes to hide exported features inherited from their parents and allows covariant redefinition of argument types of routines appears to be designed to accommodate incremental programming techniques and means that child classes need not implement behavioural subtypes of their parents. For some time it was felt that the Eiffel type system was unsafe [Cook, 1989] but it is now established it is in fact safe (but not modular), although this is not widely known.

The theory developed in this chapter may suggest that the Eiffel type rules are in fact perfectly consistent with behavioural subtyping. After all, we have displayed examples of child and parent classes that are behavioural subtypes and satisfy the Eiffel type rules. However, further analysis shows that the Eiffel type system may reject valid code that exploits the behavioural subtyping of these examples:

Example 10 *An outline Eiffel class definition:*

class C
features ...
f **is**
  local $i : INTEGER, p : MY\_INT$
  do ...$!PRIME!p.Prime(7)$;
    if $p.square$ **then** $i := p.sqrt; \ldots$
end *–class C*

where $Prime(i : INTEGER)$ *is the creation procedure for class* $PRIME$.

When the Eiffel compiler performs system-level type checking of this class, it will determine that $PRIME$ is a member of the dynamic type set of p whence, since $PRIME$ does not export $sqrt$, the code will be rejected. In fact,the Eiffel type rules allow inheritance to be used to implement *as-a* relationships [Maung *et al.*, 1994] and to represent the relationship between an actual parameter (subclass) and the constraints on a formal generic parameter (superclass), and this is in general, how the export and covariant argument rules have been exploited in the Eiffel class libraries. In principle, there is no reason why *static* assertion checking could not be used by the Eiffel compiler to allow subtypes such as those given earlier. Note that most assertions will evaluate statically to an unknown value but examples like $p.square$ above will evaluate to $false$.

Note also that if $PRIME$ was respecified as exporting *sqrt* (but with precondition *false*) then the current Eiffel compiler would allow the above example. However, it would also allow unguarded calls such as:

$$\textbf{local}\ i : INTEGER, p : MY\_INT \ldots !PRIME!p.Prime(7);\ i := p.sqrt; \ldots$$

which will be flagged at run-time as an assertion violation. So if no static assertion checking is done by the compiler, there is a tension when defining the specification of a class to, on the one hand, allow valid subtype relationships, and on the other, to reject invalid code.

13.3.3 Smalltalk-80

An untyped programming language like Smalltalk-80 could be used for formal software development. Static type checking, like run-time assertion checking, is totally superfluous if formal verification of code has been completed. The only errors that could possibly be detected are compiler (or hardware or operating system) errors, but if the compiler is unreliable then we cannot trust its type checking anyway!

If either formal verification or dynamic type and assertion checking are used, we believe that our definition is more theoretically satisfactory for Smalltalk-80 than the alternative approach of first (signature) typing a Smalltalk system and then applying a behavioural subtype definition appropriate to a statically typed OOPL.

13.3.4 Objective-C

Objects in Objective-C are untyped as in Smalltalk-80. In fact, Objective-C is even closer to our formulation since it distinguishes object identities from basic types such as Integers and Characters. Thus, by the argument given for Smalltalk-80, our definition is more appropriate (theoretically) for Objective-C inclusion polymorphism than previous definitions.

13.4 Related Work

The origin of this chapter is a model-theoretic account of behavioural subtyping [Maung, 1993a] that resulted from the author's dissatisfaction with existing formalizations of the concept [Maung, 1993b]. In [Maung, 1993a], the definitions given here are developed in terms of a detailed semantic model and shown to ensure substitutability of instances for a simple OOPL. The much simpler account given here has benefitted greatly from the theoretical framework developed by Fiadeiro and Maibaum [Fiadeiro and Maibaum, 1991]. In particular, the use of a modal logic, separate class and instance theories and a sort of identities for objects were very helpful. Fiadeiro and Maibaum have generally worked in an object-based rather than class-based framework and so have only sketched a formalization of subtypes. Kent [Kent, 1995] has applied this framework to the semantics of an object-oriented specification language and given a simple definition of subtyping. Significant contributions to the theory of behavioural subtyping include [America, 1991], [Leavens and Weihl, 1990] and [Liskov and Wing, 1994]. A comprehensive bibliography appears in [Maung, 1993b].

13.5 Conclusions

A proof-theoretic formulation has been given for the new model-theoretic definition of behavioural subtyping given in [Maung, 1993a]. In the latter article, it is shown that this new definition (unlike previous definitions [Maung, 1993b]) gives both necessary and sufficient conditions for substitutability of instances of a subtype for those of its supertype in a simple OOPL. The present work casts this definition in the very widely known setting of programming logics, and represents a definitive formulation of semantic inclusion polymorphism, a cornerstone of object technology, in that framework.

Our formulation has also led to a number of interesting possibilities - including static assertion checking, and dynamic polymorphism (including run-time checks on reassignments - a generalization of the Eiffel check on reverse assignment attempts). The latter is possible, when for example, a $Union$ instance can be attached to a $MyString$ reference so long as its $kind$ attribute has value 0.

Some issues not properly addressed here include a completely formal definition of behavioural subtyping (in particular of translation between predicates about class membership sorts and instance theories), and a proper treatment of inclusion polymorphism involving renaming and aliasing. It would also be interesting to cast the work given here into a category theoretic setting, with theories as objects and interpretations between theories as morphisms. This approach has been applied in [Kent, 1995] (but for less general cases than those given here).

Acknowledgements

The author acknowledges financial support from SERC (now EPSRC) grant number GR/H16629. Thanks to Stuart Kent, Richard Mitchell and John Howse for comments on a draft version of this chapter.

Chapter 14

Categorical Semantics for Object-Oriented Data-Specifications

Semantic data-specifications have been used for many years in the early stages of database design. The most influential example is undoubtedly the Entity-Relationship specification method ([Chen, 1976]), in which aspects of a certain reality are expressed in terms of entities, relations and attributes. The paradigm of object-oriented software development stimulated new research in the area of data-specifications. Indeed, the major goal of object-oriented analysis ([Coad and Yourdon, 1990, Van Baelen *et al.*, 1991]), is to develop a specification of the external world. Since then, many extensions to ER-like data-specifications have been suggested, either to enhance the expressiveness, or to suit them more to the needs of software engineering. However, most of the data-specification mechanisms used in practice are of an informal nature.

In this chapter, we present a categorical data-specification mechanism and its formal (categorical) semantics. Many existing data-specification methods can be translated to our categorical mechanism, thus providing them with a formal semantics. In fact, our specification mechanism is currently being used to provide a formal semantics for EROOS, described in [Van Baelen *et al.*, 1991].

We investigate two interesting properties of this specification mechanism. The first property, is the uniqueness property, which was proved in [Piessens and Steegmans, 1994]: if certain restrictions are obeyed (i.e. if the specification is *canonical*), then equivalence of the model categories implies isomorphism of the specifications. This says in fact that if the structure of entities and relations between entities in the real world can be described by the specification mechanism, it can be in only one way, up to isomorphism. This property is of utmost importance if *reuse* of data-specifications among software engineers is ever to be put in practice ([Van Baelen *et al.*, 1991]): it ensures us that two specifications of different parts of a given reality can always be combined by identifying their overlapping parts (which must be isomorphic because of the stated property).

The second interesting property is the existence of a verification protocol for specifications. Under the assumption that the software engineer can answer certain simple questions about the real world, we prove that this verification protocol is complete for a well-defined subset of specifications.

Understanding the proofs of the theorems in this chapter requires some knowledge of basic category theory as provided for instance by [Barr and Wells, 1990] and [Adámek *et al.*, 1990]. To make the chapter as accessible as possible, we have split it in two parts. The first part (sections 1, 2 and 3) gives an informal introduction to the results of the chapter: no foreknowledge of category theory is assumed. The main results are stated in an informal way, and illustrated by examples.

The second, technical, part gives formal definitions of all the concepts we have introduced, with proofs of all the theorems (or at least, references to the proofs). This part is only meant to be read by people familiar with category theory.

14.1 A Categorical Specification Mechanism

14.1.1 Classes and Dependencies

When making an object-oriented data specification, one tries to classify the different objects existing in the real world in a number of *classes*. For instance, when specifying a library system, we could have a class of persons, a class of books, a class of libraries, a class of members, etc... Every relevant object in the real world belongs to one of these classes.

We say that a *dependency* exists between class A and class B, if an object from class A can only exist if it is associated with an object of class B. For example: in a library system, there is a dependency between the class of members and the class of libraries: every member is associated with a library. In a similar way, there is a dependency between the class of members and the class of persons: every member is associated to a person.

In a classical ER-specification, one would say that the class of members is in fact a relation between the class of persons and the class of libraries. As you will see from the semantics of our specifications, a difference with classical ER-specifications is the fact that duplicates in such relations are allowed: a person could be a member of the same library twice.

We specify classes and dependencies by means of a graph: for every class, there will be a node in the graph, and for every dependency, there will be an arrow between the two involved classes. A model of such a specification associates with every node of the graph a set (the set of objects belonging to that class), and with every arrow a function (the function taking each object to its associated object in the depending class).

To make all this clear, we will now treat an extended example.

14.1.2 An Extended Example

We will specify a simple library-system. First we introduce classes for members, persons and libraries. As noted before, the following dependencies exist:

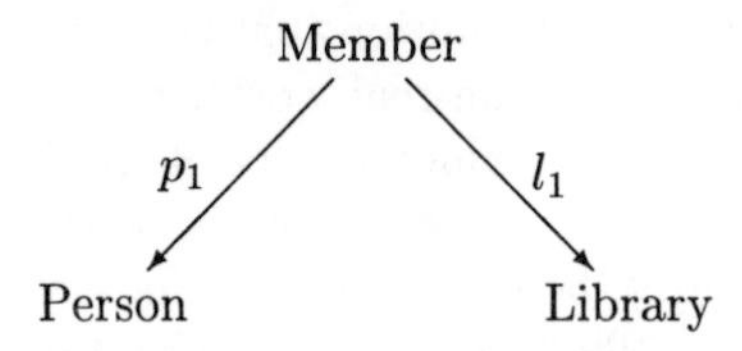

In a model, every class is taken to a set, and every arrow is taken to a function. Hence, this specification says that Member is a multi-relation over Person and Library. A person can be a member of different libraries, and can even be a member of the same library twice or more.

In a similar way, we specify the classes Book and Possession. The class Book contains all books, and the class Possession describes the possessions of the libraries: every possession is associated with a book and with a library:

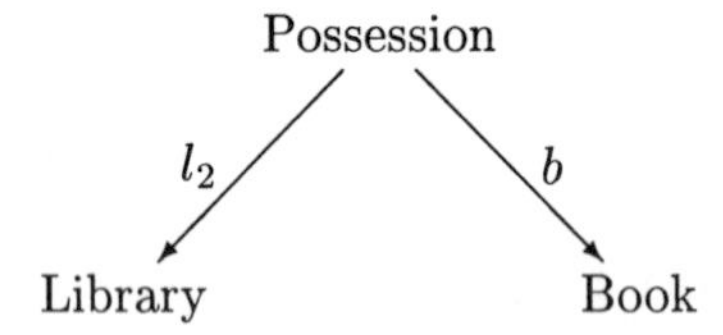

We must distinguish between physical books (specified by the class Book), and abstract books (which will be specified by the class AbstractBook). The word 'book' is used in two different ways. It can mean a physical copy of a book ("The book that is lying on my desk"), or an abstract reference ("The book [Barr and Wells, 1990] by Barr and Wells"). For these two meanings of the word book, two classes are introduced: the class Book, specifying physical books, and the class AbstractBook. That it is important to distinguish them can be seen from the following example: if a member borrows a book, he will borrow a physical book. However if a member *reserves* a book, he reserves an abstract book: he does not care which physical copy he will eventually get. Hence the following specification:

$$\text{Book} \xrightarrow{a} \text{AbstractBook}$$

This specification says that Book is actually a multiset over AbstractBook.

Finally, borrows of books can be specified in the following way:

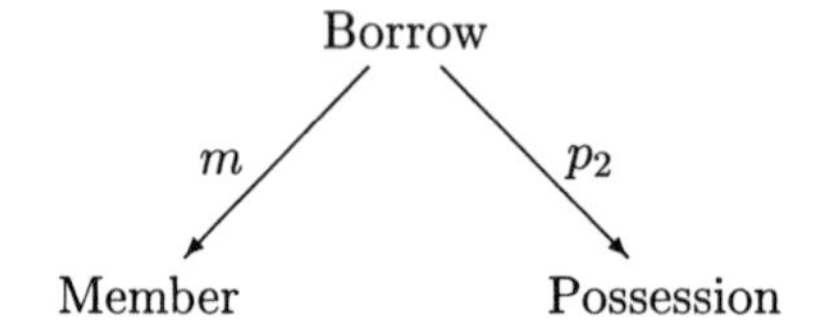

A member borrows the possession of a library.

The complete graph of our specification looks like this:

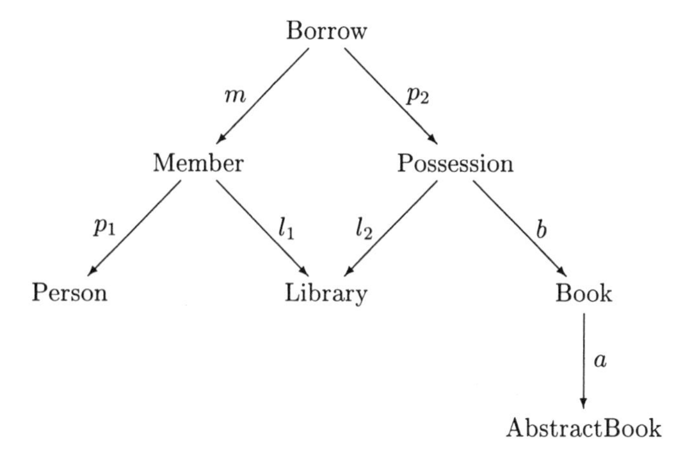

14.1.3 Constraints

As it stands, the specification of the library-system is certainly not correct: certain rules, which are always valid in the real world, can be broken in a model. Some examples:

1. In the real world, a given physical book can only be possessed by one library. In our specification however, the same book can be possessed by two different libraries. (For instance, consider a model with one Book-object B, two Possession-objects P and P' and two Library-objects L and L'. The function l_2 could take P to L and P' to L' while the function b could take P and P' to b.)

2. In the real world, a member can only borrow books from the library that he is a member of. In our specification, a member can borrow books from any library.

To incorporate such rules in the specification, one can formulate *constraints.* We distinguish two important kinds of constraints.

join constraints

With a join constraint, one can say that the composition of a given sequence of functions must be equal to the composition of another sequence of functions. Join constraints are presented graphically by means of a *diagram.* In our example specification, the following join constraint must be included:

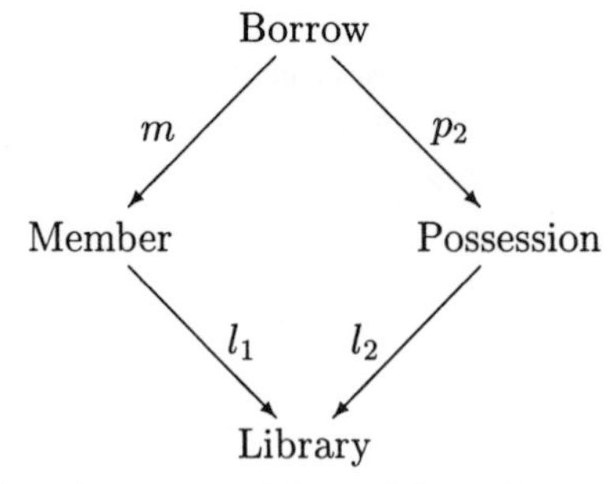

This constraint says that the composition of l_2 and p_2 must equal the composition of l_1 and m: a member can only borrow possessions of the library he is a member of.

connectivity constraints

With a connectivity constraint, we can say that a certain function must be injective in every model (In fact, connectivity constraints are more general, see the technical part of the chapter). They are called connectivity constraints, because they allow us to express the classical connectivity constraints from ER-specifications: we can say that a relation must be 1–1, 1–n or n–1 by saying that two or one of the functions coming from the relation must be injective. For instance, Possession must be a 1–n relation between Library and Book, and this is expressed by saying that the arrow b must be taken to an injective function in every model. In a similar way, the arrow p from Borrow to Possession must be injective.

14.1.4 Attributes

Attributes allow one to label the objects of a certain class with elements of a given attribute set. For example: adding an attribute SEX to the class Person specifies that every person is either male or female: we label the objects of class Person as male or female.

At first sight, the possibility to specify attributes in a data-specification mechanism seems to greatly enhance its expressivity. However, we will show in the technical part of the chapter that, in the absence of connectivity constraints, attributes can always be eliminated from a specification without changing the semantics. The price we have to pay is that the specification can become very large after the attribute elimination process. And if infinite attribute sets have to be eliminated, the specification also becomes an infinite graph. In the technical part, we only consider finite attribute sets. The results can be extended to infinite sets, but only at the expense of considerably complicating the mathematical treatment.

We give a few examples to illustrate attribute elimination. A general construction is given in Part II. As the most simple example, consider a specification consisting of one class, Person, with attribute SEX ={male,female}.

Attribute elimination gives a specification with two classes, Male and Female, with objects respectively the male persons and the female persons.

For a more substantial example, consider the following specification:

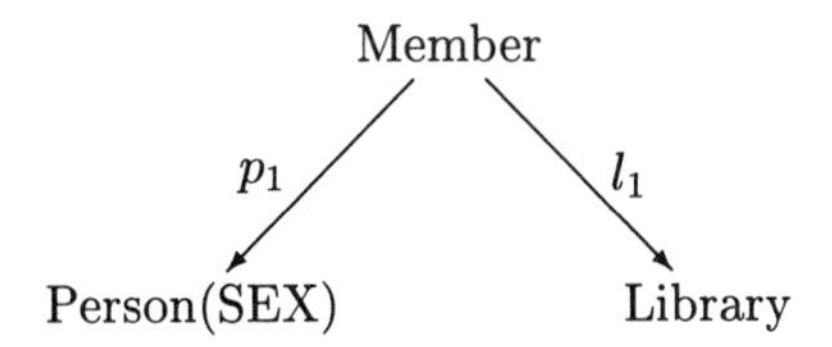

Attribute elimination leads to the following specification:

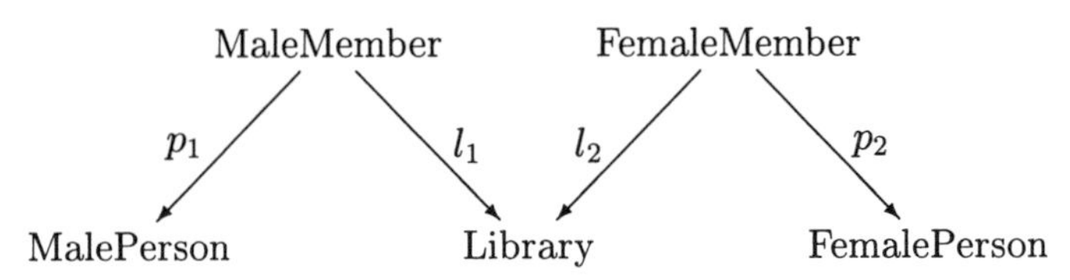

As a rule of thumb: attribute elimination splits the class with the attribute in a number of classes, and this splitting is propagated along all arrows arriving in the class.

For specifications with connectivity constraints, attribute elimination may fail to give a specification with an equivalent semantics. This problem will be considered in the technical part of the chapter.

14.2 Canonical Forms for Specifications

As attribute elimination has shown, some real-world situations can be specified in a number of different ways: eiher with attributes, or without making use of attributes. It would be desirable to have a subclass of specifications (called canonical specifications) having the following properties:

1. If two canonical specifications have the same semantics, then they are the same specification.

2. An arbitrary specification can always be converted to a canonical specification with the same semantics.

These properties greatly facilitate reuse and recombination of specifications. To find an appropriate notion of canonical specification, we must first formally define:

1. what we mean with "the same semantics"

2. what we mean with "the same specification"

The second notion is easy: two specifications are the same if they are identical up to a renaming of classes and dependencies. The first notion is a lot harder: in the technical part we will give a formal definition of the semantics of a specification. Based on this definition, it can be shown that, if attributes can be eliminated, canonical forms of specifications exist, and an algorithm to compute canonical forms of finite specifications can be given. More details can be found in the technical part of this chapter.

14.3 Verification of Specifications

Verifying specifications is very difficult, since the formal specification must be compared with the informal real world. To be able to say sensible things about the correctness of a specification, we make the following assumptions:

- Some correct specification of the real world exists: specifications have sufficient expressive power to describe the real world.

- The software engineer can answer certain questions about the real world.

The goal is now to design a protocol between a computer and the software engineer, which allows a computer to check the correctness of the specification by asking questions. Such a protocol will certainly depend on the kind of questions that may be asked.

For finite specifications with at most one dependency between two classes, we show that a finite number of simple questions suffices to prove the correctness of the specification. The questions are of the following kind: a model of the specification is presented to the software engineer and he has to decide wether this model corresponds to a possible real-world situation.

As an example, consider the following specification:

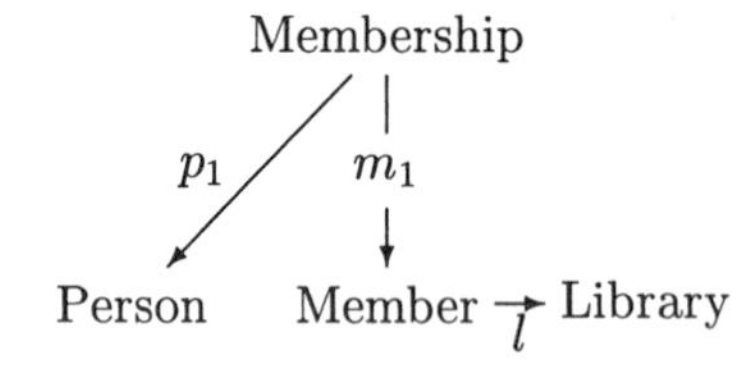

It introduces a separate class for memberships. The following model (among others) will be presented to the software engineer: one Member-object, one Library-object, no Membership-objects and no Person-objects. The software-engineer will decide that this model does not correspond to a real-world situation: it is impossible that there is a member of a library if there is no person. Hence, the specification is incorrect (a dependency between Member and Person is missing).

Another example:

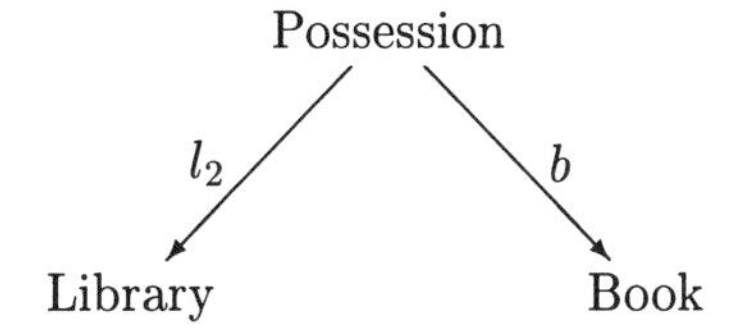

The following model (among others) will be presented to the software engineer: one Book-object B, two Possession-objects P and P' and two Library-objects L and L'. The function l_2 takes P to L and P' to L' while the function b takes P and P' to b.

The software engineer will notice that this model does not correspond to a possible real-world situation: one physical book can not be possessed by two different libraries. Hence, the specification is incorrect (it lacks a connectivity constraint).

14.4 A Note on Terminology

This section begins the technical part of the chapter.

Unfortunately, the terminology used in category-theory and in object--oriented specifications is not very compatible: many words (like object, class, model) have different meanings in the two fields. Moreover, many of the words used in object-oriented specifications do not have a formal definition.

In this section, we will try to resolve the ambiguities.

1. The word 'object' will be used as it is used in object-oriented specifications: an object is an entity, existing in the real world, with a mutable state. Examples are: a person, a book, a membership, etc ...

   For objects of a category, we will use the word 'node'. Hence, the category **FinSet** of finite sets and functions has finite sets as nodes and functions between them as arrows.

2. The word 'model' is sometimes used as a synonym for specification in object-oriented specifications. A real model is then called 'an instance'. We will use the word 'model' as it is used in category-theory and logic: the word 'instance' will not be used.

3. The word 'class' is not used in its set-theoretic sense. It has the meaning it has in object-oriented specifications: a class is a set of objects with similar characteristics.

14.5 Specifications

14.5.1 Specifications without Attributes

From the discussion in the informal part of the chapter, it should be clear that a specification without attributes is a category together with a set of arrows in the category which must be taken to mono's in every model. In fact, the definition is a little more general.

Definition 1 *A* specification *is a pair* $(\mathcal{S},\mathcal{M})$*, where* $\mathcal{S}$ *is a finite category, and* $\mathcal{M}$ *is a set of sources in* $\mathcal{S}$*.*

In practice, the category will be given as a graph and a set of diagrams. (As a linear sketch in the sense of [Barr and Wells, 1990].) The category can then be computed as the quotient of the free category on the graph by the congruence relation generated by the diagrams.

Definition 2 *A* model *of a specification* $(\mathcal{S},\mathcal{M})$ *is a functor from* $\mathcal{S}$ *to* **FinSet***, which takes every* $M \in \mathcal{M}$ *to a mono-source.*

Hence the model category is a full subcategory of the functor-category $\text{Fun}(\mathcal{S}, \mathbf{FinSet})$. For definition and properties of sources and mono-sources, see [Adámek *et al.*, 1990]. It is proved in [Piessens and Steegmans, 1994] that the model category is actually an isomorphism-closed epi-reflective subcategory of $\text{Fun}(\mathcal{S}, \mathbf{FinSet})$.

14.5.2 Specifications with Attributes

The most logical way to define the semantics of attributes is the following: consider a specification $\mathcal{F}$, and the same specification extended with an attribute set $A(X)$ on class X. The model category of the extended specification is as follows: nodes are nodes M of $\text{Mod}(\mathcal{F})$, together with a function $f : M(X) \to A(X)$ (the labelling). Arrows are natural transformations preserving the labelling. The formal definitions are:

Definition 3 *A* specification with attributes *consists of a specification* $(\mathcal{S},\mathcal{M})$*, together with a functor* $A : \mathcal{S}_0 \to \mathbf{FinSet}$*, where* $\mathcal{S}_0$ *is the discrete category containing all the nodes of* $\mathcal{S}$*.*

The functor A is essentially a function from the nodes of $\mathcal{S}$ to the finite sets. The intention is to specify that every object in class S must be labelled with a value from the set $A(S)$. If a class S is to have no attributes, $A(S)$ will be the terminal set. If it must have more than one attribute, $A(S)$ will be the product of all attribute sets.

The straightforward way to define the model category is as follows: Let $I : \mathcal{S}_0 \to \mathcal{S}$ be the inclusion, let $I^* : \text{Mod}(\mathcal{F}) \to \text{Fun}(\mathcal{S}_0, \mathbf{FinSet})$ be the functor of composition with I and let $\overline{A} : 1 \to \text{Fun}(\mathcal{S}_0, \mathbf{FinSet})$ be the functor picking out A. Then,

Definition 4 *The model category of* $(\mathcal{S},\mathcal{M},A)$ *is the comma-category* $(I^* \downarrow \overline{A})$*.*

Hence, the nodes of the model category are models of the underlying specification without attributes, together with a natural transformation from $M \circ I$ to A, which can be percieved as a labelling of the elements of the model.

14.5.3 Elimination of Attributes

Given a specification with attributes, we want to find another specification without attributes that describes the same reality. To be able to treat this mathematically, we must first define what we mean with "describing the same reality". We make the following claim:

> Two specifications describe the same reality iff their model categories are equivalent.

Of course, this claim cannot be proved, since "describing the same reality" is an intuitive notion. But in all cases we have met, the criterium is satisfactory.

In [Piessens and Steegmans, 1994], the following theorem is proved:

Theorem 1 *Let $(\mathcal{S},\mathcal{M},A)$ be a specification with attributes, and let $A^! : \mathcal{S} \to$* **FinSet** *be the right Kan extension of A along the inclusion $I : \mathcal{S}_0 \to \mathcal{S}$. If $A^!$ is a model of $(\mathcal{S},\mathcal{M})$, then $(\mathcal{S},\mathcal{M},A)$ is equivalent to a specification $(\mathcal{S}',\mathcal{M}')$ without attributes.*

Hence, we see that attribute elimination only works if $A^!$ satisfies all the connectivity constraints. If there are no connectivity constraints, attribute elimination always succeeds. In the presence of connectivity constraints, it may be impossible to find an equivalent specification without attributes.

In the proof of this theorem, the construction of the new specification $(\mathcal{S}',\mathcal{M}')$ is given: $\mathcal{S}'$ is the category of elements of $A^!$, and $\mathcal{M}'$ is the inverse image of $\mathcal{M}$ under the canonical projection form the category of elements of $A^!$ on $\mathcal{S}$. Hence, this theorem provides us with an algorithm to eliminate attributes when possible.

14.6 Canonical Forms for Specifications

Certain real-world situations can be specified in a number of different (non-isomorphic) ways, even when all attributes have been eliminated.

Example: The specifications:

$$P \circlearrowright^{r} \quad \text{with} \quad r \circ r = r \tag{14.1}$$

$$P \underset{i}{\overset{r}{\rightleftarrows}} A \quad \text{with} \quad r \circ i = \mathrm{Id}_A \tag{14.2}$$

are non-isomorphic specifications of the same real-world situation. P could be a class of persons, and r could associate with every person a 'responsible person': for a minor, the responsible person would be one of his parents. An adult is responsible for himself. Hence r is idempotent, as is expressed in the equation $r \circ r = r$. The second specification specifies the same, using however a separate class for adults. In fact, the second specification is a Cauchy-completion of the first one in the sense of [Borceux and Dejean, 1986].

Example: The specifications:

$$A \underset{g}{\overset{f}{\rightrightarrows}} B \xrightarrow{h} C \quad \text{with} \quad \begin{array}{l} h \circ f = h \circ g \\ h(x) = h(y) \Rightarrow x = y \end{array} \tag{14.3}$$

$$A \xrightarrow{f} B \xrightarrow{h} C \quad \text{with} \quad h(x) = h(y) \Rightarrow x = y \tag{14.4}$$

are another example of two non-isomorphic specifications describing the same reality: the constraints in the first specification are such that in every model, f and g must be the same. Hence, there is no need to distinguish them in the specification.

Why do we care about uniqueness of specifications? Suppose two software engineers both specify a part of a large system, and these specifications overlap partly. If it would be guaranteed that the overlapping part is specified isomorphically in the two specifications, it would be easy to merge them together: just identify the isomorphic overlapping parts. However, the examples above show that, in general, this cannot be guaranteed.

Hence, we want to define a notion of *canonical* specification with the following properties.

1. For every specification, there must exist a canonical specification with an equivalent model category. Moreover, we would like to have an algorithm to convert a specification to its canonical form.

2. Two canonical specifications have equivalent model categories iff they are isomorphic.

Definition 5 *A* source without doubles *is a source in which no arrow occurs twice or more.*

Since a source with doubles is a mono-source iff the same source with all doubles removed is a mono-source, only sources without doubles must be considered for specifications.

Definition 6 *A specification $\mathcal{F} = (\mathcal{S}, \mathcal{M})$ is* canonical *iff:*

1. *In the category $\mathcal{S}$, every idempotent arrow is split.*

2. *In the category $\mathcal{S}$, no two different nodes are isomorphic.*

3. *All representable functors from $\mathcal{S}$ to* **FinSet** *are models.*

4. *The set $\mathcal{M}$ contains all sources without doubles in $\mathcal{S}$ which are taken to mono-sources by every model.*

In [Piessens and Steegmans, 1994], the following properties of canonical specifications are proved:

Theorem 2 *For every specification $\mathcal{F} = (\mathcal{S}, \mathcal{M})$, there exists a canonical specification with an equivalent model category.*

Again, the proof of this theorem is constructive, and hence, an algorithm to compute this canonical form is given.

Theorem 3 *Let $\mathcal{F}_1 = (\mathcal{S}_1, \mathcal{M}_1)$ and $\mathcal{F}_2 = (\mathcal{S}_2, \mathcal{M}_2)$ be two canonical specifications. Then $\mathcal{F}_1$ and $\mathcal{F}_2$ are isomorphic iff their model categories are equivalent.*

This theorem is the main theorem about canonical forms: it shows that a certain real-world situation (corresponding to a specific model category) can be specified in only one way by a canonical specification.

As a consequence of these two theorems, we have a decision algorithm to decide if two specifications describe the same reality: just compute the two normal-forms and check if they are isomorphic.

14.7 Verification of Specifications

We want to design a protocol which allows a computer to verify the correctness of a specification by asking a finite number of questions at the software engineer. As pointed out in the introduction, such a protocol will of course depend on the kind of questions that can be asked.

Example: If we assume that the software engineer can construct the free model on the underlying typed set of objects, verification of specifications without connectivity constraints is very easy: construct the free models on the one-element typed sets. The specification is the full subcategory of the model category spanned by these free models.

However, we don't think it is realistic to assume that the software engineer can answer this type of question. We will make the very weak assumption that the software engineer has the following capabilities:

- We assume that the classes which are present in the specification correspond to classes in the real world: equivalently, the nodes in the graph of the specification are assumed to be correct.

- He can check wether a given typed set of objects can be the underlying typed set of objects of a model. For instance: a typed set with only one object of type Member cannot be the underlying set of a model, since in any model a Member is always associated with a Person.

- If we know the underlying category of a specification is correct, we assume he can check if a **FinSet**-valued functor from the category corresponds to a possible real-world situation.

Based on these assumptions, we develop a verification procedure, which is complete for thin specifications (to be defined later), but which also works well in practice for other specifications.

14.7.1 Verification of the Underlying Category

If we want to verify the underlying category of a specification, we must find a characteristic of the model category which is only dependent upon the underlying category. Moreover, we must be able to determine this characteristic by asking simple questions at the software engineer. We will see that the lattice of subobjects of the terminal node fits the bill.

Proposition 1 *A subobject of the terminal node of* $\mathrm{Fun}(\mathcal{S}, \mathbf{FinSet})$ *belongs to all epi-reflective isomorphism-closed subcategories of* $\mathrm{Fun}(\mathcal{S}, \mathbf{FinSet})$.

Proof: It is easy to see that the terminal node 1 belongs to all epi-reflective subcategories: all arrows out of 1 are monic, the reflection arrow of 1 must be epi and hence regular epi (all epis are regular). Thus the reflection arrow must be an isomorphism.

Now, suppose $m : s \to 1$ is a subobject of 1, and suppose $\eta_s : s \to s'$ is a reflection-arrow. Then there exists a unique m' making the following diagram commute:

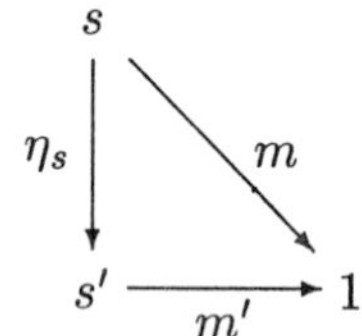

Now, since $m = m' \circ \eta_s$, and m is mono, η_s must be mono. But since η_s is a reflection arrow, it must also be epi. Hence it is an isomorphism (since all epis are regular in $\mathrm{Fun}(\mathcal{S}, \mathbf{FinSet})$). □

Hence, the lattice of subobjects of 1 is determined only by the underlying category $\mathcal{S}$. Conversely, for thin categories, the category is determined up to equivalence, by the lattice of subobjects of 1:

Proposition 2 *Suppose* $\mathcal{C}$ *and* $\mathcal{D}$ *are two thin categories.* $\mathcal{C}$ *and* $\mathcal{D}$ *are equivalent iff the lattices of subobjects of* 1 *of* $\mathrm{Fun}(\mathcal{C}, \mathbf{FinSet})$ *and* $\mathrm{Fun}(\mathcal{D}, \mathbf{FinSet})$ *are isomorphic.*

Proof: We can prove this using the uniqueness theorem for canonical specifications (theorem 3). A source with an empty set of arrows (consisting of just an object X) will be called the empty source at X. The empty source at X is a mono-source iff all arrows with codomain X are equal. In **FinSet** the empty source at a set S is a mono-source iff S contains one or zero elements.

Consider the following two specifications:

1. $\mathcal{S}_\mathcal{C}$ is the skeleton of $\mathcal{C}$ and $\mathcal{M}_\mathcal{C}$ contains all the sources without doubles in $\mathcal{S}_\mathcal{C}$.

2. $\mathcal{S}_\mathcal{D}$ is the skeleton of $\mathcal{D}$ and $\mathcal{M}_\mathcal{D}$ contains all the sources without doubles in $\mathcal{S}_\mathcal{D}$.

It is easy to see that these specifications are canonical. Moreover, in every model, every class may contain at most one object, because the empty source at that class must be a mono-source. Hence, their model categories are equivalent to the lattices of subobjects of 1 of $\text{Fun}(\mathcal{C}, \mathbf{FinSet})$ and $\text{Fun}(\mathcal{D}, \mathbf{FinSet})$.

By theorem 3, we may conclude that these lattices are isomorphic iff $\mathcal{C}$ and $\mathcal{D}$ are equivalent. □

These two propositions show that verifying the underlying category of a thin specification (i.e. a specification with a thin underlying category) can be done by verifying the lattice of subobjects of 1 of their model categories. Verifying these lattices can be done by asking a finite number of simple yes/no questions. The verification procedure goes as follows:

Let $\mathcal{F} = (\mathcal{S}, \mathcal{M})$ be a thin specification, and let $U : \text{Mod}(\mathcal{F}, \mathbf{FinSet}) \to \text{Fun}(\mathcal{S}_0, \mathbf{FinSet})$ be the functor associating to each model the underlying typed set (typed by the classes: $\mathcal{S}_0$ is the discrete category containing all nodes of $\mathcal{S}$). Present to the software enigineer all typed sets (nodes of $\text{Fun}(\mathcal{S}_0, \mathbf{FinSet})$) that are subobjects of 1 in $\text{Fun}(\mathcal{S}_0, \mathbf{FinSet})$, and ask if these typed sets underly a possible state of the real world. If the software-engineer answers yes for every typed set that is in the image of U and no for every typed set that is not in the image of U, then the category $\mathcal{S}$ of the specification is correct.

14.7.2 Verification of Specifications

Once we know that the underlying category is correct, it is rather easy to verify the complete specification. We just have to check which sources in the category must be taken to mono-sources. We have seen that it is sufficient to consider sources without doubles. Since only a finite number of sources without doubles exists for a finite category, we proceed as follows: First, we compute the canonical form of the given specification. Then, for every source without doubles:

1. We present a model of the underlying linear specification in which the source is not taken to a mono-source to the software engineer.

2. The source must be in the set of mono-sources iff the software engineer rejects the model.

The specification is correct iff the set of mono-sources given in the canonical specification is the same as the set of mono-sources obtained by the previous protocol.

14.8 Conclusion

We have defined a categorical object-oriented data specification mechanism, and its semantics. It is not so hard to extend these definitions to make the specification mechanism very expressive. However, in this chapter we concentrated on proving useful properties of a specification mechanism with only moderate expressive power.

The first interesting property of our specifications is the existence of canonical forms such that no two non-isomorphic canonical specifications have the same semantics. The second interesting property is the existence of a verification protocol for a subset of specifications.

This chapter clearly describes work that is still in progress. A number of interesting questions remain, and are the subject of further research. The most important one is: how much more expressive can specifications be, while still allowing computable canonical forms?

Acknowledgments

We wish to thank Michael Barr for his excellent diagram macro package, which was used during the preparation of this chapter.

Chapter 15

A Type-Theoretic Basis for an Object-Oriented Refinement Calculus

15.1 Introduction

A characteristic feature of object-oriented program development is its uniform way of structuring all stages of the development by *classes*: Classes describe objects found during problem analysis, and classes emerge during design and implementation. As classes serve different purposes during the development, they have to be defined in appropriate notations: in a problem-oriented way during analysis and in an efficiently implementable language for the implementation.

For covering the different stages of the development, we present an *object-oriented design notation.* Besides imperative concepts, it features objects with attributes and methods, encapsulation of attributes, subtype-polymorphism, parametric polymorphism, classes, and inheritance. Furthermore it includes *specification statements* . These are non-executable statements, standing for parts which have to be *refined* to concrete (executable) statements [Back and Wright, 1989, Morgan, 1988, Morris, 1987] . Statements containing specification statements are called *abstract.* Abstract statements can appear in class definitions, resulting in *abstract classes* .

By this extension of the programming language the same structuring principles can be applied equally to programs and specifications. The benefit of such a *wide-spectrum* notation is that transitions between notations for different stages can be avoided and that development steps can (hopefully) be more easily verified.

The contribution of this paper is a novel way of giving a formal semantics to such an object-oriented design notation, and to show its use. The semantics is given by a definitional extension of the type system $F_{\leq}$. The type system $F_{\leq}$ itself is an extension of the simple typed λ-calculus with parametric polymorphism (functions parameterized by types) and subtype-polymorphism (functions operating on values of a type as well as on values of all subtypes) [Cardelli and Wegner, 1985, Cardelli *et al.*, 1991, Curien and Ghelli, 1992]. It was developed to study flexible but decidable typing in simple, functional setting. Here, it is used for an imperative setting: Statements are defined by predicate transformers in $F_{\leq}$ by viewing predicates as Boolean valued functions and predicate transformers as higher-order functions from predicates to predicates.

A feature of the notation contributing to a flexible and decidable typing is the distinction of object types from classes. The type of an object only determines its syntactic interface. The class of an object determines its behaviour, which is given by its methods and its attributes. Subtyping is a structurally defined relation on types. By contrast, inheritance is a means for constructing classes from existing ones, and is completely independent of subtyping. Classes serve as templates for the creation of objects . Objects of different classes may have the same type.

The structure of the paper is as follows. The next section gives the semantics of statements in a subset of $F_{\leq}$. Section 3 introduces objects with attributes and methods, encapsulation of attributes, subtype-polymorphism, parametric polymorphism, classes and inheritance. Section 4 shows the use of the notation by a couple of examples. Section 5 relates the approach to other approaches and draws some conclusions. The appendix describes the type system $F_{\leq}$.

15.2 Statements as Predicate Transformers

Following various refinement calculi, concrete (executable) statements are embedded into a much richer space where statements may be demonically nondeterministic, angelically nondeterministic and miraculous. (The exact distinction of abstract and concrete statements is left open, as the calculus does not depend on it.)

Statements are typed such that the type of a statement determines the program variables (and their types) on which the statement operates. The definitions in this section are given in the fragment of the type system $F_{\leq}$ which corresponds to the simple typed λ-calculus (see also appendix).

Values and Types. Each value has a type. Types are simple types like *Int* or *Bool* (with values *true* and *false*), or are composed types like *set of* T, the type of sets with values of type T, or $S \to T$, the type of functions from values of type S to values of type T. Value e has type T is written as $e : T$. For the examples we assume that for the types used the usual operations (and constants) are defined.

Of special relevance are record types. A record is a finite association of labels to values, written $(l_1 \mapsto e_1, \ldots, l_n \mapsto e_n)$. The record type $(l_1 : T_1, \ldots, l_n : T_n)$ is the type of records with fields $l_1, \ldots, l_n$ of types $T_1, \ldots, T_n$. The empty record is written as (). It is of the empty record type, also written as (). The value of field l is selected from a record r by $r.l$.

For notational convenience, we use additional operations on records and record types. If $r : R$ and $s : S$ are records, then the overwrite $r \oplus s$ adds the fields of s to those of r, replacing fields with same label. The result is of type R overwritten by S, denoted by $R \oplus S$. If record t is of type $R \oplus S$, then the restriction $t \mid S$, removes those fields from r which are not in S. Similarly, if $T = R \oplus S$, then $T \mid S$ removes those fields from T which are not in S.

The substitution $[e/y]f$ stands for value expression f in which every free occurrence of y is replaced by value expression e. The value declaration introduces a shorthand for a parameterized value:

$$(\mathit{val}\ y\ (x : T) = e \cdot f) \quad = \quad [(\lambda\ x : T \cdot\ e)/y]f$$

Similarly, the substitution $[T/A]f$ stands for value expression f with each free occurrence of type variable A replaced by type expression T. The type declaration introduces a shorthand for a type expression:

$$(\mathit{type}\ A = T \cdot f) \quad = \quad [T/A]f.$$

States and State Predicates. A state assigns values to program variables. A state is given by a record. Program variables correspond to names of record fields. A state space is a record type.

State predicates, or predicates for short, are functions from states to *Bool*. The type *Pred R* is the type of predicates over state space (record type) R:

$$\mathit{Pred}\ R \quad = \quad R \to \mathit{Bool}$$

The Boolean operators $\neg, \wedge, \vee, \Rightarrow, \Leftarrow, \Leftrightarrow$ on state predicates are defined by pointwise extension of the corresponding operators on the Boolean values, e. g. for predicates b, c of type *Pred R* we have:

$$b \wedge c \quad = \quad (\lambda\ r : R \cdot\ b\ r \wedge c\ r)$$

Universal implication and equivalence (equality) of predicates b, c are denoted by $b \leq c$ and $b = c$, respectively. For predicates b, c of type *Pred R* we define:

$$\begin{array}{lll} b = c & \text{iff} & (\forall\ r : R \cdot\ b\ r \Leftrightarrow c\ r) \\ b \leq c & \text{iff} & (\forall\ r : R \cdot\ b\ r \Rightarrow c\ r) \end{array}$$

State predicates are conveniently written as Boolean expressions. For example, $x + c = y$ is a Boolean expression which can be understood as a predicate. If x and y are the program variables in the expression and c is a constant (logical variable), then the corresponding predicate is

$$(\lambda\ s : S \cdot\ s.x + c = s.y) \quad : \quad \mathit{Pred}\ S \quad ,$$

where $S = (x : \mathit{Int}, y : \mathit{Int})$. Hence the interpretation of a Boolean expression as a predicate depends on the *context*, which determines the program variables and constants.

Statements. If S and T are state spaces, then a statement p with initial state space S and final state space T is identified with a function (predicate transformer) mapping predicates (postconditions) over T to predicates (preconditions) over S. For a predicate c over T, the application $p\ c$ is the weakest precondition of p such that, operationally, the execution of p terminates and ends in a state satisfying c. In Dijkstra's original notation this is written as $wp(p, c)$ [Dijkstra, 1976]. The type *Tran* S T is the type of statements with initial state space S and final state space T. Note that the typing of statements allows the final state space to differ from the initial state space.

$$Tran\ S\ T \quad = \quad Pred\ T \to Pred\ S$$

Two statements p and q of type *Tran* S T are equal, written $p = q$, means that their weakest preconditions for establishing a certain postcondition are equal. Statement p is refined by statement q, written $p \sqsubseteq q$ means that the weakest precondition for p to establish a certain postcondition is stronger that the weakest precondition for q to establish the same postcondition.

$$\begin{array}{lll} p = q & \text{iff} & (\forall\ c : Pred\ T \cdot\ p\ c = q\ c) \\ p \sqsubseteq q & \text{iff} & (\forall\ c : Pred\ T \cdot\ p\ c \leq q\ c) \end{array}$$

The refinement relation is reflexive ($p \sqsubseteq p$), transitive ($p \sqsubseteq q$ and $q \sqsubseteq r$ implies $p \sqsubseteq r$), and antisymmetric ($p \sqsubseteq q$ and $q \sqsubseteq p$ implies $p = q$). Transitivity is necessary for the stepwise refinement of statements.

Composing Statements. The sequential composition of two statements p : *Tran* R S and q : *Tran* S T corresponds to their functional composition.

$$p\ ;\ q \quad = \quad (\lambda\ b : Pred\ T \cdot\ p\ (q\ b))$$

The sequential composition is associative, i. e. for appropriate statements p, q, and r we have $(p\ ;\ q)\ ;\ r = p\ ;\ (q\ ;\ r)$. Sequential composition is monotonic in both operands, meaning that if $p \sqsubseteq p'$ and $q \sqsubseteq q'$ then $p\ ;\ q \sqsubseteq p'\ ;\ q'$.

The demonic choice $q \sqcap r$ of statements q : *Tran* S T and r : *Tran* S T establishes a postcondition only if both statements do so. For the angelic choice $q \sqcup r$ to establish a postcondition it suffices that either q or r does establish it.

$$\begin{array}{lll} q \sqcap r & = & (\lambda\ b : Pred\ T \cdot\ (q\ b) \wedge (r\ b)) \\ q \sqcup r & = & (\lambda\ b : Pred\ T \cdot\ (q\ b) \vee (r\ b)) \end{array}$$

Both demonic and angelic choice are symmetric and associative and are monotonic with respect to refinement. The binary demonic and angelic choice can be generalized to a choice between an arbitrary number of alternatives. Let I be an arbitrary set and let q_i be statements indexed by $i \in I$.

$$\begin{array}{lll} (\sqcap\ i \in I \cdot\ q_i) & = & (\lambda\ b : Pred\ T \cdot\ (\forall\ i \in I \cdot\ q_i\ b)) \\ (\sqcup\ i \in I \cdot\ q_i) & = & (\lambda\ b : Pred\ T \cdot\ (\exists\ i \in I \cdot\ q_i\ b)) \end{array}$$

Primitive Statements. For a predicate $c : Pred\ T$, the guard $[c]$ and assertion $\{c\}$ are of type $Tran\ T\ T$.

$$\begin{array}{lll}[c] & = & (\lambda\ b : Pred\ T \cdot\ c \Rightarrow b) \\ \{c\} & = & (\lambda\ b : Pred\ T \cdot\ c \wedge b)\end{array}$$

Let $h : S \rightarrow T$ be a function from states to states. The update operator $\langle h \rangle$ lifts the state transformer h to a predicate transformer of type $Tran\ S\ T$.

$$\langle h \rangle \quad = \quad (\lambda\ b : Pred\ T \cdot\ (\lambda\ s : S \cdot\ b\ (h\ s)))$$

The multiple assignment $v := e$, with a list of program variables v and a list of expressions e, corresponds to an update statement with identical initial and final state space. For example, if $x + y - c$ is an expression with program variables x and y, then

$$x := x + y - c \quad = \quad \langle \lambda\ s : S \cdot\ (x \mapsto s.x + s.y - c, y \mapsto s.y) \rangle$$

where $S = (x : Int, y : Int)$. The type of this assignment is $Tran\ S\ S$.

The variable introduction *enter* $v : V := e$ extends the state space by program variables v with initial values e. The variable elimination *exit* $v : V$ removes the variables v from the state space. For example, with $S = (x : Int)$ we have:

$$\begin{array}{lll}(enter\ x : Int := 7) & = & \langle \lambda\ s : () \cdot\ (x \mapsto 7) \rangle \\ (exit\ x : Int) & = & \langle \lambda\ s : S \cdot\ () \rangle\end{array}$$

The type of *enter* $x : Int := 7$ is $Tran\ ()\ S$, the type of *exit* $x : Int$ is $Tran\ S\ ()$. We omit the types of the introduced or eliminated variables if they are clear from the context.

Embedding Statements. The initial and final state space of a statement comprise only those program variables on which it operates. Frequently, it is convenient to view a statement as operating on an enlarged state space and leaving the extra program variables unmodified. If q is of type $Tran\ S\ T$ and the program variables of state space R are disjoint from those of S and T, the embedding $q \uparrow R$ is of type $Tran\ RS\ RT$, where $RS = R \oplus S$ and $RT = R \oplus T$.

$$\begin{array}{l} q \uparrow R \quad = \\ \quad (\lambda\ c : Pred\ RT \cdot\ (\lambda\ a : RS \cdot\ (q\ (\lambda\ t : T \cdot\ c\ ((a \mid R) \oplus t)))\ (a \mid S)))\end{array}$$

For example, consider the assignment $x := 3$ of type $Tran\ (x : Int)\ (x : Int)$. With $R = (y : Int)$, we have:

$$(x := 3) \uparrow R \quad = \quad x, y := 3, y$$

Making the embedding implicitly, this allows us to write:

$$x := 3 \quad = \quad x, y := 3, y$$

Without the implicit embedding, this expression would not be well-typed. Implicit embedding is also useful for the composition with *abort*, *miracle*, and *skip*. The statement *abort* establishes no postcondition, the statement *miracle* establishes any postcondition, and the statement *skip* does nothing. All three statements are of type *Tran* () ().

$$\begin{array}{lcl} abort & = & (\lambda\ b : Pred\ () \cdot false) \\ miracle & = & (\lambda\ b : Pred\ () \cdot true) \\ skip & = & (\lambda\ b : Pred\ () \cdot b) \end{array}$$

We have that $abort = \{false\}$, $miracle = [false]$, and $skip = \{true\} = [true] = \langle id \rangle$, where *id* is the identity function on the empty record type. (With *Id* as defined in the appendix, we have that $id = Id\ ()$.)

Implicit embedding makes it possible to compose these statements in a natural way. For example, we may state that *skip* is unit of sequential composition, i. e. for statement $p : Tran\ S\ T$ we have:

$$skip\ ;\ p \quad = \quad p\ ;\ skip \quad = \quad p$$

Making the embedding explicit, this stands for:

$$skip \uparrow S\ ;\ p \quad = \quad p\ ;\ skip \uparrow T \quad = \quad p$$

Using implicit embedding, we have that both *abort* and *miracle* are left zero of sequential composition, i. e. for any statement $p : Tran\ S\ S$ we have $miracle\ ;\ p = miracle$ and $abort\ ;\ p = abort$. Furthermore, *abort* is the least element of the refinement ordering and *miracle* is the greatest, i. e. $abort \sqsubseteq p$ and $p \sqsubseteq miracle$ for any statement $p : Tran\ S\ S$.

Derived Statements. The conditional is defined by a demonic choice with guarded alternatives.

$$if\ b\ then\ q\ else\ r\ end \quad = \quad (([b]\ ;\ q) \sqcap ([\neg b]\ ;\ r))$$

Because sequential composition and demonic choice are monotonic in both operands, so is the above conditional monotonic in both q and r.

Let F be a monotonic function from statements to statements (w. r. t. refinement). Recursion is defined explicitly by a generalized demonic choice.

$$(\mu\ r \cdot F\ r) \quad = \quad (\sqcap\ q \in \{r \mid r = F\ r\} \cdot q)$$

Because F is monotonic, F has by the Knaster-Tarski fixpoint theorem a fixpoint in the lattice of statements, which is indeed $(\mu\ r \cdot F\ r)$. Iteration is defined as a special case of recursion. It is well-defined because both sequential composition and conditional are monotonic with respect to refinement.

$$while\ b\ do\ q\ end \quad = \quad (\mu\ r \cdot if\ b\ then\ q\ ;\ r\ else\ skip\ end)$$

The multiple variable declaration $(var\ v : V := e \cdot p)$ introduces new variables $v : V$ with initialization e, executes the body p of the declaration, and eliminates the variables v again.

$$(var\ v : V := e \cdot p) \quad = \quad (enter\ v : V := e\ ;\ p\ ;\ exit\ v : V)$$

The specification statement $v : [v' \cdot b]$ chooses nondeterministically (demonically) values v' such that predicate b holds and assigns them to the variables v. If no such values exist, it behaves as *miracle*.

$$v : [v' \cdot b] \quad = \quad (\sqcap\, v' \in T \cdot [b]\,;\, v := v')$$

15.3 Objects as Packed Records

In the view of objects as records, attributes and methods of an object correspond to record fields [Cardelli, 1984]. Subtyping is a relation which is defined on the structure of types. For record types, the subtype may have more fields than the supertype and the types of the common fields are also in a subtype relation.

The view of objects as records leads to a certain cyclic dependency because methods operate on records of which they are part of. This dependency is usually resolved by recursive types. Here, we follow the alternative way of using existentially quantified types instead, as proposed in [Pierce and Turner, 1994], which avoids the complexity of recursive types, but makes *packing* and *opening* of records necessary. A slight additional complication is introduced because in this encoding scheme an object is represented by a pair of records (one for the attributes and one for the methods), rather than a single record.

Objects and Object Types. Consider points in a two-dimensional plane. For example, the coordinates of a point can be given by a record $coord0$ of type $Coord0$.

$$\begin{array}{lcl} coord0 & = & (x \mapsto 9, y \mapsto 3) \\ Coord0 & = & (x : Int, y : Int) \end{array}$$

A point object encapsulates its coordinates and makes them accessible only through the methods *set* with values parameters $newx : Int, newy : Int$, method *get* with result parameters $curx : Int, cury : Int$, and parameterless method *mirror*. Parameter passing is done by enlarging and reducing the state space. That is, the body of the method *set* can access the variables $newx, newy$ from its initial state space and has to remove them at the end. Dually, the body of the method *get* adds variables $curx, cury$ to the state space. For example, $meth0$ is a record of type $Meth0$ with such method bodies:

$$\begin{array}{lcll} meth0 & = & (set & \mapsto x, y := newx, newy\,;\, exit\ newx, newy, \\ & & \ \ get & \mapsto enter\ curx, cury\,;\, curx, cury := x, y, \\ & & \ \ mirror & \mapsto x, y := -x, -y) \end{array}$$

$$\begin{array}{lcll} Meth0 & = & (set & : Tran\ (Coord0 \oplus (newx : Int, newy : Int)\ Coord0, \\ & & \ \ get & : Tran\ Coord0\ (Coord0 \oplus (curx : Int, cury : Int)), \\ & & \ \ mirror & : Tran\ Coord0\ Coord0) \end{array}$$

A point object is basically a pair with coordinates and methods, for example represented by the record $(attr \mapsto coord0, meth \mapsto meth0)$. However, the type of a point shall only reveal its method interface, but not its representation by coordinates. Hence, for defining the type *Point* of points, we replace the type *Coord*0 in *Meth*0 above by a new type variable *Attr*. Then, using existential quantification, we express that objects of type *Point* consists of attributes of type *Attr* and methods operating on these attributes, for some type *Attr*:

$$Point \;=\; (\Sigma\ Attr \cdot (attr : Attr, meth : PointMeth))$$

where

$$\begin{aligned} PointMeth \;=\; (&set &&: Tran\ (Attr \oplus (newx : Int, newy : Int))\ Attr,\\ &get &&: Tran\ Attr\ (Attr \oplus (curx : Int, cury : Int)),\\ &mirror &&: Tran\ Attr\ Attr) \end{aligned}$$

Types like *Point* are called object types. For object types we introduce a more convenient notation, which does not mention the type of the private attributes (it is a bound variable which can be given an arbitrary name anyway). For example, an equivalent definition of *Point* is:

$$\begin{aligned} Point \;=\; &object\\ &\quad set\ (newx : Int, newx : Int),\\ &\quad get\ ()\ curx : Int, curx : Int,\\ &\quad mirror\ ()\\ &end \end{aligned}$$

Now we consider colored points, where the color shall be a public attribute . This is simply achieved by requiring that the private attribute type is a subtype of the public part.

$$\begin{aligned} Col &= (color : Int)\\ ColPoint &= (\Sigma\ Attr <: Col \cdot (attr : Attr, meth : PointMeth)) \end{aligned}$$

In a more convenient notation, an equivalent definition of *ColPoint* is:

$$\begin{aligned} ColPoint \;=\; &object\\ &\quad color : Int,\\ &\quad set\ (newx : Int, newx : Int),\\ &\quad get\ ()\ curx : Int, curx : Int,\\ &\quad mirror\ ()\\ &end \end{aligned}$$

For the rest of the paper, let *Pub* stand for $(pub : P)$, where *pub* is a list of field names and *P* is a list of types. Let *Attr* be a type variable. Let *Meth* stand for

$$\begin{aligned} &(m1 : Tran\ (Attr \oplus (v1 : V1))\ (Attr \oplus (r1 : R1)),\\ &\ldots,\\ &mi : Tran\ (Attr \oplus (vi : Vi))\ (Attr \oplus (ri : Ri)))\quad , \end{aligned}$$

where $v1, \ldots, vi, r1 \ldots, ri$ are lists of labels and $V1, \ldots, Vi, R1, \ldots, Ri$ are lists of types. Using this convention, the general encoding scheme for object types is:

$$\begin{aligned} &object\ pub : P, m1\ (v1 : V1)\ r1 : R1, \ldots, mi\ (vi : Vi)\ ri : Ri\ end \;=\\ &\qquad (\Sigma\ Attr <: Pub \cdot (attr : Attr, meth : Meth)) \end{aligned}$$

Creating Objects. A value of an existentially quantified type is created by packing. For example, an object with attributes *coord*0 of type *Coord*0 and methods *meth*0 is created by:

$$pt0 \quad = \quad pack\ (attr \mapsto coord0, meth \mapsto meth0)\ by\ Coord0\ as\ Point$$

Here, the record $(attr \mapsto coord0, meth \mapsto meth0)$ is the contents of the object, *Coord*0 is the hidden type, and *Point* is the type of the resulting value. We can introduce a shorthand for this construction, writing simply:

$$pt0 \quad = \quad \langle\!|\, coord0 \;\cdot\; meth0 \,|\!\rangle$$

Here the dot separates the public from the private part. The types of the hidden attributes and the resulting object are determined by the constituents *coord*0 and *meth*0.

An object of type *ColPoint* with a public attribute is created by:

$$\begin{aligned} pt1 \quad = \quad & pack\ (attr \mapsto (coord0 \oplus (color \mapsto green), meth \mapsto meth0) \\ & by\ Coord0 \oplus (color : Int)\ as\ ColPoint \end{aligned}$$

Note that in this example we make use of implicit embedding: the method bodies in *meth*0 have to be embedded into a state space with additional program variable *color* : *Int*. As a shorthand we write:

$$pt1 \quad = \quad \langle\!|\, coord0 \;\cdot\; (color \mapsto green)\ ;\ meth0 \,|\!\rangle$$

Here the semicolon is used for separating the public attributes from the methods.

For an object like *pt*0 above, a private attribute cannot be selected because the type *Point* hides them. Hence the expression $pt0.x$ is not well-typed. However, neither can the methods be selected this way. Although the names of the methods are known by the type *Point*, the type of the methods depends on the type of the attributes, which is hidden in *Point*. Hence an expression like $pt1.mirror$ cannot be assigned a type. The only possible operation with a packed value is opening it, as shown in the next paragraphs.

Attribute Selection. A public attribute like *color* of *pt*1 above can be selected by first opening *pt*1. In the expression

$$\begin{aligned} & open\ pt1\ as\ Attr <: Col\ by\ cont : (attr : Attr, meth : PointMeth) \\ & in\ cont.attr.color \end{aligned}$$

the type variable *Attr* is introduced for the hidden type that can be used in the scope following *by*. The name *cont* is introduced for the contents of *pt*1 that can be used in the scope following *in*. Then the public attribute is selected, yielding *green* of type *Int* as result. The general form of *open* is:

$$open\ e\ as\ A <: S\ by\ x : T\ in\ f$$

Overloading the dot notation, we write the selection of a public attribute *p* of an object *e* by $e.p$. If *p* is in *pub* (see above), the general definition is:

$$\begin{aligned} e.p \quad = \quad & (open\ e\ as\ Attr <: Pub\ by\ cont : (attr : Attr, meth : Meth) \\ & in\ cont.attr.p) \end{aligned}$$

Method Calls. The definition of method calls is complicated because a method call does not yield a simple result, but changes the value of the object and the values of the result parameters. As an intermediate step, we define the construct $access\ v\ as\ A <: S\ by\ x : T\ in\ p$. It selects the value of program variable v from the initial state space and opens it with the scope of statement p. Let p be of type $Tran\ Q\ R$.

$$(access\ v\ as\ A <: S\ by\ x : T\ in\ p) \;=\; (\lambda\ b : Pred\ R \cdot (\lambda\ s : Q \cdot (open\ s.v\ as\ A <: S\ by\ x : T\ in\ (p\ b)s)))$$

For a program variable v of type Obj, the parameterless method call $v.m()$ selects the value of v from the initial state space, opens it, extends the state space by its attributes, executes the method m in the enlarged state space, and removes the modified attributes from the state space, packing them into the new value of v:

$$\begin{aligned} v.m() \;=\; & (access\ v\ as\ Attr <: Pub\ by\ cont : (attr : Attr, meth : Meth) \\ & in\ pushattr\ ;\ cont.meth.m\ ;\ popattr) \end{aligned}$$

where

$$\begin{aligned} pushattr &= \langle \lambda\ oldobj : (v : Obj) \cdot cont.attr \rangle \\ popattr &= \langle \lambda\ newattr : Attr \cdot \\ & \qquad (v \mapsto pack\ (attr \mapsto newatt, meth \mapsto cont.meth) \\ & \qquad\qquad by\ Attr\ as\ Obj)\rangle \end{aligned}$$

Statement $pushattr$ of type $Tran\ (v : Obj)\ Attr$ removes the program variable v from and adds the attributes of v to the state space. Dually, $popattr$ of type $Tran\ Attr\ (v : Obj)$ removes the attributes from the state space and adds the program variable v again.

A parameterized method call is reduced to a parameterless method call by adding the value parameters to the state space before the call and removing the result parameters from the state space after the call. For example, if pt is a program variable of type $Point$, then:

$$\begin{aligned} pt.set(6,4) &= (enter\ newx, newy := 6,4\ ;\ pt.set()) \\ a, b := pt.get() &= (pt.get()\ ;\ a, b := curx, cury\ ;\ exit\ curx, cury) \end{aligned}$$

Subtyping of Statements and Objects. The subtyping relation of $F_{\leq}$ (see also appendix) leads to a subtyping of statements and objects. Subtyping of statements is, similarly to subtyping of functions, contravariant in the first argument and covariant in the second argument.

$$\frac{E \vdash S' \leq S \qquad E \vdash T \leq T'}{E \vdash Tran\ S\ T \leq Tran\ S'\ T'}$$

This rule follows from the definition of $Tran$ and the repeated application of *sub-function*.

Two object types are in a subtype relation, if (1) all public attributes of the supertype are also public attributes of the subtype, (2) the types of the common attributes are in a subtype relation, (3) all methods of the supertype are methods of the subtype with same parameters, (4) the corresponding value parameters are in supertype relation, and (5) the corresponding result parameters are in subtype relation.

For lists S and T of types, we write $S \leq T$ if the corresponding elements of the lists are in a subtype relation. Using this convention, the subtyping rule for objects types reads as follows:

$$\frac{E \vdash Q \leq P \qquad E \vdash V1 \leq W1 \quad \ldots \quad E \vdash Vi \leq Wi \qquad E \vdash S1 \leq R1 \quad \ldots \quad E \vdash Si \leq Ri}{\begin{array}{l} E \vdash \mathit{object}\ \mathit{pub} : Q, \mathit{pub}' : Q', m1\ (v1 : W1)\ r1 : S1, \ldots, \\ \qquad mi\ (vi : Wi)\ ri : Si, \ldots, mj\ (vj : Wj)\ rj : Sj\ \mathit{end} \leq \\ \quad \mathit{object}\ \mathit{pub} : P, m1\ (v1 : V1)\ r1 : R1, \ldots, mi\ (vi : Vi)\ ri : Ri\ \mathit{end} \end{array}}$$

For example, we have $\mathit{ColPoint} \leq \mathit{Point}$.

The *subsumption* rule states that every value of a type is also a value of a supertype of that type. Therefore, $pt1$ is a value of type *ColPoint* as well as of type *Point*. This implies that, for a program variable pt of type *Point*, the assignment

$$pt := pt1$$

is indeed well-typed. However, selecting the attribute *color* from pt after the assignment is a type error, as pt does not change its type by this assignment.

Type Parameters. The type abstraction $(\Lambda\ A <: S \cdot e)$ denotes a function which takes a subtype of S as parameter and binds it to A. The result is a value of a type which may depend on S. The value declaration is extended appropriately:

$$(\mathit{val}\ y[A <: S](x : T) = e \cdot f) \quad = \quad [(\Lambda\ A <: S \cdot (\lambda\ x : T \cdot e))/y]f$$

An example of its use is (omitting the scope of the declaration):

$$\begin{array}{lll} \mathit{type} & \mathit{Colored} & = \ \mathit{object}\ \mathit{color} : \mathit{Int}\ \mathit{end} \\ \mathit{val} & \multicolumn{2}{l}{\mathit{Darker}\ [C <: \mathit{Colored}]\ (cl0 : C, cl1 : C) \ =} \\ & \multicolumn{2}{l}{\qquad \mathit{if}\ cl0.\mathit{color} \leq cl1.\mathit{color}\ \mathit{then}\ cl0\ \mathit{else}\ cl1\ \mathit{end}} \end{array}$$

If $pt2$ is an object of type *ColPoint* such that $pt2.\mathit{color} = \mathit{black}$, then we have that:

$$\mathit{Darker}\ \mathit{ColPoint}\ (pt1, pt2) \quad = \quad pt2$$

Classes. A class describes objects with same behavior, i. e. with same method bodies and attribute types. Classes are used as templates for creating objects and classes can be constructed from existing classes by inheritance. The first role implies that a class determines the initial value of objects created from that class, and the second role implies that a class must not hide the private attributes. Hence a class is given by a record of initial privates attributes, a record of initial public attributes and a record of methods. For example, the class of green points at the origin may be defined by:

$$
\begin{array}{lll}
\mathit{GreenPoint} & = & (\mathit{private} \mapsto (x \mapsto 0, y \mapsto 0), \\
& & \mathit{public} \quad \mapsto (\mathit{color} \mapsto \mathit{green}), \\
& & \mathit{methods} \mapsto \\
& & \quad (\mathit{set} \quad \mapsto x, y := \mathit{newx}, \mathit{newy}\ ;\ \mathit{exit}\ x, y, \\
& & \quad \mathit{get} \quad \mapsto \mathit{enter}\ \mathit{curx}, \mathit{cury}\ ;\ \mathit{curx}, \mathit{cury} := x, y, \\
& & \quad \mathit{mirror} \mapsto x, y := -x, -y))
\end{array}
$$

A more conventional notation for this class is:

$$
\begin{array}{lll}
\mathit{GreenPoint} & = & \mathit{class} \\
& & \quad \mathit{private}\ x : \mathit{Int} := 0, \\
& & \quad \mathit{private}\ y : \mathit{Int} := 0, \\
& & \quad \mathit{color} : \mathit{Int} := \mathit{green}, \\
& & \quad \mathit{set}\ (\mathit{newx} : \mathit{Int}, \mathit{newy} : \mathit{Int}) \ = \\
& & \quad\quad x, y := \mathit{newx}, \mathit{newy} \\
& & \quad \mathit{get}\ ()\ \mathit{curx} : \mathit{Int}, \mathit{cury} : \mathit{Int} \ = \\
& & \quad\quad \mathit{curx}, \mathit{cury} := x, y, \\
& & \quad \mathit{mirror}\ () \ = \\
& & \quad\quad x, y := -x, -y \\
& & \mathit{end}
\end{array}
$$

If $C : (\mathit{private} : \mathit{CPriv}, \mathit{public} : \mathit{CPub}, \mathit{methods} : \mathit{CMeth})$ is a class and Obj is the type of objects of class C, the function $\mathit{create}\ C$ creates an object of type Obj:

$$
\begin{array}{l}
\mathit{create}\ C \ = \\
\quad \mathit{pack}\ (\mathit{attr} \mapsto C.\mathit{private} \oplus C.\mathit{public}, \mathit{methods} \mapsto C.\mathit{methods}) \\
\quad \mathit{by}\ \mathit{CPriv} \oplus \mathit{CPub}\ \mathit{as}\ \mathit{Obj}
\end{array}
$$

When applying $\mathit{create}\ C$, we assume that the type of the objects of class C can be inferred from the context. For example,

$$\mathit{gp} := \mathit{create}\ \mathit{GreenPoint}$$

assigns an object of type $\mathit{ColPoint}$ to gp.

Inheritance on classes is, in its simplest form, expressed by record overwriting. If C, M are classes then the result of inheriting from C and modifying by M is

$$
\begin{array}{l}
(\mathit{private} \mapsto C.\mathit{private} \oplus M.\mathit{private}, \\
\mathit{public} \quad \mapsto C.\mathit{public} \oplus M.\mathit{public}, \\
\mathit{methods} \mapsto C.\mathit{methods} \oplus M.\mathit{methods})
\end{array}
$$

In a more conventional notation, an equivalent definition of *GreenPoint* is:

$$
\begin{array}{lll}
ColorlessPoint & = & class \\
 & & \quad private\ x : Int := 0, \\
 & & \quad private\ y : Int := 0, \\
 & & \quad set\ (newx : Int, newy : Int)\ = \\
 & & \qquad x, y := newx, newy \\
 & & \quad get\ ()\ curx : Int, cury : Int\ = \\
 & & \qquad curx, cury := x, y, \\
 & & \quad mirror\ ()\ = \\
 & & \qquad x, y := -x, -y \\
 & & end
\end{array}
$$

$$
\begin{array}{lll}
GreenPoint & = & class\ (ColorlessPoint) \\
 & & \quad color : Int := green \\
 & & end
\end{array}
$$

Note that here we make use of implicit embedding: The methods inherited from *ColorlessPoint* operate in an enlarged state space in *GreenPoint*.

15.4 Using Abstract and Concrete Classes

As classes are values (of a certain structure), classes can be parameterized by both values and types. By contrast, programming languages like Eiffel treat classes as types, allowing only parameterization by types in order to make type-checking decidable. By separating classes and object types, more flexibility is gained. The following examples show the use of type and value parameters in class definitions. They also show the use of abstract statements for method definitions.

We make use of the type *bag*. An empty bag is written as $\prec\succ$, a bag containing only e by $\prec e \succ$. The addition of bags b, c is written as $b + c$, the subtraction as $b - c$. The test $e\ in\ b$ determines if e is contained in the bag b and $\#b$ determines the number of elements in b.

For a buffer, the basic operations are storing an element in the buffer and taking an element from the buffer. A buffer does not guarantee any order of incoming and outgoing elements. Let *error* be a suitable error handling procedure:

$$
\begin{array}{l}
val\ Buffer[Item] = \\
\quad class \\
\qquad private\ q : bag\ of\ Item := \prec\succ, \\
\qquad put\ (i : Item) = q := q + \prec i \succ, \\
\qquad get\ ()\ i : Item = \\
\qquad\quad if\ q \neq \prec\succ\ \ then\ i : [i' \cdot i'\ in\ q]\ ;\ q := q - \prec i \succ\ \ else\ error\ end, \\
\qquad empty\ ()\ e : Bool = e := (q = \prec\succ) \\
\quad end
\end{array}
$$

A priority queue is a buffer in which the elements are removed according to their priority. The priority of an element is given by a public attribute of type *Int*.

$$\mathit{type}\ \mathit{OrderedItem} = \mathit{object}\ \mathit{key} : \mathit{Int}\ \mathit{end}$$

$$\begin{array}{l}
\mathit{val}\ \mathit{PriorityQueue}\ [\mathit{Item} <: \mathit{OrderedItem}] = \\
\quad \mathit{class}\ (\mathit{Buffer}\ \mathit{Item}) \\
\quad\quad \mathit{get}\ ()\ i : \mathit{Item} = \\
\quad\quad\quad \mathit{if}\ q \neq \prec\succ\ \mathit{then} \\
\quad\quad\quad\quad i :[i' \cdot (i'\ \mathit{in}\ q) \wedge (\forall\ x\ \mathit{in}\ q \cdot x.\mathit{key} \leq i'.\mathit{key})]\ ; \\
\quad\quad\quad\quad q := q - \prec i \succ \\
\quad\quad\quad \mathit{else}\ \mathit{error}\ \mathit{end} \\
\quad \mathit{end}
\end{array}$$

In this example, a public attribute *key* of type *Int* is selected in a Boolean expression. This could not have been expressed if *key* were a method, because method calls are statements and not simple values.

A bounded priority queue is a priority queue which has a limit for the number of elements stored in it. Hence it is parameterized by the type of the elements and their number.

$$\begin{array}{l}
\mathit{val}\ \mathit{BoundedPriorityQueue}\ [\mathit{Item} <: \mathit{OrderedItem}]\ (\mathit{max} : \mathit{Int}) = \\
\quad \mathit{class}\ (\mathit{PriorityQueue}\ \mathit{Item}) \\
\quad\quad \mathit{put}\ (i : \mathit{Item}) = \{\#q < \mathit{max}\}\ ;\ q := q + \prec i \succ \\
\quad \mathit{end}
\end{array}$$

A bounded priority queue can be simply implemented by storing the elements in an array. We omit the initialization of the array because it is irrelevant.

$$\begin{array}{l}
\mathit{val}\ \mathit{ArrayPriorityQueue}\ [\mathit{Item} <: \mathit{OrderedItem}]\ (\mathit{max} : \mathit{Int}) = \\
\quad \mathit{class} \\
\quad\quad \mathit{private}\ a : \mathit{array}\ \mathit{max}\ \mathit{of}\ \mathit{Item}, \\
\quad\quad \mathit{private}\ n : \mathit{Int} := 0, \\
\quad\quad \mathit{put}\ (i : \mathit{Item}) = a[n] := i\ ;\ n := n + 1, \\
\quad\quad \mathit{get}\ ()\ i : \mathit{Item} = \\
\quad\quad\quad \mathit{if}\ n \neq 0\ \mathit{then} \\
\quad\quad\quad\quad \mathit{var}\ p, i := 0, 1\ \cdot \\
\quad\quad\quad\quad \mathit{while}\ i < n\ \mathit{do}\ \mathit{if}\ a[p].\mathit{key} < a[i].\mathit{key}\ \mathit{then}\ p := i\ \mathit{end}\ \mathit{end}\ ; \\
\quad\quad\quad\quad i := a[p]\ ;\ n := n - 1\ ;\ a[p] := a[n] \\
\quad\quad\quad \mathit{else}\ \mathit{error}\ \mathit{end}, \\
\quad\quad \mathit{empty}\ ()\ e : \mathit{Bool} = e := (n = 0) \\
\quad \mathit{end}
\end{array}$$

For a type $\mathit{Item} \leq \mathit{OrderedItem}$ and integer *max*, objects of classes

- *Buffer Item*,
- *PriorityQueue Item*,

- *BoundedPriorityQueue Item max*,
- *ArrayPriorityQueue Item max*

all have the same type. This is independent of the use of inheritance for defining the classes.

If classes are viewed as abstract data types, the technique of data refinement can be used to establish refinement relationships between classes . In this example, *PriorityQueue Item* refines *Queue Item* and *ArrayPriorityQueue Item max* refines *BoundedPriorityQueue Item max* [Sekerinski, 1994]. This is related to the approaches of [America, 1990, Lano and Haughton, 1992, Leavens, 1991].

15.5 Related Work and Conclusions

This work started as an approach to formalize the idea of abstract and concrete classes in an analogy to the amalgamation of abstract and concrete statements in the refinement calculus. The motivation is to overcome the limitations when specifying classes by pre- and postconditions: in this approach, a method call cannot appear in a pre- or postconditions (because method calls in general change program variables). Hence, objects and classes cannot be used for specifying other classes (only for implementing them), which means that the approach does not scale up.

Another approach in the same direction is presented in [Utting, 1992]. Statements are defined by predicate transformers e. g. as in [Morgan, 1988], rather than in a typed approach as here. This leads to a simpler semantics without the need for type-theory, but disallows for technical reasons specification statements in methods. Another typed approach is reported in [Naumann, 1994], also defining statements by predicate transformer and subtyping on record types. The semantics is rich enough to model the features of Oberon, but it does not include encapsulation and parametric polymorphism.

The approach taken here has mostly been influenced by the development of the refinement calculus in higher order logic (a variant of the simple typed λ-calculus) in [Back, 1993]. The difference is that states are encoded here by records rather than tuples. The motivation for this is that inheritance is more naturally defined by records, which are used for modeling the state of objects. This has a number of minor consequences, e. g. a different treatment of variable declarations, of expressions (in guards, assertions and assignments), and of embedding. However, nested declarations of the same program variables cannot be considered in this model.

Our encoding of object types by existential types is inspired by that of [Pierce and Turner, 1994] in $F^{\omega}_{\leq}$, an extension of $F_{\leq}$ by higher-order polymorphism (functions from types to types). Here, we restrict ourselves to $F_{\leq}$, at the cost of treating some definitions less elegantly.

Correct typing of expressions in $F_{\leq}$ is known to be decidable, with some minor restrictions as reported in [Pierce, 1994]. This implies that the correct typing of expressions constructed by operators of this paper is also decidable, although we did not derive (efficient) type-checking algorithms. Such type-checking algorithms could be used as part of a compiler or for type-checking specifications.

The idea of distinguishing classes from object types has already been given a type-theoretic account in [Bruce and Gent, 1993, Pierce and Turner, 1994]. The motivation for this distinction is that decidable properties should be separated from undecidable properties. For example, the equivalence of two types, the subtyping of two types and the containment of a value in a type is decidable. By contrast, it is undecidable whether two classes are equivalent (which would involve comparing the method bodies for equivalence), and it is undecidable whether an object belongs to a class (in the sense that is belongs to the set of reachable values of the class).

A number of object-oriented concepts have not been treated, for example recursive types (like points with method $distance(other : Point)$) and object names as values. Also, the treatment of inheritance is simplified by not considering mutually dependent methods. These concepts remain the subject of further work.

Acknowledgments. The comments of Michael Butler on an earlier version were very helpful. Discussions with David Naumann lead to further insight and a number of improvements.

Appendix: The Type System $F_{\leq}$

The type system $F_{\leq}$ was developed to study the problems of sound, decidable typing of object-oriented languages in a simple, functional setting. It provides a basis for flexible typing but still guarantees the absence of "message not understood" errors of untyped languages.

$F_{\leq}$ is an extension of the simple typed λ-calculus by type parameters and subtyping. Type parameters allow to express parametric polymorphism, i. e. functions operating on values of different types. Subtyping between two types leads to the idea that a value of a type is also a value of a supertype of that type. This allows to express subtype polymorphism, i. e. functions operating on values of a type and, consequently, on values of all its subtypes.

In this appendix, we present $F_{\leq}$ with record types and existential types, which corresponds to the language *Fun* in [Cardelli and Wegner, 1985]. Existential types allow to express encapsulation. Furthermore we use in the examples basic types like *Int* and *Bool* as well products $T \times S$ and sets $set\ of\ T$, arrays $array\ of\ T$ and bags $bag\ of\ T$. For these types we assume that suitable operations (and constants) are available. Both record types and existential types, as well as other types, can also be encoded in pure $F_{\leq}$ (see for example [Curien and Ghelli, 1992]). The reader is referred to [Cardelli and Wegner, 1985] and [Pierce and Turner, 1994] for tutorial introductions.

Syntax. The syntactic categories are that of values (e), types (T), variables (x), type variables (A), and labels (l).

T	$::=$	$Bool \mid Int \mid \ldots$	basic types
	$\mid$	A	type variable
	$\mid$	Top	top type
	$\mid$	$(l_1 : T_1, \ldots, l_n : T_n)$	record type
	$\mid$	$T \to T$	function type
	$\mid$	$(\Pi\ A <: T \cdot T)$	universally quantified type
	$\mid$	$(\Sigma\ A <: T \cdot T)$	existentially quantified type
e	$::=$	$false \mid true \mid 0 \mid 1 \mid \ldots$	constants
	$\mid$	x	variable
	$\mid$	$(l_1 \mapsto e_1, \ldots, l_n \mapsto e_n)$	record construction
	$\mid$	$e.l$	field selection
	$\mid$	$(\lambda\ x : T \cdot e)$	abstraction
	$\mid$	$e\ e$	application
	$\mid$	$(\Lambda\ A <: T \cdot e)$	type abstraction
	$\mid$	$e\ T$	type application
	$\mid$	$pack\ e\ by\ T\ as\ T$	packing
	$\mid$	$(open\ e\ as\ A <: T\ by\ x : T\ in\ e)$	opening

Value e has type T is written as $e : T$. Type S is a subtype of T is written as $S \leq T$. For example, the type *Top* is the supertype of all types, i. e. for any type S, $S \leq Top$. Furthermore we have that any values e is (also) of type *Top*, $e : Top$. When the type bound in an universally quantified type, in an existentially quantified type, in a type abstraction or in an opening is omitted, it is taken to be *Top*:

$$
\begin{aligned}
(\Pi\ A \cdot T) &= (\Pi\ A <: Top \cdot T) \\
(\Sigma\ A \cdot T) &= (\Sigma\ A <: Top \cdot T) \\
(\Lambda\ A \cdot e) &= (\Lambda\ A <: Top \cdot e) \\
(open\ e\ as\ A\ by\ x : T\ in\ e) &= (open\ e\ as\ A <: Top\ by\ x : T\ in\ e)
\end{aligned}
$$

Examples of Quantified Types. Universally quantified types are functions with a type parameter and a value result. For example

$$Id = (\Lambda\ A \cdot (\lambda\ x : A \cdot x))$$

is the polymorphic identity function. Its type is $(\Pi\ A \cdot A \to A)$. It can be applied to any type. For example, applying it to *Int* yields the identity function on *Int*:

$$Id\ Int = (\lambda\ x : Int \cdot x)$$

The type of *Id Int* is $Int \to Int$.

Existentially quantified types express encapsulation. Intuitively, a value of the type $(\Sigma\ R \cdot S)$ can be thought of as a pair with first component a type

R and the second component a value of type S, where S depends on R. For example, the type

$$T \quad = \quad (\Sigma\ R\ \cdot\ R \times (R \rightarrow \mathit{Bool}))$$

is the type of "objects" which consist of an "attribute" of the hidden type R and a "method" from the attribute type to Bool. By

$$o \quad = \quad \mathit{pack}\ (5, (\lambda\ x : \mathit{Int}\ \cdot\ \mathit{odd}\ x))\ \mathit{by}\ \mathit{Int}\ \mathit{as}\ T$$

an object of type T is created with attribute 5, method $(\lambda\ x : \mathit{Int}\ \cdot\ \mathit{odd}\ x)$ and hidden attribute type Int. As the type of the object o is T, its attribute type is not visible, we only know that one exists. By

$$b \quad = \quad (\mathit{open}\ o\ \mathit{as}\ A\ \mathit{by}\ \mathit{cont} : A \times (A \rightarrow \mathit{Bool})\ \mathit{in}\ (\mathit{snd}\ \mathit{cont})(\mathit{fst}\ \mathit{cont}))$$

the object is opened, meaning that a name is given to the hidden type and the content. Then the method is extracted from the content and applied to the attribute, resulting in $b = \mathit{true}$.

Subtyping Rules. The inference rules given below allow to decide whether the subtype relation between to types holds. However, this depends on the environment, i. e. the enclosing declarations of the types and values. More formally, an environment is a list whose individual components have the form $x : T$ or $A <: T$, where x is a variable, A is a type variable, and T is a type expression. Each variable and type variable must occur at most once on the left hand side of $x : T$ or $A <: T$. A subtyping judgment of the form $E \vdash S \leq T$ means that from the environment E it follows that S is a subtype of T.

$$E \vdash T \leq T \qquad \text{sub-reflexive}$$

$$\frac{E \vdash S \leq T \qquad E \vdash T \leq U}{E \vdash S \leq U} \qquad \text{sub-transitive}$$

$$E, A <: T, F \vdash A \leq T \qquad \text{sub-var}$$

$$E \vdash T \leq \mathit{Top} \qquad \text{sub-top}$$

$$\frac{E \vdash S_1 \leq T_1 \quad \ldots \quad E \vdash S_m \leq T_m}{E \vdash (l_1 : S_1, \ldots, l_m : S_m, \ldots, l_n : S_n) \leq (l_1 : T_1, \ldots, l_m : T_m)} \qquad \text{sub-record}$$

$$\frac{E \vdash S' \leq S \qquad E \vdash T \leq T'}{E \vdash S \rightarrow T \leq S' \rightarrow T'} \qquad \text{sub-function}$$

$$\frac{E \vdash S' \leq S \qquad E, A <: S' \vdash T \leq T'}{E \vdash (\Pi\ A <: S\ \cdot\ T) \leq (\Pi\ A <: S'\ \cdot\ T')} \qquad \text{sub-universal}$$

$$\frac{E \vdash S' \leq S \qquad E, A <: S' \vdash T \leq T'}{E \vdash (\Sigma\ A <: S\ \cdot\ T) \leq (\Sigma\ A <: S'\ \cdot\ T')} \qquad \text{sub-existential}$$

Typing Rules. The inference rules given next allow to infer the minimal type of a value expression. (Every well-formed value expression is of type *Top*, so inferring an arbitrary type for a value expression is trivial.) A typing judgment of the form $E \vdash e : T$ means that from the environment E the type T for the expression e can be inferred.

$$\frac{E \vdash e : S \quad E \vdash S \leq T}{E \vdash e : T} \qquad \text{subsumption}$$

$$E, x : T, F \vdash x : T \qquad \text{var-type}$$

$$\frac{E \vdash e_1 : T_1, \quad \ldots \quad E \vdash e_n : T_n}{E \vdash (l_1 \mapsto e_1, \ldots, l_n \mapsto e_n) : (l_1 : T_1, \ldots, l_n : T_n)} \qquad \text{record-cons-type}$$

$$\frac{E \vdash e : (l_1 : T_1, \ldots, l_n : T_n)}{E \vdash e.l_i : T_i} \qquad \text{field-selection-type}$$

$$\frac{E, x : S \vdash e : T}{E \vdash (\lambda\ x : S \cdot e) : S \to T} \qquad \text{abstraction-type}$$

$$\frac{E \vdash e : S \to T \qquad E \vdash f : S}{E \vdash e\ f : T} \qquad \text{application-type}$$

$$\frac{E, A <: S \vdash e : T}{E \vdash (\Lambda\ A <: S \cdot e) : (\Pi\ A <: S \cdot T)} \qquad \text{type-abstraction-type}$$

$$\frac{E \vdash e : (\Pi\ A <: S \cdot T) \qquad E \vdash S' \leq S}{E \vdash e\ S' : [S'/A]T} \qquad \text{type-application-type}$$

$$\frac{E \vdash S' <: S \qquad E \vdash e : [S'/A]S}{E \vdash (\mathit{pack}\ e\ \mathit{by}\ S'\ \mathit{as}\ (\Sigma\ A <: S \cdot T)) : (\Sigma\ A <: S \cdot T)} \qquad \text{packing-type}$$

$$\frac{E \vdash e : (\Sigma\ A <: S \cdot T) \quad E, A <: S, x : T \vdash f : U}{E \vdash (\mathit{open}\ e\ \mathit{as}\ A <: S\ \mathit{by}\ x : T\ \mathit{in}\ f) : U} \qquad \text{opening-type}$$

Bibliography

[Abadi and Cardelli, 1995] M. Abadi and L. Cardelli. An imperative object calulus. In P. D. Mosses, M. Nielsen, and M. I. Schwartzbach, editors, *TAPSOFT '95*, volume 915 of *Lecture Notes in Computer Science*. Springer-Verlag, May 1995.

[Abrial, 1991] J. R. Abrial. A refinement case study (using the Abstract Machine Notation). In *4th Refinement Workshop*, Workshops in Computing. Springer-Verlag, 1991.

[Abrial, 1995 to appear] J. R. Abrial. *Assigning Programs to Meaning.* Cambridge University Press, 1995 (to appear).

[Aceto and Hennessy, 1989] L. Aceto and M. Hennessy. Towards action refinement in process algebras. In *Proc. 4th LICS, IEEE*, pages 138–145, 1989.

[Adámek *et al.*, 1990] Jiří Adámek, Horst Herrlich, and George E. Strecker. *Abstract and Concrete Categories.* Pure and Applied Mathematics. John Wiley & Sons, Inc., New York, 1990.

[Adámek, 1981] Jiri Adámek. Observability and Nerode equivalence in concrete categories. In Ferenc Gécseg, editor, *Fundamentals of Computation Theory.* Springer-Verlag Lecture Notes in Computer Science 117, 1981.

[Agha and Hewitt, 1987] G. Agha and C. Hewitt. Concurrent programming using actors. In A. Yonezawa and M. Tokoro, editors, *Object Oriented Concurrent Programming.* MIT Press, 1987.

[Agoulmine, 1994] M. Agoulmine. *Une methode a multi-formalismes pour specification et la validation de systemes distribues.* PhD thesis, Paris VI, January 1994.

[Alcatel ISR., 1995] Alcatel ISR. *SPOKE - Reference Manual, Version 3.0*, 1995.

[Alencar and Goguen, 1991] A. J. Alencar and J. A. Goguen. OOZE: An object-oriented Z environment. In P. America, editor, *ECOOP '91 Proceedings*, volume 512 of *Lecture Notes in Computer Science*, pages 180–199. Springer-Verlag, July 1991.

[America, 1987] P. America. Pool-t: a parallel object-oriented language. In *Object Oriented Concurrent Programming.* MIT Press, 1987.

[America, 1990] Pierre America. A Behavioral Approach to Subtyping in Object-Oriented Programming Languages. In J. W. de Bakker, W. P. de Roever, and G. Rozenberg, editors, *REX School/Workshop on the Foundations of Object-Oriented Languages*, Lecture Notes in Computer Science 489, pages 60–90, Noordwijkerhout, The Netherlands, 1990. Springer-Verlag.

[America, 1991] P. America. A behavioural approach to subtyping in object-oriented programming languages. In M. Lenzerini, D. Nardi, and M. Simi, editors, *Inheritance Hierarchies in Knowledge Representation and Programming Languages.* John Wiley and Sons, 1991.

[Andrews, 1993] D. J. Andrews. *Information Technology Programming Languages – VDM-SL ; First Committee Draft Standard: CD 13817-1*, document iso/iec jtc1/sc22/wg19 n-20 edition, November 1993.

[Arbib and Manes, 1974] Michael Arbib and Ernest Manes. Machines in a category: an expository introduction. *SIAM Review*, 16:163–192, 1974.

[Armstrong *et al.*, 1992] J. M. Armstrong, J. Howse, I. Maung, and R. Mitchell. The role of inheritance as an import/export mechanism. In W. J. Taylor, editor, *Ada in Transition*, pages 170–181. IOS Press, 1992. Studies in Computer Communications Systems (4).

[Armstrong, 1991] J. M. Armstrong. *The Roles of Inheritance in Software Development.* PhD thesis, Council for National Academic Awards, 1991.

[Atkinson *et al.*, 1991] W. D. Atkinson, J. P. Booth, and W. J. Quirk. Modal action logic for the specification and validation of safety. In *Mathematical Structures for Software Engineering.* The Institute of Mathematics and its Applications Conference Series 27, Clarendon Press, 1991.

[Atkinson, 1991] C. Atkinson. *Object-Oriented Reuse, Concurrency and Distribution.* ACM Press: Addison-Wesley, 1991.

[Austin and Parkin, 1993a] S. Austin and G. I. Parkin. Benefits, Limitations and Barriers to Formal Methods. Technical report, Division of Information Technology and Computing, National Physical Laboratory (NPL), United Kingdom, 1993.

[Austin and Parkin, 1993b] S. Austin and G. I. Parkin. Formal Methods: A Survey. Technical report, National Physical Laboratory, Queens Road, Teddington, Middlesex, TW11 0LW, March 1993.

[Back and Wright, 1989] R. J. R. Back and J. von Wright. Refinement Calculus, Part I: Sequential Nondeterministic Programs. In J. W. deBakker, W-P. deRoever, and G. Rozenberg, editors, *REX Workshop on Stepwise Refinement of Distributed Systems - Models, Formalisms, Correctness*, Lecture

Notes in Computer Science 430, pages 42–66, Mook, The Netherlands, 1989. Springer-Verlag.

[Back, 1993] Ralph J. R. Back. Refinement Calculus, Lattices and Higher Order Logic. In M. Broy, editor, *Program Design Calculi*, NATO ASI Series, pages 53–72. Springer-Verlag, 1993.

[Bar-David, 1992] T. Bar-David. Practical Consequences of Formal Definitions of Inheritance. *Journal of Object-Oriented Programming*, 5(4):43–49, July/August 1992.

[Barr and Wells, 1990] Michael Barr and Charles Wells. *Category Theory for Computing Science.* Prentice-Hall International Series in Computer Science. Prentice-Hall International, New York, 1990.

[Barringer *et al.*, 1985] H. Barringer, R. Kuiper, and A. Pnueli. A compositional temporal approach to a csp-like language. In E. Neuhold and G. Chroust, editors, *Formal Models in Programming*, pages 207–227. North-Holland, 1985.

[Barroca *et al.*, 1995] L. M. Barroca, J. S. Fitzgerald, and L. Spencer. The architectural specification of an avionic subsystem. In *IEEE Workshop on Industrial-strength Formal Specification Techniques*, pages 17–29. IEEE Press, 1995.

[Bastide *et al.*, 1993] R. Bastide, P. Palanque, and C. Sibertin-Blanc. Cooperative objects: A concurrent, petri-net based, object-oriented language. In *IEEE International Conference on Systems, Man and Cybernetics, Le Touquet, FRANCE, Vol. III*, pages 286–291, 1993.

[Battiston and cindio, 1993] E. Battiston and F. De cindio. Net formalisms for representing objects, classes and inheritance in concurrent object-oriented languages. Technical report, CNR, 1993. Prog. Finalizzato "Sistemi Informatici e Calcolo Parallelo", Tech. Report.

[Bauer *et al.*, 1989] F. Bauer, B. Möller, H. Partsch, and P. Pepper. Formal program construction by transformations—intuition-guided programming. *IEEE Transactions on Software Engineering*, 15(2), February 1989.

[Berthelot and Roucairol, 1980] G. Berthelot and G. Roucairol. *Reductions of nets and parallel programs*, pages 277–290. 1980.

[Birtwistle *et al.*, 1979] G. Birtwistle, O-J. Dahl, B. Myhrhaug, and K. Nygard. *Simula Begin.* Lund, Sweden: Studentlitteratur, 1979.

[Bolognesi and Brinksma, 1987] T. Bolognesi and E. Brinksma. Introduction to the ISO Specification Language LOTOS. *Computer Networks and ISDN Systems*, 14(1):25–59, 1987.

[Booch and Bryan, 1994] G. Booch and D. Bryan. *Software Engineering in Ada (4th Edition).* Benjamin/Cummings, 1994.

[Booch, 1987] G. Booch. *Software Engineering With Ada.* Benjamin/Cummings, second edition, 1987.

[Booch, 1991] Grady Booch. *Object Oriented Design with Applications.* Menlo Park, CA : Benjamin Cummings, 1991.

[Borba and Goguen, 1994] Paulo Borba and Joseph Goguen. An operational semantics for FOOPS. In Roel Wieringa and Remco Feenstra, editors, *Working papers of the International Workshop on Information Systems—Correctness and Reusability, IS-CORE'94*, Technical Report IR-357. Vrije Universiteit, Amsterdam, September 1994. A longer version appeared as Technical Monograph PRG-115, Oxford University, Computing Laboratory, Programming Research Group, November, 1994.

[Borba, 1995] Paulo Borba. *Semantics and Refinement for a Concurrent Object Oriented Language.* PhD thesis, Oxford University, Computing Laboratory, Programming Research Group, July 1995.

[Borceux and Dejean, 1986] Francis Borceux and Dominique Dejean. Cauchy completion in category theory. *Cahiers de Topologie et Géométrie Différentielle Catégorique*, XXVII(2):133–146, 1986.

[Borning and Ingalls, 1982] A. H. Borning and D. H. H. Ingalls. A type declaration and inference system for Smalltalk. In *POPL*, pages 133–139, 1982.

[Bowen and Stavridou, 1993] J. Bowen and V. Stavridou. Safety-critical systems, formal methods and standards. *Software Engineering Journal*, 8(4):189–209, July 1993.

[Bracha and Griswold, 1993] Gilad Bracha and David Griswold. Strongtalk: Typechecking Smalltalk in a Production Environment. In *OOPSLA*, pages 45–56, 1993.

[Brinksma (ed), 1988] E. Brinksma (ed). *Information Processing Systems — Open Systems Interconnection — LOTOS — A Formal Description Technique Based on the Temporal Ordering of Observation Behaviour, ISO 8807*, 1988.

[Bruce and Gent, 1993] Kim B. Bruce and Robert van Gent. TOIL: A new Type-safe Object-oriented Imperative Language. Technical report, Department of Computer Science, Williams College, October 13 1993.

[Bruce *et al.*, 1993] Kim B. Bruce, Jon Crabtree, Thomas Murtagh, Robert van Gent, Allyn Dimock, and Robert Muller. Safe and decidable type checking in an object oriented language. In *OOPSLA*, pages 29–46, 1993.

[Bruce, 1991] Kim B. Bruce. The equivalence of two semantic definitions for inheritance in object-oriented languages. In *Mathematical Foundations of Programming Semantics*, LNCS. Springer, June 1991.

[Bruce, 1992] Kim B. Bruce. A paradigmatic object-oriented programming language: Design, static typing and semantics. Technical Report CS-92-01, Williams College, January 1992.

[Canning *et al.*, 1989] Peter S. Canning, William R. Cook, Walter L. Hill, and Walter G. Olthoff. Interfaces for Strongly-Typed Object Oriented Programming Languages. In *Proceedings OOPSLA*, pages 457–467, 1989.

[Cardelli and Mitchell, 1989] Luca Cardelli and John Mitchell. Operations on records. Technical report, DEC SRC, 1989.

[Cardelli and Wegner, 1985] L. Cardelli and P. Wegner. On Understanding Types, Data Abstraction and Polymorphism. *ACM Computing Surveys*, 17(4):471–522, 1985.

[Cardelli *et al.*, 1991] L. Cardelli, J. C. Mitchell, S. Martini, and A. Scedrov. An extension of system F with subtyping. In *Theoretical Aspects of Computer Software*, Lecture Notes in Computer Science 526, Sendai, Japan, 1991. Springer-Verlag.

[Cardelli, 1984] Luca Cardelli. A Semantics of Multiple Inheritance. In G. Kahn, D. MacQueen, and G. Plotkin, editors, *International Symposium on the Semantics of Data Types*, Lecture Notes in Computer Science 173, pages 51–67. Springer-Verlag, 1984.

[Carrington *et al.*, 1989] D. Carrington, D. Duke, R. Duck, P. King, G. Rose, and G. Smith. *Object-Z: an object-oriented extension to Z.* North-Holland, 1989.

[Carrington *et al.*, 1990] D. Carrington, D. Duke, R. Duke, P. King, G. A. Rose, and G. Smith. Object-Z: An object-oriented extension to Z. In *Formal Description Techniques, II (FORTE'89)*, pages 281–296. North-Holland, 1990.

[Castagna *et al.*, 1992] Giuseppe Castagna, Giorgio Ghelli, and Giuseppe Longo. A Calculus for Overloaded Functions with Subtyping. In *ACM Conference on LISP and Functional Programming*, 1992.

[Castagna, 1994] Cuiseppe Castagna. Covariance and contravariance: Conflict wihout a cause. Technical report, Ecole Normale Superieure, October 1994.

[Chapront, 1992] P. Chapront. Vital coded processor and safety related software design. In H. H. Frey, editor, *Safety of Computer Control Systems 1992 (SAFECOMP '92), Computer Systems in Safety Critical Applications, Proc IFAC Symp.*, pages 141 – 145. Pergamon Press, 1992.

[Chen, 1976] P. P. Chen. The entity-relationship model – towards a unified view of data. *ACM Transactions on Database Systems*, 1(1):9–36, 1976.

[Cîrstea, 1995] Corina Cîrstea. A distributed semantics for FOOPS. Technical Report PRG-TR-20-95, Programming Research Group, University of Oxford, 1995.

[Clark and Gregory, 1986] K. L. Clark and S. Gregory. Parlog : Parallel programming in logic. *ACM transactions on Programming Languages and Systems*, 8(1):1–49, 1986.

[Clark and Wang, 1994] K. L. Clark and T. I Wang. Distributed object oriented logic programming. In *ICOT Fifth Generation Computer System Workshop on Heterogeneous Cooperative Knowledge-Bases*. 1994.

[Coad and Yourdon, 1990] P. Coad and E. Yourdon. *Object-Oriented Analysis*. Yourdon Press, New Jersey, 1990.

[Coad and Yourdon, 1991] P. Coad and E. Yourdon. *Object Oriented Analysis*. Yourdon Press, Prentice-Hall, 2nd edition, 1991.

[Coleman *et al.*, 1992] D. Coleman, F. Hayes, and S. Bear. Introducing objectcharts or how to use statecharts in object-oriented design. *IEEE Transactions on Software Engineering*, 18(1), January 1992.

[Coleman *et al.*, 1994] D. Coleman, P. Arnold, S. Bodoff, C. Dollin, H. Gilchrist, F. Hayes, and P. Jeremaes. *Object-oriented Development: The FUSION Method*. Prentice Hall Object-oriented Series, 1994.

[Cook and Daniels, 1994] S. Cook and J. Daniels. *Designing Object Systems: Object-Oriented Modelling with Syntropy*. Prentice Hall, Sept 1994.

[Cook and Palsberg, 1989] W. Cook and J. A. Palsberg. A denotational semantics of inheritance and its correctness. *Proceedings Of 1989 International Conference on Object-oriented Programming, Systems and Languages*, 24(10):433–443, 1989. Special issue of SIGPLAN Notices.

[Cook, 1989] W. Cook. A proposal for making eiffel type safe. *Computer Journal*, 32, 1989.

[Costa *et al.*, 1994] José Felix Costa, Amílcar Sernadas, and Cristina Sernadas. Object inheritance beyond subtyping. *Acta Informatica*, 5:5–26, 1994.

[Cox, 1986] B. J. Cox. *Object Oriented Programming : An Evolutionary Approach*. Addison-Wesley, 1986.

[Craigen *et al.*, 1993] D. Craigen, S. Gerhart, and T. Ralston. An International Survey of Industrial Applications of Formal Methods. Technical Report NISTGCR 93/626, U.S. Department of Commerce, Technology Administration, National Institute of Standards and Technology, Computer Systems Laboratory, Gaithersburg, MD 20899, March 1993.

[Crammond *et al.*, 1993] J. Crammond, A. Davison, A. Burt, M. Huntbach, M. Lamand, Y. Cosmadopoulos, and D. Chu. The parallel parlog user manual. Technical report, Dept. of Computing, Imperial College, 1993.

[Curien and Ghelli, 1992] Pierre-Louis Curien and Giorgio Ghelli. Coherence of Subsumption, Minimum Typing and Type-Checking in $F_{\leq}$. *Mathematical Structures in Computer Science*, 2:55–91, 1992.

[Cusack, 1991] E. Cusack. Object-oriented modelling in Z. In P. America, editor, *ECOOP '91 Proceedings*, Lecture Notes in Computer Science. Springer-Verlag, 1991.

[DaSilva *et al.*, 1991] C. DaSilva, B. Dehbonei, and F. Mejia. Formal specification in the development of industrial applications: The subway speed control mechanism. In *FORTE '91*, pages 207 – 221. North-Holland, 1991.

[Davison, 1989] A. Davison. *Polka:A Parlog Object Oriented Language.* PhD thesis, Imperial College, 1989.

[Dawes, 1991] John Dawes. *The VDM-SL Reference Guide.* Pitman (London, UK), 1991.

[de Champeaux *et al.*, 1993] D. de Champeaux, D. Lea, and P. Faur. *Object-Oriented System Development.* Addison-Wesley, 1993.

[Dijkstra, 1976] E. W. Dijkstra. *A Discipline of Programming.* Prentice-Hall, 1976.

[Dijkstra, 1993] Edsger W Dijkstra. The unification of three calculi. In Manfred Broy, editor, *Program Design Calculi*, pages 197–231. Springer Verlag, 1993.

[DOD, 1983] DOD (US Department of Defense), Ada Joint Program Office. Washington D.C. : Government Printing Office. *Reference Manual for the Ada Programming Language*, 1983.

[Draper, 1993] C. Draper. Practical Experiences of Z and SSADM. In J. Bowen, editor, *Proceedings of 1992 Z User Meeting*, Workshops in Computing. Springer-Verlag, 1993.

[Drossopoulou and Karathanos, 1993] S. Drossopoulou and S. Karathanos. Static typing for dynamic binding. In *BCS Formal Aspects of Computing, OO Special Interest Group*. Springer Verlag - to appear, December 1993.

[Drossopoulou and Karathanos, 1994] S. Drossopoulou and S. Karathanos. ST&T: Smalltalk with Types. Technical report, Imperial College, July 1994. Technical Report DOC 94/11.

[Drossopoulou and Yang, 1995] Sophia Drossopoulou and Dan Yang. $\lambda^{\&,S}$-Subtypes can be Subclasses. Technical report, Imperial College, 1995. at http://www-ala.doc.ic.ac.uk/papers/S.Drossopoulou/LambdaS, a short version presented at the BCTCS Swansea.

[Drossopoulou *et al.*, December 1995 to appear] Sophia Drossopoulou, Stephan Karathanos, and Dan Yang. Type Checking Smalltalk. *Journal of Object Oriented Programming*, December 1995 - to appear.

[Dubey, 1991] Rakesh Dubey. On a general definition of safety and liveness. Master's thesis, School of Electrical Engineering and Comp. Sci., Washington State Univ., 1991.

[Durr and Dusink, 1993] E. Durr and E. Dusink. The role of VDM^{++} in the development of a real-time tracking and tracing system. In J. Woodcock and P. Larsen, editors, FME '93, Lecture Notes in Computer Science. Springer-Verlag, 1993.

[Durr and Plat, 1994] E. H. Durr and N. Plat. *VDM^{++} Language Reference Manual.* Cap Volmac, Utrecht, Netherlands, afrodite (esprit-iii project number 6500) document afro/cg/ed/lrm/v9 edition, March 1994.

[Durr and van Katwijk, 1992] E. Durr and J. van Katwijk. Vdm++: A formal specification language for object-oriented design. In *TOOLS Europe '92*, pages 63–77, 1992.

[Eertink and Wolz, 1993] H. Eertink and D. Wolz. Symbolic Execution of LOTOS Specifications. In M. Diaz and R. Groz, editors, *Formal Description Techniques V*, pages 295–310, North-Holland, 1993.

[Ehrich *et al.*, 1991] Hans-Dieter Ehrich, Joseph Goguen, and Amílcar Sernadas. A categorial theory of objects as observed processes. In J. W. de Bakker, Willem de Roever, and Gregorz Rozenberg, editors, *Foundations of Object Oriented Languages.* Springer-Verlag Lecture Notes in Computer Science 489, 1991.

[Ehrig and Mahr, 1985] H. Ehrig and B. Mahr. *Fundamentals of Algebraic Specification 1.* Springer-Verlag, 1985.

[Ehrig and Mahr, 1990] H. Ehrig and B. Mahr. *Fundamentals of Algebraic Specification 2.* Springer-Verlag, 1990.

[Enderton, 1972] H. B. Enderton. *A Mathematical Introduction to Logic.* Academic press, 1972.

[Eva, 1992] M. Eva. *SSADM Version 4: A User's Guide.* International Series in Software Engineering. McGraw-Hill, 1992.

[Fiadeiro and Maibaum, 1991] J. Fiadeiro and T. Maibaum. Towards object calculi. Technical report, Imperial College, 1991.

[Fiadeiro and Maibaum, 1992] J. Fiadeiro and T. Maibaum. Temporal theories as modularisation units for concurrent system specification. *Formal Aspects of Computing*, 4:239–272, 1992.

[Fiadeiro and Maibaum, 1994] J. Fiadeiro and T. Maibaum. Sometimes 'tomorrow' is 'sometime': action refinement in a temporal logic of objects. In D. M. Gabbay and H. J. Ohlbach, editors, *Temporal Logic, LNAI 827*. Springer-Verlag, 1994.

[Fiadeiro and Maibaum, 1995] J. Fiadeiro and T. Maibaum. Interconnecting formalisms: supporting modularity, reuse and incrementality. In *Proc. 3rd Symposium on Foundations of Software Engineering*. ACM Press, 1995.

[France, 1992] R. France. Semantically extended data flow diagrams: A formal specification tool. *IEEE Transactions on Software Engineering*, 18(4), April 1992.

[George *et al.*, 1992] C. George, P. Haff, K. Havelund, A. E. Haxthausen, R. Milne, C. B. Nielsen, S. Prehn, and K. R. Wagner. *The RAISE Specification Language*. Prentice Hall, 1992.

[Ghelli, 1991] Giorgio Ghelli. A Static Type System for Message Passing. In *OOPSLA*, pages 129–145, 1991.

[Goguen and Burstall, 1992] Joseph Goguen and Rod Burstall. Institutions: Abstract model theory for specification and programming. *Journal of the Association for Computing Machinery*, 39(1):95–146, 1992.

[Goguen and Diaconescu, 1994] Joseph Goguen and Răzvan Diaconescu. Towards an algebraic semantics for the object paradigm. In Hartmut Ehrig and Fernando Orejas, editors, *Recent Trends in Data Type Specification*. Springer-Verlag Lecture Notes in Computer Science 785, 1994.

[Goguen and Ginali, 1978] J. Goguen and S. Ginali. A categorical approach to general systems theory. In G. Klir, editor, *Applied General Systems Research*, pages 257–270. Plenum, 1978.

[Goguen and Malcolm, 1994] Joseph Goguen and Grant Malcolm. Proof of correctness of object representations. In A. W. Roscoe, editor, *A Classical Mind: essays dedicated to C.A.R. Hoare*, chapter 8, pages 119–142. Prentice-Hall International, 1994.

[Goguen and Meseguer, 1987] Joseph Goguen and José Meseguer. Unifying functional, object-oriented and relational programming, with logical semantics. In Bruce Shriver and Peter Wegner, editors, *Research Directions in Object-Oriented Programming*, pages 417–477. MIT, 1987.

[Goguen and Winkler, 1988] Joseph Goguen and Timothy Winkler. Introducing OBJ3. Technical Report SRI-CSL-88-9, SRI International, Computer Science Lab, August 1988. Revised version to appear with additional authors José Meseguer, Kokichi Futatsugi and Jean-Pierre Jouannaud, in *Applications of Algebraic Specification using OBJ*, edited by Joseph Goguen.

[Goguen and Wolfram, 1991] Joseph Goguen and David Wolfram. On types and FOOPS. In Robert Meersman, William Kent, and Samit Khosla, editors, *Object Oriented Databases: Analysis, Design and Construction*, pages 1–22. North Holland, 1991. Proceedings, IFIP TC2 Conference, Windermere, UK, 2–6 July 1990.

[Goguen *et al.*, 1978] J. A. Goguen, J. Thatcher, and E. Wagner. An initial algebra approach to the specification, correctness and implementation of abstract data types. In Raymond Yeh, editor, *Trends in Programming Methodology IV*, pages 80–149. Prentice Hall, 1978.

[Goguen, 1970] Joseph Goguen. Mathematical representation of hierarchically organised systems. In E. O. Attinger, editor, *Global Systems Dynamics*, pages 111–129. S. Karger, 1970.

[Goguen, 1972a] Joseph Goguen. Minimal realization of machines in closed categories. *Bulletin of the American Mathematical Society*, 78(5):777–783, 1972.

[Goguen, 1972b] Joseph Goguen. Systems and minimal realization. In *Proceedings, 1971 IEEE Conf. on Decision and Control*, pages 42–46, 1972.

[Goguen, 1975a] Joseph Goguen. Discrete-time machines in closed monoidal categories. *Journal of Computer and System Sciences*, 10:1–43, 1975.

[Goguen, 1975b] Joseph Goguen. Objects. *International Journal of General Systems*, 1:237–243, 1975.

[Goguen, 1986] J. Goguen. *Reusing and Interconnecting Software Components*, volume 19, pages 16–28. 1986.

[Goguen, 1989] Joseph Goguen. Principles of parameterized programming. In Ted Biggerstaff and Alan Perlis, editors, *Software Reusability, Volume I: Concepts and Models*, pages 159–225. Addison Wesley, 1989.

[Goguen, 1991a] Joseph Goguen. A categorical manifesto. *Mathematical Structures in Computer Science*, 1(1):49–67, 1991.

[Goguen, 1991b] Joseph Goguen. Types as theories. In George Michael Reed, Andrew William Roscoe, and Ralph F. Wachter, editors, *Topology and Category Theory in Computer Science*, pages 357–390. Oxford University Press, 1991.

[Goguen, 1992] Joseph Goguen. Sheaf semantics for concurrent interacting objects. *Mathematical Structures in Computer Science*, 11:159–191, 1992.

[Goguen, 1994] Joseph Goguen. *Theorem Proving and Algebra.* October 1994. Lecture Notes, Oxford University Computing Laboratory.

[Goldberg and Robson, 1983] A. Goldberg and D. Robson. *Smalltalk-80 : The Language and its Implementation.* Addison-Wesley, 1983.

[Goldblatt, 1982] R. Goldblatt. *Axiomatizing the logic of computer programming.* Springer-Verlag, 1982.

[Goldblatt, 1987] R. Goldblatt. *Logics of Time and Computation.* CSLI, 1987.

[Goldsack and Lano, 1996] S.J. Goldsack and K. Lano. Refinement of object structures in vdm++. In *Imperial College Report*, 1996.

[Graver, 1989] Justin Owen Graver. *Type-Checking and Type-Inference for Object-Oriented Programming Languages.* PhD thesis, University of Illinois at Urbana-Champaign, 1989.

[Gray, 1980] John Gray. Fragments of the history of sheaf theory. In M. P. Fourman, C. J. Mulvey, and D. S. Scott, editors, *Applications of Sheaves.* Springer-Verlag Lecture Notes in Mathematics 753, 1980.

[Gregory, 1987] S. Gregory. *Parallel logic programming in PARLOG.* Addison Wesley, Reading, MA, 1987.

[Guttag, 1980] J. Guttag. Abstract data types and the development of data structures. In *Programming Language Design.* Los Alamitos, CA. : IEEE Computer Society Press, 1980.

[Halbert and O'Brien, 1987] D. C. Halbert and P. D. O'Brien. Using types and inheritance in object-oriented programming. *I.E.E.E. Software*, pages 71–79, September 1987.

[Hall, 1994] A. Hall. Specifying and Interpreting Class Hierarchies in Z. In *8th Z User Meeting*, Workshops in Computing. Springer-Verlag, 1994.

[Harel, 1987] D. Harel. Statecharts: A visual formalism for complex systems. *Science of Computer Programming*, (8):231–274, 1987.

[Hares, 1990] J. S. Hares. *SSADM for the Advanced Practitioner.* Wiley Series in Software Engineering Practice, 1990.

[Harris, 1991] W. Harris. Contravariance for the rest of us. *Journal of Object-oriented Programming*, November 1991.

[Haughton and Lano, 1994] H. Haughton and K. Lano. Testing and safety analysis of B specifications. In *6th Refinement Workshop.* Springer-Verlag, 1994.

[Haughton and Lano, 1995] H. Haughton and K. Lano. *B Abstract Machine Notation: A Reference Manual.* McGraw-Hill, 1995.

[Hayes and Mahony, 1992] I. J. Hayes and B. Mahony. A case study in timed refinement: A mine pump. *IEEE Software*, 18(9), September 1992.

[Hennessy, 1988] Matthew Hennessy. *Algebraic Theory of Processes.* The MIT Press, 1988.

[Hense, 1993] A. V. Hense. Denotational semantics of an object-oriented language with explicit wrappers. *BCS Formal Aspects of Computing*, 5:181–207, 1993.

[Hildebrand and Treves, 1990] Th. Hildebrand and N. Treves. S-cort: A method for the development of an electronic payment system. In *Advances in Petri Nets 1989*, number 424 in LNCS, pages 262–280. Springer Verlag, 1990.

[Hill, 1991] J. V. Hill. *Microprocessor Based Protection Systems*. Elsevier, 1991.

[Hoare *et al.*, 1987] C.A.R. Hoare, J. Michael Spivey, Ian Hayes, Jifeng He, Carroll Morgan, A. William Roscoe, Jeff Sanders, Ib Sorenson, and Bernard Sufrin. Laws of programming. *Communications of the Association for Computing Machinery*, 30(8):672–686, 1987.

[Hoare, 1972] C. A. R. Hoare. Proof of correctness of data representation. *Acta Informatica*, 1:271–281, 1972.

[Hoare, 1974] C. A. R. Hoare. Monitors : An operating system structuring concept. *Communications of the ACM*, 17, 1974.

[Hoare, 1985a] C. A. R. Hoare. *Communicating Sequential Processes*. Prentice Hall, 1985.

[Hoare, 1985b] C. A. R. Hoare. *Communicating Sequential Processes*. Prentice Hall, 1985.

[Hogg, 1991] J. Hogg. Islands: Aliasing protection in object-oriented languages. In *OOPSLA '91 Proceedings*. Springer-Verlag, 1991.

[Houston, 1994] I. Houston. Formal Specification of the OMG Core Object Model. Technical report, IBM UK, Hursely Park, 1994.

[Iachini and Giovanni, 1990] P. L. Iachini and R. Di Giovanni. HOOD and Z for the development of complex software systems. In *VDM and Z, VDM 90*, volume 428 of *Lecture Notes in Computer Science*, pages 262–289. Springer-Verlag, 1990.

[IEC, 1995] IEC. *Software for Computers in the Application of Industrial Safety-Related Systems*, 1995. IEC 65A (Secretariat) 122.

[Jacobson, 1992] I. Jacobson. *Object-Oriented Software Engineering*. Addison-Wesley, 1992.

[Jarvinen *et al.*, 1990] H-M. Jarvinen, R. Kurki-Suonio, M. Sakkinen, and K. Systa. Object-oriented specification of reactive systems. In *Proc. 12th Int. Conf. on Software Enginnering*, pages 63–71. IEEE Computer Society Press, 1990.

[Jensen, 1987] K. Jensen. Coloured petri nets. In *Advances in Petri Nets 1986, LNCS 255, Part I*, pages 248–299. Springer Verlag, 1987.

[Jeremaes *et al.*, 1986] P. Jeremaes, S. Khosla, and T. S. E. Maibaum. A modal [action] logic for the specification of embedded real-time systems. In *Software Engineering 86*, Southampton UK, 1986.

[Jifeng *et al.*, 1986] He Jifeng, C. A. R. Hoare, and Jeff Sanders. Data refinement refined. volume 213 of *Lecture Notes in Computer Science*. Springer-Verlag, 1986.

[Johnson *et al.*, 1988] Ralph E. Johnson, Justin O. Graver, and Lawrence W. Zurawski. An optimizing compiler for Smalltalk. In *OOPSLA*, pages 18–26, 1988.

[Johnson, 1993] Ralph Johnson. Typechecking Smalltalk. In *OOPSLA*, pages 315–321, 1993.

[Jones, 1983] Cliff Jones. Specification and design of (parallel) programs. In R.E.A. Mason, editor, *Information Processing'83*, pages 321–332. North-Holland, 1983.

[Jones, 1990] C. B. Jones. *Systematic Software Development using VDM*. Prentice Hall International, (2nd edition) edition, 1990.

[Jones, 1992] Cliff Jones. An object-based design method for concurrent programs. Technical Report UMCS-92-12-1, Department of Computer Science, University of Manchester, 1992.

[Jones, 1993] Cliff Jones. Process-algebraic foundations for an object-based design notation. Technical Report UMCS-93-10-1, Department of Computer Science, University of Manchester, 1993.

[Kahn *et al.*, 1987] Kenneth M. Kahn, Eric D. Tribble, Mark S. Miller, and Daniel G. Bobrow. Vulcan: Logical concurrent objects. In P. Shriver and P. Wegner, editors, *Research Directions in Object-Oriented Programming*. MIT Press, 1987.

[Kamin, 1988] S. Kamin. Inheritance in Smalltalk-80: a denotational definition. In *Proceedings of the Fifteenth Annual ACM-SIGACT Symposium on the Principles of Programming Languages*, pages 80–87. ACM Press, 1988.

[Karathanos, 1993] Stephan Karathanos. Type checking and type inference for Smalltalk, September 1993. Imperial College, M.Sc. Project.

[Kent, 1995] S. J. H. Kent. Structured theory semantics for vdm++. in preparation, 1995.

[Lakos and Keen, 1991] C. A. Lakos and C.D. Keen. Loopn - language for object-oriented petri nets. In *Proceedings of SCS Multiconference on Object-Oriented Simulation, Anaheim, California*, pages 22–30, 1991.

[Lambek and Scott, 1986] Joachim Lambek and Philip J. Scott. *Introduction to Higher Order Categorical Logic.* Cambridge University Press, 1986. Cambridge Studies in Advanced Mathematics, Volume 7.

[Lamport, 1983] L. Lamport. Specifying concurrent program modules. *ACM TOPLAS*, 6(2), 1983.

[Lamport, 1994] L. Lamport. The temporal logic of actions. *ACM TOPLAS*, 16:872–923, 1994.

[Lane and Moerdijk, 1992] Saunders Mac Lane and Ieke Moerdijk. *Sheaves in Geometry and Logic.* Springer-Verlag, 1992.

[Lane, 1971] Saunders Mac Lane. *Categories for the Working Mathematician*, volume 5 of *Graduate Texts in Mathematics.* Springer Verlag, 1971.

[Lano and Goldsack, 1994] Kevin Lano and Stephen Goldsack. Refinement and subtyping in formal object-oriented development. In Roel Wieringa and Remco Feenstra, editors, *Working papers of the International Workshop on Information Systems—Correctness and Reusability, IS-CORE'94*, number IR-357. Vrije Universiteit, Amsterdam, September 1994.

[Lano and Haughton, 1992] K. Lano and H. Haughton. Reasoning and Refinement in Object-Oriented Specification Languages. In O. Lehrmann Madsen, editor, *European Conference on Object-Oriented Programming '92*, Lecture Notes in Computer Science 615. Springer-Verlag, 1992.

[Lano and Haughton, 1993] K. Lano and H. Haughton. *Object-Oriented Specification Case Studies (First Edition).* Prentice Hall, 1993.

[Lano, 1991] K. Lano. Z^{++}, an object-oriented extension to Z. In J. Nicholls, editor, *Z User Meeting, Oxford, UK*, Workshops in Computing. Springer-Verlag, 1991.

[Lano, 1994] K. Lano. Refinement in object-oriented specification languages. In D. Till, editor, *6th Refinement Workshop.* Springer-Verlag, 1994.

[Lano, 1995] K. Lano. Reactive system specification and refinement. In *TAPSOFT '95*, volume 915 of *Lecture Notes in Computer Science.* Springer-Verlag, 1995.

[Leavens and Weihl, 1990] G. T. Leavens and W. E. Weihl. Reasoning about object-oriented programs that use subtypes. In *OOPSLA'90 Proceedings.* ACM Press, 1990.

[Leavens, 1991] Gary T. Leavens. Modular Specification and Verification of Object-Oriented Programs. *IEEE Software*, 8(4):72–80, 1991.

[LeBlanc, 1995] P. LeBlanc. VENUS User Manual; Combined Use of OMT and VDM^{++}. Technical Report afro/verilog/plb/um/v2.3, Verilog, 1995.

[Lee *et al.*, 1991] M. K. O. Lee, P. N. Scharbach, and I. H. Sørensen. Engineering real software using formal methods. In *4th Refinement Workshop*, Workshops in Computing. Springer-Verlag, 1991.

[Lilius, 1993] Johan Lilius. A sheaf semantics for Petri nets. Technical Report A23, Dept. of Computer Science, Helsinki University of Technology, 1993.

[Lin, 1994] T. Lin. A Formal Semantics for MooZ, PhD Thesis. Technical report, DI/UFPE, Recife/PE, Brazil, 1994.

[Lin, 1995] H. Lin. Complete inference systems for weak bisimulation equivalences in the π-calculus. In *TAPSOFT '95*, volume 915 of *Lecture Notes in Computer Science*. Springer-Verlag, May 1995.

[Liskov and Wing, 1993] B. Liskov and J. Wing. Family values: A behavioral notion of subtyping. Technical Report CMU-CS-93-187, School of Computer Science, Carnegie Mellon University, 1993.

[Liskov and Wing, 1994] B. Liskov and J. Wing. Family values: A behavioural notion of subtyping. *ACM TOPLAS*, 16(6), November 1994.

[Liskov and Zilles, 1977] B. Liskov and S. Zilles. An introduction to formal specifications of data abstractions. In *Current Trends in Programming Methodology, vol 1*, page 18. Englewood Cliffs, N.J.: Prentice Hall, 1977.

[Liskov, 1988] B. Liskov. Data Abstraction and Hierarchy. In *OOPSLA '87 (Addendum to Proceedings): ACM SIGPLAN Notices*, 23(5):17-34, May 1988.

[Mac Lane and Moerdijk, 1992] Saunders Mac Lane and Ieke Moerdijk. *Sheaves in Geometry and Logic.* Universitext. Springer-Verlag, 1992.

[MacQueen, 1986] David MacQueen. Using Dependent Types to Express Modular Structure. In *POPL*, 1986.

[Magnusson, 1991] B. Magnusson. Code reuse considered harmful. *The Journal of Object-oriented Programming*, 8, November/December 1991. Guest Editorial.

[Malcolm and Goguen, 1994] Grant Malcolm and Joseph Goguen. Proving correctness of refinement and implementation. Technical Monograph PRG-114, Programming Research Group, Oxford University, 1994.

[Manna and Pnueli, 1991] Z. Manna and A. Pnueli. *The Temporal Logic of Reactive and Concurrent Systems.* Springer-Verlag, 1991.

[Maung *et al.*, 1994] I. Maung, J. R. Howse, and R. J. Mitchell. Towards a formalization of programming by difference. In *FME'94 Proceedings.* Springer-Verlag, 1994.

[Maung, 1993a] I. Maung. Simulation, subtyping and substitutability. Technical report, University of Brighton, September 1993. revised version accepted for publication in Formal Aspects of Computing.

[Maung, 1993b] I. Maung. A survey of behavioural subtyping. Technical report, University of Brighton, September 1993.

[May, 1990] D. May. Use of formal methods by a silicon manufacturer. In C. A. R. Hoare, editor, *Developments in Concurrency and Communication*, chapter 4, pages 107–129. Addison-Wesley, 1990.

[Mazurkiewicz, 1984] Antoni Mazurkiewicz. Traces, histories, graphs: instances of a process monoid. In M. P. Chytil and V. Koubek, editors, *Mathematical Foundations of Computer Science*. Springer-Verlag Lecture Notes in Computer Science 176, 1984.

[Meira and Cavalcanti, 1991] S. R. L. Meira and A. L. C. Cavalcanti. Modular object-oriented Z specifications. In *Z User Meeting 1990*, Workshops in Computing, pages 173–192. Springer-Verlag, 1991.

[Meyer, 1988] B. Meyer. *Object-oriented Software Construction*. Prentice-Hall, 1988.

[Meyer, 1990] B. Meyer. Lessons from the design of the Eiffel libraries. *Communications of the ACM*, 33(9), September 1990.

[Meyer, 1992] B. Meyer. *Eiffel : The Language*. Prentice-Hall, 1992.

[Milner *et al.*, 1989] Robin Milner, Joachim Parrow, and David Walker. A calculus of mobile processes. Technical Report ECS-LFCS-89-85, 86, Laboratory for Foundations of Computer Science, Edinburgh University, 1989.

[Milner, 1971] Robin Milner. An algebraic definition of simulation between programs. Technical Report CS-205, Stanford University, Computer Science Department, 1971.

[Milner, 1989] Robin Milner. *Communication and Concurrency*. Prentice Hall, 1989.

[Milner, 1991] R. Milner. *Communication and Concurrency*. Prentice Hall, 1991.

[Milner, 1992] R. Milner. The polyadic π-calculus: A tutorial. In M. Broy, editor, *Logic and Algebra of Specification*. Springer-Verlag, 1992.

[Moller-Pedersen *et al.*, 1987] B. Moller-Pedersen, D. Belsnes, and H. P. Dahle. Relational and tutorial on osdl: An object-oriented extension of sdl. *Computer Networks and ISDN Systems*, 13(2):97–117, 1987.

[Monteiro and Pereira, 1986] Luis Monteiro and Fernando Pereira. A sheaf-theoretic model of concurrency. In *Proc. Logic in Computer Science (LICS '86)*. IEEE Press, 1986.

[Moreira and Clark, 1993] A. M. D. Moreira and R. G. Clark. Rigorous Object-Oriented Analysis. Technical Report CSM-109, Computing Science Department, University of Stirling, Scotland, 1993.

[Moreira and Clark, 1994] A. M. D. Moreira and R. G. Clark. Combining Object-Oriented Analysis and Formal Description Techniques. In M. Tokoro and R. Pareschi, editors, *8th European Conference on Object-Oriented Programming: ECOOP '94*, LNCS 821, pages 344–364. Springer-Verlag, July 1994.

[Morgan, 1988] Carroll C. Morgan. The Specification Statement. *ACM Transactions on Programming Languages and Systems*, 10(3):403–419, 1988.

[Morgan, 1990] C. Morgan. *Programming from Specifications.* Prentice-Hall, 1990.

[Morgan, 1994] Carroll Morgan. *Programming from Specifications.* Prentice Hall, second edition, 1994.

[Morris, 1987] J. M. Morris. A Theoretical Basis for Stepwise Refinement and the Programming Calculus. *Science of Computer Programming*, 9(3), 1987.

[Mosses, 1992] Peter Mosses. *Action Semantics.* Tracts in Theoretical Computer Science. Cambridge University Press, 1992.

[Naumann, 1994] David A. Naumann. Predicate Transformer Semantics of an Oberon-Like Language. In Ernst-R. Olderog, editor, *Programming Concepts, Methods and Calculi*, pages 460–480, San Miniato, Italy, 1994. International Federation for Information Processing.

[Paepcke, 1991] Andreas Paepcke, editor. *OOPSLA'91.* ACM, ACM Press, November 1991.

[Park, 1981] David Park. Concurrency and automata on infinite sequences. In *Proceedings of 5th GI Conference*, volume 104 of *Lecture Notes in Computer Science.* Springer-Verlag, 1981.

[Parnas, 1972] D. Parnas. On the criteria for decomposing a system into modules. *Communications ACM*,]December 1972.

[Pierce and Turner, 1994] Benjamin C. Pierce and David N. Turner. Simple Type-Theoretic Foundations For Object-Oriented Programming. *Journal of Functional Programming*, 4(2):207–247, 1994.

[Pierce, 1991] Benjamin C. Pierce. *Basic Category Theory for Computer Science.* MIT Press, 1991.

[Pierce, 1994] Benjamin C. Pierce. Bounded Quantification is Undecidable. In Carl A. Gunter and John C. Mitchell, editors, *Theoretical Aspects of Object-Oriented Programming.* MIT Press, Massachusetts, London, 1994.

[Piessens and Steegmans, 1994] Frank Piessens and Eric Steegmans. Canonical forms for data-specifications. In *Proceedings of Computer Science Logic 94*, number 933 in Lecture Notes in Computer Science, pages 397–411. Springer-Verlag, 1994.

[Plotkin, 1981] Gordon Plotkin. A structural approach to operational semantics. Technical Report DAIMI FN–19, Computer Science Department, Aarhus University, September 1981.

[Pnueli, 1977] A. Pnueli. The temporal logic of programs. In *Proc 18th Annual Symposium on Foundations of Computer Science*, pages 45–57. IEEE, 1977.

[Polack and Whiston, 1991] F. Polack and M. Whiston. Formal methods and system analysis. In *Methods Integration Conference*, Workshops in Computing. Springer-Verlag, 1991.

[Ponder and Bush, 1992] C. Ponder and B. Bush. Polymorphism considered harmful. *ACM Sigplan Notices*, 27(6), June 1992.

[Pyle and Josephs, 1991] D. R. Pyle and M. Josephs. Enriching a structured method with Z. In *Methods Integration Conference*, Workshops in Computing. Springer-Verlag, 1991.

[Rapanotti and Socorro, 1992] Lucia Rapanotti and Adolfo Socorro. Introducing FOOPS. Technical Report PRG-TR-28-92, Oxford University, Computing Laboratory, Programming Research Group, November 1992.

[Reisig, 1987] W. Reisig. Petri nets in software engineering. In *Advances in Petri Nets 1986*, number 225, Part II. in LNCS, 1987.

[Remy, 1993] Didier Remy. Typechecking records and variants in a natural extension of Ml. In *POPL*, pages 77–87, 1993.

[Ritchie, 1993] B. Ritchie. Proof with Mural, 1993. Rutherford Appleton Laboratory, Informatics Department, Chilton, Didcot, Oxon OX11 0QX, UK.

[Rudkin, 1992] S. Rudkin. Inheritance in LOTOS. In Parker K. R. and Rose G. A., editors, *Formal Description Techniques IV*, pages 409–423, North-Holland, 1992.

[Rumbaugh *et al.*, 1991] J. Rumbaugh, M. Blaha, W. Premerlani, F. Eddy, and W. Lorensen. *Object-Oriented Modelling and Design*. Prentice-Hall, 1991.

[Rumbaugh, 1993] J. Rumbaugh. Disinherited! examples of the misuse of inheritance. *Journal of Object-oriented Programming*, 5(9):22–24, February 1993.

[Ryan *et al.*, 1991] M. Ryan, J. Fiadeiro, and T. Maibaum. Sharing actions and attributes in modal action logic. In T. Ito and A. Meyer, editors, *Proceedings of the International Conference on Theoretical Aspects of Computer Software (TACS'91)*. Springer Verlag, 1991.

[Sakkinen, 1989] M. Sakkinen. Disciplined inheritance. In S Cook, editor, *ECOOP '89, Proceedings of the Third European Conference on Object-oriented Programming*, pages 39–56. British Computer Society Workshop Series, 1989.

[Sampaio and Meria, 1990] A. Sampaio and S. Meria. Modular extensions to Z. In *VDM and Z*, volume 428 of *Lecture Notes in Computer Science*. Springer-Verlag, 1990.

[Sannella and Tarlecki, 1988] D. Sannella and A. Tarlecki. Building specifications in an arbitrary institution. *Information and Control*, 76:165–210, 1988.

[Sekerinski, 1994] Emil Sekerinski. *Verfeinerung in der Objektorientierten Programmkonstruktion.* Doctoral thesis, University of Karlsruhe, 1994.

[Semmens *et al.*, 1992] L. Semmens, R. France, and T. Docker. Integrated structured analysis and formal specification techniques. *The Computer Journal*, 35(6), 1992.

[Sernadas *et al.*, 1990] A. Sernadas, H-D. Ehrich, and J-F. Costa. From processes to objects. *The INESC Journal of Research and Development*, pages 7–27, 1990.

[Shaffert *et al.*, 1986] C. Shaffert, T. Cooper, B. Bullis, M. Killian, and C. Wilpolt. An introduction to Trellis/Owl. *Proceedings Of 1986 International Conference on Object-oriented Programming, Systems and Languages*, 21(11):9–16, 1986. Special issue of SIGPLAN Notices.

[Shapiro and Takeuchi, 1983] E. Y. Shapiro and A. Takeuchi. Object oriented programming in concurrent prolog. In *New Generation Computing*. 1983.

[Shibayama and Yonezawa, 1987] E. Shibayama and A. Yonezawa. Distributed computing in abcl/1. In A. Yonezawa and M. Tokoro, editors, *Object Oriented Concurrent Programming*. MIT Press, 1987.

[Smith and Smith, 1977] J. M. Smith and D. C. P. Smith. Database Abstractions: Aggregation. *Communications of the ACM*, 20(6):405–413, June 1977.

[Smith, 1995] G. Smith. A logic for Object-Z. In *Z User Meeting '95*, Lecture Notes in Computer Science. Springer Verlag, 1995.

[Socorro, 1993] Adolfo Socorro. *Design, Implementation, and Evaluation of a Declarative Object Oriented Language.* PhD thesis, Oxford University, Computing Laboratory, Programming Research Group, December 1993.

[Spivey, 1992] M. Spivey. *The Z Notation: A Reference Manual.* Prentice Hall, 2nd edition, 1992.

[Stroustrup, 1991] B. Stroustrup. *The C++ Programming Language.* Addison-Wesley, second edition, 1991.

[Suzuki, 1989] Norisha Suzuki. Inferring types in Smalltalk. In *8th POPL*, 1989.

[Swatman, 1993] P. A. Swatman. Using formal specification in the acquisition of information systems: Educating information systems professionals. In *Z User Meeting 1992*, Workshops in Computing. Springer-Verlag, 1993.

[Tennison, 1975] B. R. Tennison. *Sheaf Theory*, volume 20 of *London Mathematical Society Lecture Notes.* Cambridge University Press, 1975.

[Thomas and Kirkwood, 1995] M. Thomas and C. Kirkwood. Experiences with specification and verification in LOTOS. In *Industrial-Strength Formal Specification Techniques.* IEEE Press, 1995.

[Treves, 1992] N. Treves. A survey on ca nets. Technical report, ESPRIT Project EP 5342 PROOFS, 1992. Technical Report SLI-TR-038-V1.0-WP1.

[Utting, 1992] M. Utting. *An Object-Oriented Refinement Calculus with Modular Reasoning.* Doctoral thesis, University of New South Wales, Kensington, 1992.

[Van Baelen *et al.*, 1991] Stefan Van Baelen, Johan Lewi, Eric Steegmans, and Helena Van Riel. EROOS: An entity-relationship based object-oriented specification method. In G. Heeg, B. Magnusson, and B. Meyer, editors, *Technology of Object-Oriented Languages and Systems TOOLS 7*, pages 103–117. Prentice Hall, 1991.

[van den Enden and Verhoeckx, 1989] A.W.M. van den Enden and N.A.M Verhoeckx. Discrete Signal processing: an Introduction. Prentice Hall International, 1989.

[van der Aalst *et al.*, 1993] W. van der Aalst, K. M. van Hee, N. Treves, and R. di Giovanni. Proofs: Formalisms and methods, June 1993. ESPRIT Project EP 5342 'PROOFS'.

[van Hee *et al.*, 1990] K. M. van Hee, L. J. Somers, and M. Voorhoeve. *Executable specification for distributed information systems.* North-Holland, 1990.

[Verheijen and van Backkum, 1982] G. M. A. Verheijen and J. van Backkum. Niam: an information analysis method. In *Information System Design Methodologies: A Comparative Review*, pages 537–589. IFIP, 1982.

[Wang, 1995] T. I. Wang. *Distributed Object-oriented Logic Programming.* PhD thesis, Imperial College, 1995.

[Ward, 1989] P. T. Ward. How to integrate object orientation with structured analysis and design. *IEEE Software*, pages 74–82, March 1989.

[Wegner and Zdonik, 1988] P. Wegner and S. Zdonik. Inheritance as an incremental modification mechanism or what like is and isn't like. In *Proceedings of ECOOP88.* New York, NY: Springer Verlag, 1988.

[Wegner, 1987] P. Wegner. The object oriented classification paradigm. In *Research Directions in Object Oriented Programming.* New York, NY.: Springer Verlag, 1987.

[Wieringa *et al.*, 1993] R. Wieringa, W. de Jonge, and P. Spruit. Roles and dynamic subclasses: A model logic approach. Technical Report IS-CORE report, Faculty of Mathematics and Computer Science, Vrije Universiteit Amsterdam, 1993.

[Wieringa *et al.*, 1994] R. Wieringa, W. de Jonge, and P. Spruit. Roles and dynamic subclasses: a modal logic approach. In *ECOOP'94 Proceedings.* Springer-Verlag, 1994.

[Wilde and Huitt, 1991] N. Wilde and R. Huitt. Maintenance support for object-oriented programs. In *Proceedings of Conference on Software Maintenance.* IEEE Computer Society Press, 1991.

[Wills, 1991] A. Wills. Capsules and types in Fresco: Program verification in Smalltalk. In P. America, editor, *ECOOP '91 Proceedings*, volume 512 of *Lecture Notes in Computer Science*, pages 59–76. Springer-Verlag, 1991.

[Wirth, 1988] N. Wirth. *Programming in Modula-2.* Springer-Verlag New York, fourth edition, 1988.

[Wolczko, 1989] M. Wolczko. Semantics of Smalltalk-80. In J. Bezivin, J. Hullot, P. Cointe, and H. Lieberman, editors, *Proceedings of the 1987 European Conference on Object-oriented Programming*, pages 108–120. Hiedelberg: Springer-Verlag, 1989.

[Wolfram and Goguen, 1992] David A. Wolfram and Joseph A. Goguen. A sheaf semantics for FOOPS expressions (extended abstract). In M. Tokoro, O. Nierstrasz, P. Wegner, and A. Yonezawa, editors, *Proceedings of the ECOOP'91 Workshop on Object-Based Concurrent Computing*, pages 81–98. Springer-Verlag Lecture Notes in Computer Science 612, 1992.

[Yang, 1994] Y. Yang. *Integration de l'approche orientee objets dans un contexte de multi-formalismes pour la modelisation d'applications distribuees.* PhD thesis, PARIS VI, November 1994.

[Yoshida and Chikayama, 1988] K. Yoshida and T. Chikayama. A'um-a stream-based concurrent object-oriented language. In *Proceeding of FGCS'88.* ICOT, 1988.

[Yourdon, 1989] E. Yourdon. *Modern Structured Analysis.* Englewood Cliffs, N.J.: Prentice Hall, 1989.

Index